HAZARDOUS WASTE
AND
SOLID WASTE

HAZARDOUS WASTE AND SOLID WASTE

Edited by
David H.F. Liu
Béla G. Lipták

Paul A. Bouis
Special Consultant

LEWIS PUBLISHERS
Boca Raton London New York Washington, D.C.

Library of Congress Cataloging-in-Publication Data

Catalog record is available from the Library of Congress.

The material in this book is taken from, Environmental Engineers' Handbook, Second Edition, edited by David H. F. Liu and Béla G. Lipták, Paul A Bouis, special consultant.

© 2000 by CRC Press LLC
Lewis Publishers is an imprint of CRC Press LLC

No claim to original U.S. Government works
International Standard Book Number 1-56670-512-6
Printed in the United States of America 1 2 3 4 5 6 7 8 9 0
Printed on acid-free paper

Foreword

Hazardous Waste and Solid Waste covers in depth the interrelated factors and principles which affect our environment and how we have dealt with them in the past, how we are dealing with them today, and how we might deal with them in the future. Although this book is clearly aimed at the environmental professional, it is written and structured in a way that will allow others outside the field to educate themselves about our environment, and what can and must be done to continue to improve the quality of life on spaceship earth. The book covers in detail the ongoing global transition among the cleanup of the re-mains of abandoned technology and the prevention of pollution from existing technology. *Hazardous Waste and Solid Waste* tries to strike a balance between the danger of hazardous wastes and the low probability that a dangerous environmental event will occur because of these wastes.

A great deal of effort has gone into providing as much information as possible in easy-to-use tables and figures. We have chosen to use schematic diagrams rather than actual pictures of equipment, devices, or landscapes to explain or illustrate technology and techniques used in var-

Preface

Dr. David H.F. Liu passed away prior to the preparation of this book.
He will be long remembered by his coworkers,
and the readers of this handbook will carry his memory into the 21st Century

Engineers respond to the needs of society with technical innovations. Their tools are the basic sciences. Some engineers might end up working *on* these tools instead of working *with* them. Environmental engineers are in a privileged and challenging position, because their tools are the totality of man's scientific knowledge, and their target is nothing less than human survival through making man's peace with nature.

Because the environment is a complex web, the straining of some of the strands affects the entire web. This book recognizes this integrated nature of our environment, where the various forms of pollution are interrelated symptoms and therefore can not be treated separately.

The contributors to this book came from all continents and their backgrounds cover not only engineering, but also legal, medical, agricultural, meteorological, biological and other fields of training. In addition to discussing the causes, effects, and remedies of pollution, this book also emphasizes reuse, recycling, and recovery. Nature does not cause pollution; by total recycling, *nature makes resources out of all wastes.* Our goal should be to learn from nature in this respect.

To the best of our knowledge today, life in the universe exists only in a ten-mile-thick layer on the 200-million-square-mile surface of this planet. During the 5 million years of human existence, we lived in this thin crust of earth, air, and water. Initially man relied only on inexhaustible resources. The planet appeared to be without limits and the laws of nature directed our evolution. Later we started to supplement our muscle power with exhaustible energy sources (coal, oil, uranium) and to substitute the routine functions of our brains by machines. As a result, in some respects we have "conquered nature" and today we are directing our own evolution. Today, our children grow up in man-made environments; virtual reality

or cyberspace is more familiar to them than the open spaces of meadows.

While our role and power have changed, our consciousness did not. Subconsciously we still consider the planet inexhaustible and we are still incapable of thinking in timeframes which exceed a few lifetimes. These human limitations hold risks, not only for the planet, nor even for life on this planet, but for our species. Therefore, it is necessary to pay attention not only to our physical environment but also to our cultural and spiritual environment.

It is absolutely necessary to bring up a new generation which no longer shares our deeply rooted subconscious belief in continuous growth: A new generation which no longer desires the forever increasing consumption of space, raw materials, and energy.

It is also necessary to realize that, while as individuals we might not be able to think in longer terms than centuries, as a society we must. This can and must be achieved by developing rules and regulations which are appropriate to the time-frame of the processes which we control or influence. The half-life of plutonium is 24,000 years, the replacement of the water in the deep oceans takes 1000 years. For us it is difficult to be concerned about the consequences of our actions, if those consequences will take centuries or millennia to evolve. Therefore, it is essential that we develop both an educational system and a body of law which would protect our descendants from our own shortsightedness.

Protecting life on this planet will give the coming generations a unifying common purpose. The healing of environmental ills will necessitate changes in our subconscious and in our value system. Once these changes have occurred, they will not only guarantee human survival, but will also help in overcoming human divisions and thereby change human history.

Nature never produces anything that it can not decompose and return into the pool of fresh resources. Man does. Nature returns organic wastes to the soil as fertilizer. Man often dumps such wastes in the oceans, buries them in landfills, or burns them in incinerators. Man's deeply rooted belief in continuous growth treats nature as a commodity, the land, oceans, and atmosphere as free dumps. There is a subconscious assumption that the planet is inexhaustible. In fact the dimensions of the biosphere are fixed and the planet's resources are exhaustible.

In addition to resource depletion and the disposal of toxic, radioactive, and municipal wastes, the natural environment is also under attack from strip mining, clear cutting, noise, and a variety of other human activities. In short, there is a danger of transforming the diverse and stable ecosystem into an unstable one which consists only of man and his chemically sustained food factory.

When man started to supplement his muscle energy with outside sources, these sources were all renewable and inexhaustible. The muscle power of animals, the burning of wood, the use of hydraulic energy were man's external energy sources for millions of years. Only during the last couple of centuries have we started to use exhaustible energy sources, such as coal, oil, gas, and nuclear. This change in energy sources not only resulted in pollution but has also caused uncertainty about our future because we can not be certain if the transition from an exhausted energy source to the next one can be achieved without major disruptions.

Today, as conventional energy use increases, pollution tends to rise exponentially. As the population of the U.S. has increased 50% and our per capita energy consumption has risen 25%, the emission of pollutants has soared by 2000%. While the population of the world doubles in about 50 years, energy consumption doubles in about 20 and electric energy use even faster. In addition to chemical pollution, thermal pollution also rises with fossil energy consumption, because for each unit of electricity generated, two units of heat energy are discharged into the environment.

It is time to redirect our resources from the military—whose job it is to protect dwindling oil resources—and from deep sea drilling—which might cause irreversible harm to the ocean's environment—and use these resources to develop the new, permanent, and inexhaustible energy supplies of the future.

Protecting the global environment, protecting life on this planet, must become a single-minded, unifying goal for all of us. The struggle will overshadow our differences, will give meaning and purpose to our lives and, if we succeed, it will mean survival for our children and the generations to come.

Béla G. Lipták

Contributors

Richard C. Bailie
BSChE, MSChE, PhDChE;
Professor of Chemical Engineering,
West Virginia University

Paul A. Bouis
BSCh, PhDCh; Assistant Director, Research &
Development, Mallinckrodt-Baker, Inc.

Jerry L. Boyd
BSChE; Chief Process Application Engineer, Eimco
Corp.

Mary Anna Evans
BS, MS, PE; Senior Engineer, Water and Air Research,
Inc.

Jess W. Everett
BSE, MS, PhD, PE; Assistant Professor, School of Civil
Engineering and Environmental Engineering, University
of Oklahoma

Lloyd H. Ketchum, Jr.
BSCE, MSE, MPH, PhD, PE; Associate Professor,
Civil Engineering and Geological Sciences,
University of Notre Dame

David H.F. Liu
PhD, ChE; Principal Scientist, J.T. Baker, Inc. a division
of Procter & Gamble

Béla G. Lipták
ME, MME, PE; Process Control and Safety Consultant,
President, Liptak Associates, P.C.

F. Mack Rugg
BA, MSES, JD, Environmental Scientist,
Project Manager, Camp Dresser & McKee Inc.

Michael S. Switzenbaum
BA, MS, PhD; Professor,
Environmental Engineering Program,
Department of Civil and Environmental Engineering,
University of Massachusetts, Amherst

William C. Zegel
ScD, PE, DEE; President, Water and Air Research, Inc.

Contents

10 Treatment and Disposal 212

Hazardous Waste

Paul A. Bouis | Mary A. Evans | Lloyd H. Ketchum, Jr. | David H.F. Liu | William C. Zegel

1

1
Sources and Effects

1.1
HAZARDOUS WASTE DEFINED

Purpose and Scope

Hazardous waste is often defined as waste material that everyone wants picked up but no one wants put down. The legal and scientific definitions have become more complex as more compounds are found and more is learned about the toxicity of compounds and elements. The Resource Conservation and Recovery Act (RCRA) hazardous waste regulations (40 CFR §261 1987) provide the legal definition of hazardous waste. This definition is not always clear because the regulations are written in language general enough to apply to all possible situations, including unusual terminology, several exemptions, and exclusions.

The purpose of this section is to present the various definitions of hazardous waste in a manner useful to the environmental engineer. To be a hazardous waste, material must first conform to the definition of *waste*; second, it must fit the definition of *solid waste*; and third, it must fit the definition of *hazardous waste*. The environmental engineer must test the material against each of these definitions. This section assumes that the generator can demonstrate whether the material is indeed a waste.

Definition of Solid Waste

Solid waste need not literally be a solid. It may be a solid, a semisolid, a liquid, or a contained gaseous material. In accordance with RCRA regulations, a solid waste is any discarded material that is not specifically excluded by the regulation or excluded by granting of a special variance by the regulatory agency. Discarded material is considered abandoned, recycled, or inherently wastelike. Materials are considered abandoned if they are disposed of, burned or incinerated, or accumulated, stored, or treated (but not recycled) before being abandoned.

Materials are considered recycled if they are recycled or accumulated, stored, or treated before recycling. However, materials are considered solid waste if they are used in a manner constituting disposal, burned for energy recovery, reclaimed, or accumulated speculatively. Table 1.1.1 presents various classes of materials and general situations in which they would be considered solid wastes.

Inherently wastelike materials are solid wastes when they are recycled in any manner. This includes:

- Certain wastes associated with the manufacturing of tri-, tetra-, or pentachlorophenols or tetra-, penta-, or hexachlorobenzenes (for listed wastes F020, F021, F022, F023, F026, and F028, see the following section for an explanation of F designations
- Secondary materials that, when fed to a halogen acid furnace, exhibit characteristics of hazardous waste or are listed as hazardous waste
- Other wastes that are ordinarily disposed of, burned, or incinerated
- Materials posing a substantial hazard to human health and the environment when they are recycled.

For a material to be considered recycled and not a solid waste, the material must be used or reused in making a product without reclamation. The material is also considered recycled if it is used as an effective substitute for commercial products or returned to the process from which it was generated without reclamation. In this latter case, the material must be a substitute for raw material feedstock, and the process must use raw materials as its principal feedstocks.

The process for determining whether a waste is a solid waste is summarized in Figure 1.1.1.

Definition of Hazardous Waste

A solid waste is classified as a hazardous waste and is subject to regulation if it meets any of the following four conditions:

The waste is a characteristic hazardous waste, exhibiting any of the four characteristics of a hazardous waste: ignitability, corrosivity, reactivity, or toxicity (see Section 2.1 Hazardous Waste Characterization).

The waste is specifically listed as hazardous in one of the four tables in Part 261, Subpart D of the RCRA regulations: Hazardous Wastes From Nonspecific Sources, Hazardous Wastes From Specific Sources, Acute

TABLE 1.1.1 CONDITIONS UNDER WHICH COMMON MATERIALS ARE SOLID WASTES

Material	Use Constituting Disposal*	Energy Recovery Fuel†	Reclamation‡	Speculative Accumulation§
Spent Materials	Solid Waste	Solid Waste	Solid Waste	Solid Waste
Sludge	Solid Waste	Solid Waste	Solid Waste	Solid Waste
Sludge Exhibiting Characteristics of Hazardous Waste	Solid Waste	Solid Waste	NOT a Solid Waste	Solid Waste
By-products	Solid Waste	Solid Waste	Solid Waste	Solid Waste
By-products Exhibiting Characteristics of Hazardous Waste	Solid Waste	Solid Waste	NOT a Solid Waste	Solid Waste
Commercial Chemical Products	Solid Waste	Solid Waste	NOT a Solid Waste	NOT a Solid Waste
Scrap Metal	Solid Waste	Solid Waste	Solid Waste	Solid Waste

*Use constituting disposal includes application to or placement on the land, and use in the production of (or incorporation in) products that are applied to or placed on the land. Exceptions are made for materials that are applied to the land in ordinary use.

†Energy recovery fuel includes direct burning, use in producing a fuel, and incorporation in a fuel. However, selected commercial chemical products are not solid wastes if their common use is fuel.

‡Reclamation includes materials processed to recover useable products, or regenerated. Examples are recovery of lead from old automobile batteries or used wheel weights and regeneration of spent catalysts or spent solvents.

§Speculative accumulation refers to materials accumulated before the precise mechanism for recycle is known. This designation can be avoided if: the material is potentially recyclable; a feasible means for recycle is available; and during each calendar year the amount of material recycled or transferred to another site for recycling equals at least 75% of the material accumulated at the beginning of the period.

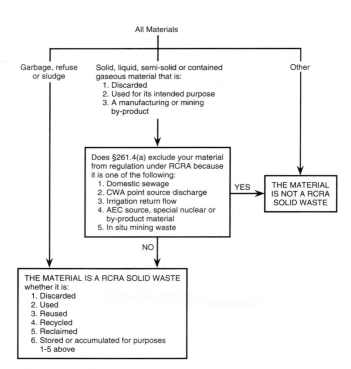

FIG. 1.1.1 Definition of a solid waste.

Hazardous Wastes, or Toxic Wastes.

The waste is a mixture of a listed hazardous waste and a nonhazardous waste.

The waste is declared hazardous by the generator of the waste. This is true even if the waste is not hazardous by any other definition and was declared hazardous in error.

The environmental engineer is referred to Section 261.3 of the RCRA regulations (40 CFR §261.3) for more information on exceptions to these criteria. A hazardous waste must be a solid waste and thus may be in the form of a solid, semisolid, liquid, or contained gas.

The EPA developed *listed wastes* by examining different types of wastes and chemical products to see if they exhibited one of the characteristics of a hazardous waste, then determining whether these met the statutory definition of hazardous waste, were acutely toxic or acutely hazardous, or were otherwise toxic. The following series letters denote the origins of such wastes.

F Series includes hazardous wastes from nonspecific sources (e.g., halogenated solvents, nonhalogenated solvents), electroplating sludges, cyanide solutions from plating batches). These are generic wastes commonly produced by manufacturing and industrial processes.

K Series is composed of hazardous waste from specific sources (e.g., brine purification muds from the mercury cell process in chlorine production where separated, purified brine is not used and API separator sludges). These are wastes from specifically identified industries, such as wood preserving, petroleum refining and organic chemical manufacturing.

P Series denotes acutely hazardous waste of specific commercial chemical products (e.g., potassium silver cyanide, toxaphene, or arsenic oxide) including discarded and off-specification products, containers, and spill residuals.

U Series includes toxic hazardous wastes that are chemical products, (e.g., xylene, DDT, and carbon tetrachloride) including discarded products, off-specification products, containers, and spill residuals.

Acute hazardous wastes are defined as fatal to humans in low doses, or capable of causing or contributing to serious irreversible, or incapacitating reversible illness. They are subject to more rigorous controls than other listed hazardous wastes.

Toxic hazardous wastes are defined as containing chemicals posing substantial hazards to human health or the environment when improperly treated, stored, transported, or disposed of. Scientific studies show that they have toxic, carcinogenic, mutagenic, or teratogenic effects on humans or other life forms.

The environmental engineer needs to understand when a waste becomes a hazardous waste, since this change initiates the regulatory process. A solid waste that is not excluded from regulation (see previous sections) becomes a hazardous waste when any of the following events occur:

- For listed wastes—when the waste first meets the listing description
- For mixtures of solid waste and one or more listed wastes—when a listed waste is first added to the mixture
- For other wastes—when the waste first exhibits any of the four characteristics of a hazardous waste

After a waste is labeled hazardous, it generally remains a hazardous waste forever. Some characteristic hazardous wastes may be declared no longer hazardous if they cease to exhibit any characteristics of a hazardous waste. However, wastes that exhibit a characteristic at the point of generation may still be considered hazardous even if they no longer exhibit the characteristic at the point of land disposal.

Figures 1.1.2 and 1.1.3 summarize the process used to determine whether a solid waste is a hazardous waste and whether it is subject to special provisions for certain hazardous wastes.

EXCLUSIONS

The regulations allow several exemptions and exclusions when determining whether a waste is hazardous. These exclusions center on recycled wastes and several large-vol-

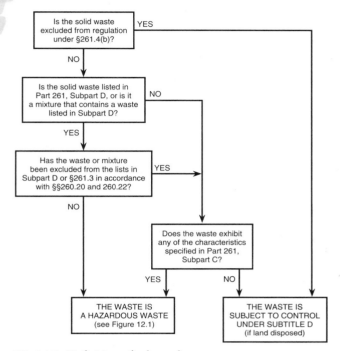

FIG. 1.1.2 Definition of a hazardous waste.

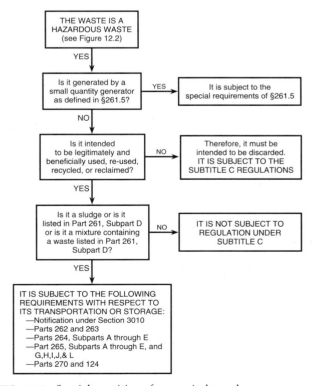

FIG. 1.1.3 Special provisions for certain hazardous waste.

ume or special-interest wastes. Wastes specifically excluded from regulation include industrial wastewater discharges, nuclear materials, fly ash, mining overburden, drilling fluids, and ore processing wastes. A major exemption is also granted to small-quantity generators of hazardous wastes (i.e., those generating less than 100 kg/month [220 lb/month] of hazardous wastes).

The exclusions cover materials that are not solid wastes, solid wastes that are not hazardous wastes, hazardous wastes that are exempt from certain regulations, and samples associated with chemical and physical testing or treatability studies. For regulatory purposes, the following are not considered solid wastes:

Domestic sewage, or any mixture of domestic sewage and other wastes, passing through a sewer system to a publicly-owned treatment works
Industrial wastewater point discharges regulated under Section 402 of the Clean Water Act
Irrigation return flows
Source, special nuclear, or by-product material as defined by the Atomic Energy Act of 1954, as amended
Materials subject to in situ mining techniques but not removed from the ground as part of the extraction process
Pulping liquids that are reclaimed in a pulping liquor recovery furnace and reused in the pulping process
Spent sulfuric acid used to produce virgin sulfuric acid
Secondary materials that are reclaimed and, with certain restrictions, returned to their original generation process(es) and reused in the production process
Spent wood-preserving solutions that are reclaimed and reused for their original intended purpose
Wastewaters from the wood-preserving process that are reclaimed and reused to treat wood
Listed hazardous wastes from coking and coke by-products processes that are hazardous only because they exhibit toxicity characteristics when, after generation, they are (1) recycled to coke ovens, (2) recycled to the tar recovery process as a feedstock to produce coal tar, or (3) mixed with coal tar prior to the tar's sale or refining
Nonwastewater splash condenser dross residue resulting from treating emission control dust and sludge in high-temperature metals-recovery units in primary steel production (a listed waste)

The following solid wastes are not considered hazardous by the RCRA regulations:

Household wastes, including garbage, trash, and sanitary wastes in septic tanks
Solid wastes generated in growing and harvesting agricultural crops or raising animals; this includes animal manures that are returned to the soil as fertilizers
Mining overburden returned to the mine site
Fly ash waste, bottom ash waste, slag waste, and flue gas emission control waste, generated from coal or other fossil fuels combustion

Drilling fluids, produced waters, and other wastes associated with the exploration, development, or production of crude oil, natural gas, or geothermal energy
Waste that could be considered hazardous based on the presence of chromium *if* it can be demonstrated that the chromium is not in the hexavalent state. Such a demonstration is based on information showing only trivalent chromium in the processing and handling of the waste in a non-oxidizing environment, or a specific list of waste sources known to contain only trivalent chromium.
Solid waste from extracting, beneficiating, and processing of ores and minerals
Cement kiln dust waste, unless the kiln is used to burn or process hazardous waste

Before an environmental engineer concludes a company or concern is not subject to regulation under RCRA, the engineer should confirm this conclusion via the RCRA Hotline (1-800-424-9346). Preferably, the decision should also be confirmed by an attorney or other qualified professional familiar with RCRA regulations.

SMALL-QUANTITY GENERATORS (40 CFR §261.5)

A small-quantity generator is conditionally exempt if it generates no more than 100 kg of hazardous waste in a calendar month. In determining the quantity of hazardous waste generated in a month, the generator does not need to include hazardous waste removed from on-site storage, only waste generated that month. Also excluded is waste that is counted more than once. This includes hazardous waste produced by on-site treatment of already-counted hazardous waste, and spent materials that are generated, reclaimed, and subsequently reused on site, so long as such spent materials have been counted once.

The limits on generated quantities of hazardous waste are different for acute hazardous waste (P list). The limit is equal to the total of one kg of acute hazardous waste or a total of 100 kg of any residue or contaminated soil, waste, or other debris resulting from the clean-up of any spilled acute hazardous wastes.

With exceptions, wastes generated by conditionally exempt small-quantity generators are not subject to regulation under several parts of RCRA (Parts 262 through 266, 268, and Parts 270 and 124 of Chapter 2, and the notification requirements of section 3010). The primary exception is compliance with section 262.11, hazardous waste determination. Hazardous wastes subject to these reduced requirements may be mixed with nonhazardous wastes and remain conditionally exempt, even though the mixture exceeds quantity limits. However, if solid waste is mixed with a hazardous waste that exceeds the quantity exclusion level, the mixture is subject to full regulation. If hazardous wastes are mixed with used oil and this mixture is to be burned for energy recovery, the mixture is subject to used oil management standards (Part 279 of RCRA).

RECYCLABLE MATERIALS (40 CFR §261.6)

Recycled hazardous wastes are known as recyclable materials. These materials remain hazardous, and their identification as recyclable materials does not exempt them from regulation. With certain exceptions, recyclable materials are subject to the requirements for generators, transporters, and storage facilities. The exceptions are wastes regulated by other sections of the regulations and wastes that are exempt, including: waste recycled in a manner constituting disposal; waste burned for energy recovery in boilers and industrial furnaces; waste from which precious metals are reclaimed; or spent lead-acid batteries being reclaimed. Wastes generally exempt from regulation are reclaimed industrial ethyl alcohol, used batteries or cells returned to a battery manufacturer for regeneration, scrap metal, and materials generated in a petroleum refining facility. Recycled used oil is subject to used oil management standards (Part 279 of RCRA).

CONTAINER RESIDUE (40 CFR §261.7)

Any hazardous waste remaining in a container or an inner liner removed from an empty container is not subject to regulation. The problem is determining whether a container is empty or not. RCRA regulations consider a container empty when all possible wastes are removed using common methods for that type of container, and no more than an inch (2.5 cm) of residue remains on the bottom of the container or liner. Alternately, a container with a volume of 110 gal or less can be considered empty if no more than 3% of the capacity, by weight, remains in the container or liner. Larger containers are considered empty when no more than 0.3% of capacity, by weight, remains in the container or liner. If the material in the container was a compressed gas, the container is considered empty when its pressure is reduced to atmospheric pressure.

Regarding acute hazardous waste (P list), the test for an empty container is much more stringent. The container or inner liner must be triple-rinsed using a solvent capable of removing the commercial chemical product or manufacturing chemical intermediate. Alternative cleaning methods can be used if they are demonstrated to be equivalent to or better than triple rinsing. Of course, a container can also be considered empty if a contaminated liner is removed.

—*Mary A. Evans*
William C. Zegel

References

Code of Federal Regulations. (1 July 1987): Title 40, sec. 261.
U.S. Environmental Protection Agency (EPA). 1986. *RCRA orientation manual."* Office of Solid Waste, Washington, D.C.

1.2
HAZARDOUS WASTE SOURCES

The *reported* quantities of hazardous waste generated in the U.S. remained in the range of 250–270 million metric tn per year through most of the 1980s. Figure 1.2.1 indicates which industrial sectors generate these wastes. The majority of hazardous waste is generated by the chemical manufacturing, petroleum, and coal processing industries. As Figure 1.2.2 shows, waste generation is not broadly distributed throughout these industries; instead, a few dozen facilities account for most waste generation. While it is striking that a few dozen manufacturing facilities generate most of the country's hazardous wastes, these waste generation rates must be viewed in context. Figure 1.2.3 shows that 250–270 million tn of hazardous waste generated annually are over 90% wastewater. Thus, the rate of generation of hazardous constituents in the waste is probably on

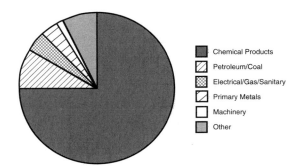

FIG. 1.2.1. Hazardous waste generation in 1986, classified by industry sector. (Reprinted from U.S. Environmental Protection Agency (EPA), 1988, *1986 national survey of hazardous waste treatment, storage, disposal and recycle facilities,* EPA/530-SW-88/035.)

the order of 10 to 100 million tons per year. In relation to the 300+ million tons of commodity chemicals produced annually and the 1000 million tons of petroleum refined annually (C&E News 1991), the mass of hazardous constituents in waste is probably less than 5% of all chemical production.

Examples of basic industries and types of hazardous wastes produced are listed in Table 1.2.1, illustrating the

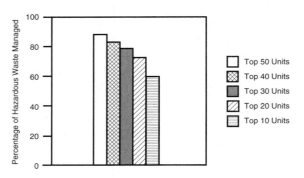

FIG. 1.2.2 Percentages of hazardous waste managed in the 50 largest facilities in 1986. (Reprinted from U.S. EPA, 1988.)

wide range and complexity of the wastes. However, these few examples do not adequately suggest the numbers and kinds of hazardous chemical constituents in hazardous wastes to be managed. There are approximately 750 listed wastes in 40 CFR Part 261, and countless more characteristic wastes. The intensity of industrial competition constantly engenders the introduction of new products, thus wastes are generated at an awesome pace.

Hazardous Waste from Specific Sources (40 CFR §261.32)

The following solid wastes are listed as hazardous wastes from a specific source unless they meet an exclusion. Except for K044, K045, and K047, which are reactive wastes, they are toxic wastes.

WOOD PRESERVATION

Bottom sediment sludge from wastewater treatment in wood-preserving processes using creosote or pentachlorophenol (K001) is a hazardous waste.

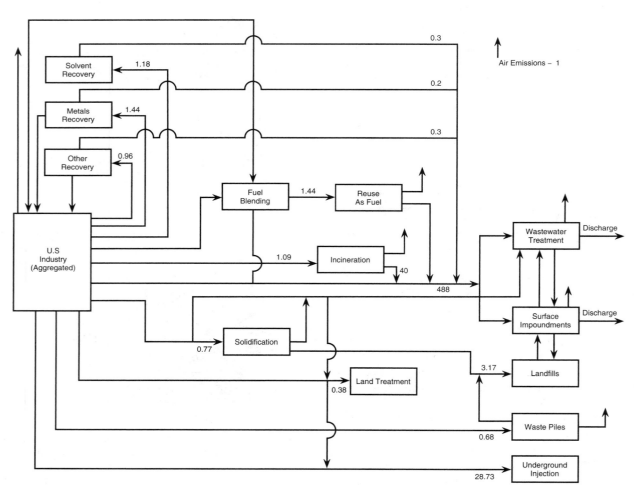

FIG. 1.2.3 Flow of industrial hazardous waste treatment operations (1986 data in tn per yr).

TABLE 1.2.1   TYPES OF HAZARDOUS WASTE

Industry	*Wastes Produced*

Chemical Manufacturing	• Spent solvents and still bottoms
	   White spirits, kerosene, benzene, xylene, ethyl benzene, toluene, isopropanol, toluene diisocyanate, ethanol, acetone, methyl ethyl ketone, tetrahydrofuran, methylene chloride, 1,1,1-trichloroethane, trichloroethylene
	• Ignitable wastes not otherwise specified (NOS)
	• Strong acid/alkaline wastes
	   Ammonium hydroxide, hydrobromic acid, hydrochloric acid, potassium hydroxide, nitric acid, sulfuric acid, chromic acid, phosphoric acid
	• Other reactive wastes
	   Sodium permanganate, organic peroxides, sodium perchlorate, potassium perchlorate, potassium permanganate, hypochlorite, potassium sulfide, sodium sulfide
	• Emission control dusts and sludges
	• Spent catalysts

Construction	• Ignitable paint wastes
	   Ethylene dichloride, benzene, toluene, ethyl benzene, methyl isobutyl ketone, methyl ethyl ketone, chlorobenzene
	• Ignitable wastes not otherwise specified (NOS)
	• Spent solvents
	   Methyl chloride, carbon tetrachloride, trichlorotrifluoroethane, toluene, xylene, kerosene, mineral spirits, acetone
	• Strong acid/alkaline wastes
	   Ammonium hydroxide, hydrobromic acid, hydrochloric acid, hydrofluoric acid, nitric acid, phosphoric acid, potassium hydroxide, sodium hydroxide, sulfuric acid

Metal Manufacturing	• Spent solvents and solvent still bottoms
	   Tetrachloroethylene, trichloroethylene, methylene chloride, 1,1,1-trichloroethane, carbon tetrachloride, toluene, benzene, trichlorofluoroethane, chloroform, trichlorofluoromethane, acetone, dichlorobenze, xylene, kerosene, white spirits, butyl alcohol
	• Strong acid/alkaline wastes
	   Ammonium hydroxide, hydrobromic acid, hydrochloric acid, hydrofluoric acid, nitric acid, phosphoric acid, nitrates, potassium hydroxide, sodium hydroxide, sulfuric acid, perchloric acid, acetic acid
	• Spent plating wastes
	• Heavy metal wastewater sludges
	• Cyanide wastes
	• Ignitable wastes not otherwise specified (NOS)
	• Other reactive wastes
	   Acetyl chloride, chromic acid, sulfides, hypochlorites, organic peroxides, perchlorates, permanganates
	• Used oils

Paper Industry	• Halogenated solvents
	   Carbon tetrachloride, methylene chloride, tetrachloroethylene, trichloroethylene, 1,1,1-trichloroethane, mixed spent halogenated solvents
	• Corrosive wastes
	   Corrosive liquids, corrosive solids, ammonium hydroxide, hydrobromic acid, hydrochloric acid, hydrofluoric acid, nitric acid, phosphoric acid, potassium hydroxide, sodium hydroxide, sulfuric acid
	• Paint wastes
	   Combustible liquid, flammable liquid, ethylene dichloride, chlorobenzene, methyl ethyl ketone, paint waste with heavy metals
	• Solvents
	   Petroleum distillates

Source: Reprinted from U.S. Environmental Protection Agency (EPA), *Does your business produce hazardous wastes?* (Office of Solid Waste and Emergency Response, (EPA/530-SW-010, Washington, D.C.)

INORGANIC PIGMENTS

Hazardous wastes include wastewater treatment sludge from the production of various metal-based pigments: chrome yellow and orange (K002), molybdate orange (K003), zinc yellow (K004), chrome green from the solvent recovery column in the production of toluene diiosocyanate via phosgenation of toluenediamine (K005), anhydrous and hydrated chrome-oxide green (K006), iron blue (K008), and oven residue from the production of chrome-oxide green (K008).

ORGANIC CHEMICALS

Numerous hazardous wastes occur in organic chemical production facilities. In the production of acetaldehyde from ethylene, distillation bottoms (K009) and distillation side cuts (K010) are hazardous wastes. In acrylonitrile production, the bottom streams from the wastewater stripper (K011), the acetonitrile column (K013), and the acetonitrile purification column (K014) are hazardous wastes. In 1,1,1-trichlorethane production, hazardous wastes include spent catalyst from the hydrochlorinator reactor (K028), waste from the product steam stripper (K029), distillation bottoms (K095), and heavy ends from the heavy end column (K096).

In the production of toluenediamine via hydrogenation of dinitrotoluene, hazardous wastes are generated in reaction by-product water from the drying column (K112) and condensed liquid light ends (K113), vicinals (K114), and heavy ends (K115) from the purification of toluenediamine.

In the production of ethylene dibromide via bromination of ethylene, hazardous wastes result from reactor vent gas scrubber wastewater (K117), spent adsorbent solids (K118), and still bottoms (K136) from purification.

Hazardous wastes are found in heavy ends or still bottoms from benzyl chloride distillation (K015), ethylene dichloride in ethylene dichloride production (K019), and vinyl chloride in vinyl chloride monomer production (K020). Heavy ends or distillation residues from carbon tetrachloride production (K016); the purification column in the production of epichlorohydrin (K017); the fractionation column in ethyl chloride production (K018); the production of phenol/acetone from cumene (K022); the production of phthalic anhydride from naphthalene (K024); the production of phthalic anhydride from ortho-xylene (K094); the production of nitro-benzene by the nitration of benzene (K025); the combined production of trichloroethylene and perchloroethylene (K030); the production of aniline (K083); and the production of chlorobenzenes (K085) are also hazardous wastes.

Other sources of hazardous wastes include distillation light ends from the production of phthalic anhydride from ortho-xylene (K093) or naphthalene (K024); aqueous spent antimony catalyst waste from fluoromethanes production (K021); stripping still tails from the production of methyl ethyl pyridines (K026); centrifuge and distillation residues from toluene diisocyanate production (K027); process residues from aniline extraction in aniline production (K103); combined wastewater streams generated from nitrobenzene/aniline production (K104); the separated aqueous stream from the reactor product washing step in the production of chlorobenzenes (K105); and the organic condensate from the solvent recovery column in the production of toluene diisocyanate via phosgenation of toluenediamine.

INORGANIC CHEMICALS

Chlorinated hydrocarbon waste from the purification step of the diaphragm cell process using graphite anodes (K073); wastewater treatment sludge from the mercury cell process (K106); and brine purification muds from the mercury cell process where separately prepurified brine is not used (K071) are hazardous wastes related to the production of chlorine.

PESTICIDES

Hazardous wastes are generated in the production of nine pesticides: MSMA and cacodylic acid, chlordane, creosote, disulfoton, phorate, toxaphene, 2,4,5–T, 2,4–D, and ethylenebisdithiocarbamic acid and its salts. In MSMA and cacodylic acid production, hazardous waste is generated as by-product salts (K031). In chlordane production, hazardous wastes include: wastewater treatment sludge (K032); wastewater and scrub water from the chlorination of cyclopentadiene (K033); filter solids from the filtration of hexachlorocyclopentadiene (K034); and vacuum stripper discharge from the chlordane chlorinator (K097). Wastewater treatment sludges generated in creosote production (K035) are also defined as hazardous waste. Hazardous wastes from the production of disulfoton are still bottoms from toluene reclamation distillation (K036), and wastewater treatment sludges (K037). Phorate production generates hazardous wastes from washing and stripping wastewater (K038), wastewater treatment sludge (K040), and filter cake from filtration of diethylphosphorodithioic acid (K039).

Wastewater treatment sludge (K041) and untreated process wastewater (K098) from toxaphene production and heavy ends, or distillation residues from tetrachlorobenzene in 2,4,5–T production (K042) are hazardous wastes. Similarly, 2,6–dichlorophenol waste (K043) and untreated wastewater (K099) from 2,4–D production are hazardous wastes.

Hazardous wastes from the production of ethylenebisdithiocarbamic acid and its salts are: process wastewaters (including supernates, filtrates, and washwaters) (K123); reactor vent scrubber water (K124); filtration, evaporation, and centrifugation solids (K125); and baghouse dust and floor sweepings in milling and packaging operations (K126).

EXPLOSIVES

Hazardous wastes from explosives production include: wastewater treatment sludges from manufacturing and processing explosives (K044) and manufacturing, formulation, and loading lead-based initiating compounds (K046); pink or red water from TNT operations (K047); and spent carbon from the treatment of wastewater-containing explosives (K045).

PETROLEUM REFINING

Dissolved air flotation (DAF) float (K048), slop oil emulsion solids (K049), heat exchanger bundle cleaning sludge (K050), API separator sludge (K051), and tank bottoms from storage of leaded fuel (K052) are hazardous wastes.

IRON AND STEEL

Emission control dust and sludges from primary steel production in electric furnaces (K061) and spent pickle liquor generated in steel finishing operations (K062) are hazardous wastes.

SECONDARY LEAD

Emission control dust and sludge (K069) and waste solution from acid leaching of emission control dust and sludge (K100) are hazardous wastes.

VETERINARY PHARMACEUTICALS

Wastewater treatment sludges generated in the production of veterinary pharmaceuticals from arsenic or organo-ar-

senic compounds (K084), distillation tar residues from the distillation of aniline-based compounds (K101), and residue from the use of activated carbon for decolorization (K102) are hazardous wastes.

INK FORMULATION

Solvent washes and sludges, caustic washes and sludges, or water washes and sludges from cleaning tubs and equipment used in ink formulation from pigments, driers, soaps, and stabilizers containing chromium and lead are hazardous wastes (K086).

COKING

Ammonia still lime sludge (K060) and decanter tank tar sludge (K087) are hazardous wastes.

Hazardous Wastes from Nonspecific Sources (40 CFR §261.31)

Hazardous wastes are also generated from nonspecific sources, depending upon the type of waste. Table 1.2.1 lists a number of these categories, although it is by no means an exhaustive listing.

—Mary A. Evans
William C. Zegel

Reference

Code of Federal Regulations. (1 July 1981): Title 40, sec. 261.3.

1.3
EFFECTS OF HAZARDOUS WASTE

It is virtually impossible to describe a "typical" hazardous waste site, as they are extremely diverse. Many are municipal or industrial landfills. Others are manufacturing plants where operators improperly disposed of wastes. Some are large federal facilities dotted with contamination from various high-tech or military activities.

While many sites are now abandoned, some sites are partially closed down or still in active operation. Sites range dramatically in size, from quarter-acre metal plating shops to 250-sq mi mining areas. The wastes they contain vary widely, too. Chief constituents of wastes in solid, liquid, and sludge forms include heavy metal, a common by-product of electroplating operations, and solvents or degreasing agents.

Human Health Hazards

Possible effects on human and environmental health also span a broad spectrum. The nearly uninhibited movement, activity, and reactivity of hazardous chemicals in the atmosphere are well established, and movement from one medium to another is evident. Hazardous wastes may enter the body through ingestion, inhalation, dermal absorption, or puncture wounds.

Human health hazards occur because of the chemical and physical nature of the waste, and its concentration and quantity; the impact also depends on the duration of exposure. Adverse effects on humans range from minor temporary physical irritation, dizziness, headaches, and nausea to long-term disorders, cancer or death. For example, the organic solvent carbon tetrachloride (CCl_4) is a central nerve system depressant as well as an irritant and can cause irreversible liver or kidney damage. Table 1.3.1 shows the potential effects of selected hazardous substances.

Site Safety

Transportation spills and other industrial process or storage accidents account for some hazardous waste releases. Such releases can result in fires, explosions, toxic vapors, and contamination of groundwater used for drinking.

Danger arises from improper handling, storage, and disposal practices (refer to Section 4.1 on Treatment, Storage, and Disposal Requirements). At hazardous waste sites, fires and explosions may result from investigative or remedial activities such as mixing incompatible contents of drums or from introduction of an ignition source, such as a spark from equipment.

A site safety plan is needed to establish policies and procedures for protecting workers and personnel during clean-up and day-to-day waste-handling activities. The minimum contents of a site safety plan are listed in Table 1.3.2.

TABLE 1.3.1 HEALTH EFFECTS OF SELECTED HAZARDOUS SUBSTANCES

Chemical	Source	Health Effects
Pesticides		
DDT	Insecticides	Cancer; damage to liver, embryos, bird eggs
BHC	Insecticides	Cancer, embryo damage
Petrochemicals		
BENZENE	Solvents, pharmaceuticals and detergents	Headaches, nausea, loss of muscle coordination, leukemia, damage to bone marrow
VINYL CHLORIDE	Plastics	Lung and liver cancer, depression of central nervous system, suspected embryotoxin
Other Organic Chemicals		
DIOXIN	Herbicides, waste incineration	Cancer, birth defects, skin disease
PCBs	Electronics, hydraulic fluid, fluorescent lights	Skin damage, possible gastro-intestinal damage, possibly cancer-causing
Heavy Metals		
LEAD	Paint, gasoline	Neurotoxic; causes headaches, irritability, mental impairment in children; brain, liver, and kidney damage
CADMIUM	Zinc, batteries, fertilizer	Cancer in animals, damage to liver and kidneys

Source: World Resources Institute and International Institute for Environment and Development, 1987; *World Resources 1987,* (New York, N.Y.: Basic Books, pp. 205–06.

TABLE 1.3.2 SITE SAFETY PLANS

- Name key personnel and alternates responsible for site safety.
- Describe the risks associated with each operation conducted.
- Confirm that personnel are adequately trained to perform their job responsibilities and to handle the specific hazardous situations they may encounter.
- Describe the protective clothing and equipment to be worn by personnel during various site operations.
- Describe any site-specific medical surveillance requirements.
- Describe the program for periodic air monitoring, personnel monitoring, and environmental sampling, if needed.
- Describe the actions to be taken to mitigate existing hazards (e.g., containment of contaminated materials) to make the work environment less hazardous.
- Define site control measures and include a site map.
- Establish decontamination procedures for personnel and equipment.
- Set forth the site's standard operating procedures for those activities that can be standardized, and where a checklist can be used.
- Set forth a contingency plan for safe and effective response to emergencies.

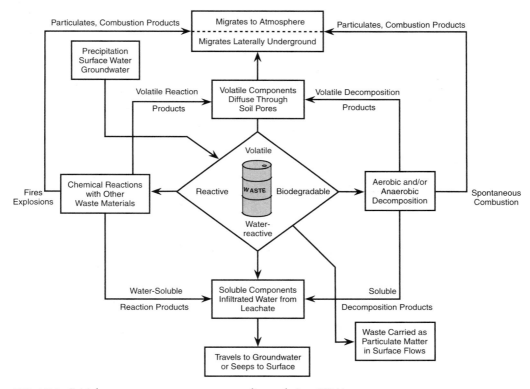

FIG. 1.3.1 Initial transport processes at waste disposal sites (EPA).

TABLE 1.3.3 ENVIRONMENTAL PERFORMANCE GUIDELINES

Prevention of adverse effects on air quality considering
1. Volume and physical and chemical characteristics of facility waste, including potential for volatilization and wind dispersal
2. Existing quality of the air, including other sources of contamination and their cumulative impact on the air
3. Potential for health risks caused by human exposure to waste constituents
4. Potential damage to wildlife, crops, vegetation, and physical structures caused by exposure to waste constituents
5. Persistence and permanence of the potential adverse effects

Prevention of adverse effects on surface water quality considering
1. Volume and physical and chemical characteristics of facility waste
2. Hydrogeological characteristics of the facility and surrounding land, including topography of the area around the facility
3. Quantity, quality, and directions of groundwater flow
4. Patterns of rainfall in the region
5. Proximity of facility to surface waters
6. Uses of nearby surface waters and any water quality standards established for those surface waters
7. Existing quality of surface water, including other sources of contamination and their cumulative impact on surface water
8. Potential for health risks caused by human exposure to waste constituents
9. Potential damage to wildlife, crops, vegetation, and physical structures caused by exposure to waste constituents
10. Persistence and permanence of the potential adverse effects

Prevention of adverse effects on groundwater quality considering
1. Volume and physical and chemical characteristics of the waste in the facility, including its potential for migration through soil or through synthetic liner materials
2. Geologic characteristics of the facility and surrounding land
3. Patterns of land use in the region
4. Potential for migration of waste constituents into subsurface physical structures
5. Potential for migration of waste constituents into the root zone of food-chain crops and other vegetation
6. Potential for health risks through human exposure to waste constituents
7. Potential damage to wildlife, crops, vegetation, and physical structures through exposure to waste constituents
8. Persistence and permanence of potential adverse effects

Environmental Contamination

Hazardous waste disposers need to understand the potential toxic effects of these wastes and realize how strictly the wastes must be contained. Dangerous chemicals often migrate from uncontrolled sites, percolating from holding ponds and pits into underlying groundwater, then flowing into lakes, streams, and wetlands. Produce and livestock in turn become contaminated, then enter the food chain. Hazardous chemicals then build up, or bioaccumulate, when plants, animals, and people consume contaminated food and water.

Most groundwater originates as surface water. Great quantities of land-deposited hazardous wastes evaporate into the atmosphere, runoff to surface waters, then percolate to groundwaters (Figure 1.3.1). Atmospheric and surface water waste releases commingle with other releases or are lost to natural processes, but groundwater contamination may remain highly concentrated, relatively lo-

calized, and persistent for decades or centuries. Although current quantities of waste are being reduced, any additional releases together with previously released materials will continue contaminating aquifers in many areas, and many groundwater supplies are now impaired.

Table 1.3.3 presents EPA guidelines for hazardous handling facilities performance with respect to human health and the environment.

—*David H.F. Liu*

References

U.S. Environmental Protection Agency (U.S. EPA). 1981. *Interim standard for owners and operators of new hazardous waste land disposal facilities*. Code of Federal Regulations. Title 40, Part 267. Washington, D.C.: U.S. Government Printing Office.
———. 1985. *Protecting health and safety at hazardous waste sites: an overview*. Technology Transfer, EPA 625/9–25/006, Cincinnati, OH.

2
Characterization, Sampling, and Analysis

2.1
HAZARDOUS WASTE CHARACTERIZATION

Criteria

The EPA applies two criteria in selecting four characteristics as inherently hazardous in any substance:

The characteristics must be listed in terms of physical, chemical, or other properties causing the waste to meet the definition of a hazardous waste in the act; and

The properties defining the characteristics must be measurable by standardized, available testing protocols.

The second criterion was adopted because generators have the primary responsibility for determining whether a solid waste exhibits any of the characteristics. EPA regulation writers believed that unless generators were provided with widely available and uncomplicated methods for determining whether their wastes exhibited the characteristics, the identification system would not work (U.S. EPA 1990).

Because of this second criterion, the EPA did not add carcinogenicity, mutagenicity, bioaccumulation potential, or phytotoxicity to the characteristics. The EPA considered the available protocols for measuring these characteristics either insufficiently developed, too complex, or too highly dependent on skilled personnel and professional equipment. In addition, given the current knowledge of such characteristics, the EPA could not confidently define the numerical threshold levels where characteristic wastes would present a substantial hazard (U.S. EPA 1990).

Characteristics

As testing protocols become accepted and confidence in setting minimum thresholds increases, more characteristics may be added. To date, waste properties exhibiting any or all of the existing characteristics are defined in 40 CFR §261.20–261.24.

CHARACTERISTIC OF IGNITABILITY

Ignitability is the characteristic used to define as hazardous those wastes that could cause a fire during transport, storage, or disposal. Examples of ignitable wastes include waste oils and used solvents.

A waste exhibits the characteristics of ignitability if a representative sample of the waste has any of the following properties:

1. It is a liquid, other than an aqueous solution containing less than 24% alcohol by volume, and has flash point less than 60°C (140°F), as determined by a Pensky-Martens Closed Cup Tester (using the test method specified in ASTM Standard D-93-79 or D-93-80) or by a Setaflash Closed Cup Tester (using the test method specified in ASTM Standard D-3278-78).
2. It is not a liquid and is capable, under standard temperature and pressure, of causing fire through friction, absorption of moisture, or spontaneous chemical changes and, when ignited, burns so vigorously and persistently that it creates a hazard.
3. It is an ignitable compressed gas as defined in the 49 Code of Federal Regulations 173.300 DOT regulations.
4. It is an oxidizer as defined in the 49 Code of Federal Regulations 173.151 DOT regulations.

A waste that exhibits the characteristic of ignitability but is not listed as a hazardous waste in Subpart D of RCRA has the EPA hazardous waste number of D001.

CHARACTERISTIC OF CORROSIVITY

Corrosivity, as indicated by pH, was chosen as an identifying characteristic of a hazardous waste because wastes with high or low pH can react dangerously with other wastes or cause toxic contaminants to migrate from certain wastes. Examples of corrosive wastes include acidic wastes and used pickle liquor from steel manufacture. Steel corrosion is a prime indicator of a hazardous waste since wastes capable of corroding steel can escape from drums and liberate other wastes.

A waste exhibits the characteristic of corrosivity if a representative sample of the waste has either of the following properties:

1. It is aqueous and has a pH less than or equal to 2 or greater than or equal to 11.5, as determined by a pH meter using an EPA test method. The EPA test method for pH is specified as Method 5.2 in "Test Methods for

the Evaluation of Solid Waste, Physical/Chemical Methods."

2. It is a liquid and corrodes steel (SAE 1020) at a rate greater than 6.35 mm (0.250 inch) per year at a test temperature of 55°C (130°F), as determined by the test method specified in NACE (National Association of Corrosion Engineers) Standard TM-01-69 and standardized in "Test Methods for the Evaluation of Solid Waste, Physical/Chemical Methods."

A waste that exhibits the characteristic of corrosivity but is not listed as a hazardous waste in Subpart D has the EPA hazardous waste number of D002.

CHARACTERISTIC OF REACTIVITY

Reactivity was chosen as an identifying characteristic of a hazardous waste because unstable wastes can pose an explosive problem at any stage of the waste management cycle. Examples of reactive wastes include water from TNT operations and used cyanide solvents.

A waste exhibits the characteristic of reactivity if a representative sample of the waste has any of the following properties:

1. It is normally unstable and readily undergoes violent change without detonating.
2. It reacts violently with water.
3. It forms potentially explosive mixtures with water.
4. When mixed with water, it generates toxic gases, vapors, or fumes in a quantity sufficient to present a danger to human health or the environment.
5. It is a cyanide- or sulfide-bearing waste which, when exposed to pH conditions between 2 and 11.5, can generate toxic gases, vapors, or fumes in a quantity sufficient to present a danger to human health or the environment.
6. It is capable of detonation or explosive reaction if subjected to a strong initiating source or if heated under confinement.
7. It is readily capable of detonation or explosive decomposition or reaction at standard temperature and pressure.
8. It is a forbidden explosive as defined in the 49 Code of Federal Regulations 173.51, or a Class A explosive as defined in the 49 Code of Federal Regulations 173.53, or a Class B explosive as defined in the 49 Code of Federal Regulations 173.88 DOT regulations.

A waste that exhibits the characteristic of reactivity but is not listed as a hazardous waste in Subpart D has the EPA hazardous waste number of D003.

TABLE 2.1.1   MAXIMUM CONCENTRATION OF CONTAMINANTS FOR RCRA TOXICITY CHARACTERISTICS

EPA Hazardous Waste Number	Contaminant	Maximum Concentration (mg/L)	EPA Hazardous Waste Number	Contaminant	Maximum Concentration (mg/L)
D004	Arsenic[a]	5.0	D036	Hexachloro-1,3-butadiene	0.5
D005	Barium[a]	100.0			
D019	Benzene	0.5	D037	Hexachloroethane	3.0
D006	Cadmium[a]	1.0	D008	Lead[a]	5.0
D022	Carbon tetrachloride	0.5	D013	Lidane[a]	0.4
D023	Chlordane	0.03	D009	Mercury[a]	0.2
D024	Chlorobenzene	100.0	D014	Methoxychlor[a]	10.0
D025	Chloroform	6.0	D040	Methyl ethyl ketone	200.0
D007	Chromium	5.0	D041	Nitrobenzene	2.0
D026	o-Cresol	200.0	D042	Pentachlorophenol	100.0
D027	m-Cresol	200.0	D044	Pyridine	5.0
D028	p-Cresol	200.0	D010	Selenium	1.0
D016	2,4-D[a]	10.0	D011	Silver[a]	5.0
D030	1,4-Dichlorobenzene	7.5	D047	Tetrachloroethylene	0.7
			D015	Toxaphene[a]	0.5
D031	1,2-Dichloroethane	0.5	D052	Trichloroethylene	0.5
D032	1,1-Dichloroethylene	0.7	D053	2,4,5-Trichlorophenol	400.0
D033	2,4-Dinitrotoluene	0.13	D054	2,4,6-Trichlorophenol	2.0
D012	Endrin[a]	0.02			
D034	Heptachlor (and its hydroxide)	0.008	D017	2,4,5-TP (Silvex)[a]	1.0
			D055	Vinyl chloride	0.2
D035	Hexachlorobenzene	0.13			

[a]Formerly EP Toxicity Contaminants.
Source: Code of Federal Regulations, Title 40, sec. 261.24.

CHARACTERISTIC OF TOXICITY

The test, toxicity characteristic leaching procedure (TCLP), is designed to identify wastes likely to leach hazardous concentrations of particular toxic constitutents into the groundwater as a result of improper management. During the TCLP, constituents are extracted from the waste to stimulate the leaching actions that occur in landfills. If the concentration of the toxic constituent exceeds the regulatory limit, the waste is classified as hazardous.

If the extract from a representative waste sample contains any of the contaminants listed in Table 2.1.1 at a concentration equal to or greater than the respective value given, the waste exhibits the toxicity characteristic. Where the waste contains less than 0.5 percent filterable solids, the waste itself is considered to be the extract. A waste that exhibits the toxicity characteristic but is not a listed hazardous waste has the EPA hazardous waste number specified in Table 2.1.1. The TCLP test replaced the EP toxicity test in September 1990 and added 25 organic compounds to the eight metals and six pesticides that were subject to the EP toxicity test.

Specific Compounds

Information about waste is needed to evaluate the health effects, determine the best method of handling, and evaluate methods of storage, treatment or disposal. Items of interest include:

- Physical properties such as density or viscosity
- Toxicity in water
- Permissible exposure limits (PELs) in the air
- Health hazards
- Precautions
- Controls
- Emergency and first aid procedures
- Disposal methods

There are a number of references that define the properties of specific compounds (Sax 1984, Sittig 1985, Weiss 1986), however, no current source defines the impact of hazardous mixtures.

—*David H.F. Liu*

References

Sax, N. 1984. *Dangerous properties of hazardous materials.* 6th ed. New York, N.Y.: Van Nostrand Reinhold.

Sittig, M. 1985. *Handbook of toxic and hazardous chemicals and carcinogens.* 2d ed. Park Ridge, N.J.: Noyes Publications.

U.S. Environmental Protection Agency (EPA). 1990. *RCRA orientation manual.* Office of Solid Waste. Washington, D.C.

Weiss, G. 1986. *Hazardous chemical data book.* 2d ed. Park Ridge, N.J.: Noyes Publications.

2.2
SAMPLING AND ANALYSIS

Safety and data quality are the two major concerns when sampling hazardous waste. Where environmental data are collected, quality assurance provides the means to determine data quality. This entails planning, documentation and records, audits, and inspections. Data quality is known when there are verifiable and defensible documentation and records associated with sample collection, transportation, sample preservation and analysis, and other management activities.

Sampling Equipment and Procedures

SAFETY

Samples must be secured to ensure the safety of the sampler, all others working in the area, and the surroundings.

If the source and nature of the hazardous waste are known, the sampler should study the properties of the material to determine the necessary safety precautions, including protective clothing and special handling precautions.

If the nature of the hazardous waste is unknown, such as at an abandoned waste disposal site, then the sampler should take additional precautions to prevent direct contact with the hazardous waste. Stored, abandoned, or suspect waste will often be containerized in drums and tanks. Such containers and materials buried under abandoned waste sites pose special safety problems (De Vera, Simmons, Stephens, Storn, 1980; EPA 1985). Care must be exercised in opening drums or tanks to prevent sudden releases of pressurized materials, fire, explosions, or spillage.

SAMPLING EQUIPMENT

Drums should be opened using a spark-proof brass bung wrench. Drums with bulged heads are particularly dangerous. The bulge indicates that the contents are under extreme pressure. To sample a bulged drum, a remotely operated drum opening device should be used, enabling the sampler to open the drum from a safe distance. Such operations should be carried out only by fully trained technicians in full personnel protective gear.

Liquid waste in tanks must be sampled in a manner that represents the contents of the tank. The EPA specifies that the *colawassa* sampler is used for such sampling. The colawassa is a long tube with a stopper at the bottom that opens or closes using the handle at the top. This device enables the sampler to retrieve representative material at any depth within the tank. The colawassa has many shortcomings, including the need for completely cleaning it and removal of all residues between each sampling. This is difficult, and it also creates another batch of hazardous waste to be managed.

A glass colawassa, which eliminates sample contamination by metals and stopper materials, is available through technical and scientific supply houses. In most situations, ordinary glass tubing can be used to obtain a representative sample, and can be discarded after use.

Bomb samplers that are lowered into a liquid waste container, then opened at the selected depth, are also useful in special situations.

Long-handled dippers can be used to sample ponds, impoundments, large open tanks, or sumps: however these devices cannot cope with stratified materials. Makeshift devices using tape or other porous or organic materials introduce the likelihood of sample contamination.

Dry solid samples may be obtained using a thief or trier, or an augur or dipper. Sampling of process units, liquid discharges, and atmospheric emissions all require specialized equipment training.

The EPA has published several guidance documents detailing hazardous waste, soil, surface water and groundwater and waste stream sampling (EPA 1985a, 1985b; De Vera et al. 1980; Evans and Schweitzer 1984).

Procedures used or materials contacting the sample should not cause gain or loss of pollutants. Sampling equipment and sample containers must be fabricated from inert materials and must be thoroughly cleaned before use. Equipment that comes into contact with samples to be analyzed for organic compounds should be fabricated of (in order of preference):

- Glass (amber glass for organics; clear glass for metals, oil, cyanide, BOD, TOC, COD, sludges, soil, and solids, and others)
- Teflon (Teflon lid liners should be inserted in caps to prevent contamination normally supplied with bottles)
- Stainless steel
- High-grade carbon steel
- Polypropylene
- Polyethylene (for common ions, such as fluoride, chloride, and sulfate)

Classic commercial analytic schedules require a sample of more than 1,500 ml. Commercial field samplers collect samples of 500 to 1,000 ml. If such volumes are insufficient, multibottle samples can be collected. Special containers may be designed to prolong sample duration.

PROCEDURES

Representative samples should be obtained to determine the nature of wastes.

If the waste is in liquid form in drums, it should be completely mixed (if this is safe) before sampling, and an aliquot should be taken from each container. Within a group of drums containing similar waste, random sampling of 20% of the drums is sufficient to characterize the wastes. If the sampler is unsure of the drum contents, each must be sampled and analyzed.

If the waste source is a manufacturing or waste treatment process solid, composite sampling and analysis are recommended. In such cases, an aliquot is periodically collected, composited, and analyzed.

If the solid waste is in a lagoon, abandoned disposal facility, tank, or similar facility, three-dimensional sampling is recommended. Although samples collected three-dimensionally are sometimes composited, they are usually analyzed individually. This process characterizes the solid waste and aids in determining whether the entire quantity of material is hazardous.

If the source and nature of the material is known, sampling and analysis are limited to the parameters of concern. When the waste is unknown, a full analysis for 129 priority pollutants is often required.

SAMPLE PRESERVATION

Aqueous samples are susceptible to rapid chemical and physical reactions between the sampling time and analysis. Since the time between sampling and analysis could be greater than 24 hours, the following preservation techniques are recommended to avoid sample changes resulting in errors: all samples except metals must be refrigerated. Refrigeration of samples to 4°C is common in fieldwork, and helps stabilize samples by reducing biological and chemical activity (EPA 1979).

In addition to refrigeration, specific techniques are required for certain parameters (see section 10.9). The preservation technique for metals is the addition of nitric acid (diluted 1:1) to adjust the pH to less than 2, which will stabilize the sample up to 6 months; for cyanide, the addition of 6N caustic will adjust the pH to greater than 12, and refrigeration to 4°C, which will stabilize the sample for up to 14 days. Little other preservation can be performed on solid samples.

Quality Assurance and Quality Control

Quality assurance has emerged significantly during the past decade. Permit compliance monitoring, enforcement, and litigation are now prevalent in the environmental arena. Only documented data of known quality will be sustained under litigation. This section focuses on two areas.

SAMPLE CUSTODY

Proper chain-of-custody procedures allow sample processing and handling to be traced and identified from the time containers are initially prepared for sampling to the final disposition of the sample. A chain-of-custody record (Figure 2.2.1) should accompany each group of samples from the time of collection to their destination at the analytical laboratory. Each person with custody of the samples must sign the chain-of-custody form, ensuring that the samples are not left unattended unless properly secured.

Within the laboratory, security and confidentiality of all stored material should always be maintained. Analysts should sign for any sample removed from a storage area for performing analyses and note the time and date of returning a sample to storage. Before releasing analytical results, all information on sample labels, data sheets, tracking logs, and custody records should be cross-checked to ensure that data are consistent throughout the record. Gummed paper custody seals or custody tape should be used to ensure that the seal must be broken when opening the container.

CHAIN OF CUSTODY RECORD

PROJECT	SAMPLERS: *(Signed)*

LAB #	STATION	DATE	TIME	SAMPLE TYPE: WATER	SEDIMENT	TISSUE	AIR	OIL	OTHER	NUMBER OF CONTAINERS	REMARKS

RELINQUISHED BY: *(Signed)*		RECEIVED BY: *(Signed)*	DATE/TIME
RELINQUISHED BY: *(Signed)*		RECEIVED BY: *(Signed)*	DATE/TIME
RELINQUISHED BY: *(Signed)*		RECEIVED BY: *(Signed)*	DATE/TIME
RELINQUISHED BY: *(Signed)*		RECV'D BY MOBILE LAB FOR FIELD ANAL.: *(Signed)*	DATE/TIME
DISPATCHED BY: *(Signed)*	DATE/TIME	RECEIVED FOR LAB BY: *(Signed)*	DATE/TIME
METHOD OF SHIPMENT:			

FIG. 2.2.1 Example chain of custody record. Distribution: Original—accompany shipment; One copy—survey coordinator-field files.

PRECISION AND ACCURACY

One of the objectives of the QA or QC plan is to ensure that there is no contamination from initial sampling through final analysis. For this reason, duplicate, field blank, and travel blank samples should be prepared and analyzed.

Duplicate sampling requires splitting one field sample into two aliquots for laboratory analysis. Typically, 10% of the samples should be collected in duplicate. Duplicates demonstrate the reproducibility of the sampling procedure.

A travel blank is a contaminant-free sample prepared in the laboratory that travels with empty sample bottles to the sampling site and returns to the laboratory with the samples. Typically, two travel blanks are prepared and shipped. Travel blanks identify contamination in the preparation of sample containers and shipping procedures.

Field blanks are empty sampling bottles prepared using contaminant-free water following general field sampling procedures for collection of waste samples. These are returned to the laboratory for analysis. Field blanks identify contamination associated with field sampling procedures.

For liquid samples, all three types of the above QA/QC samples are prepared. For soils, semi-soils, sludges, and solids, only duplicate samples are typically prepared.

The field supervisor of sample collection should maintain a bound logbook so that field activity can be completely reconstructed without relying on the memory of the field crew. Items noted in the logbook should include:

- Date and time of activity
- Names of field supervisor and team members
- Purpose of sampling effort
- Description of sampling site
- Location of sampling site
- Sampling equipment used
- Deviation(s) from standard operating procedures
- Reason for deviations
- Field observations
- Field measurements
- Results of any field measurements
- Sample identification
- Type and number of samples collected
- Sample handling, packaging, labeling, and shipping information

The logbook should be kept in a secure place until the project activity is completed, when the logbook should be kept in a secured project file.

Analysis

If the source and nature of the waste is known, sampling and analysis are limited to the parameters of concern. If the waste is unknown, a full spectrum analysis is often required, including analysis for the 129 priority pollutants. Table 2.2.1 divides priority pollutants into seven categories (EPA 1980–1988).

Table 2.2.2 presents the recommended analytical procedures for the following categories: volatile organics, acid-extractable organics, base and neutral organics, pesticides and PCBs, metals, cyanides, asbestos, and others. Typically, organic analysis is performed using gas chromatography and mass spectrometry (GC/MS). Typical sensitivity is on the order of 1–100 parts per

billion (ppb), depending on the specific organic compound and the concentration of compounds that may interfere with the analysis. This technique gives good quantification and excellent qualification about the organics in the waste.

A number of references should be consulted before determining the analytical protocols for the waste sample (EPA 1979; EPA 1977; EPA 1985a; EPA 1979a; APHA 1980).

Because analysis of hazardous waste samples is costly, it is beneficial to prepare several samples and subject them to one of several screening procedures. Depending on the data obtained, the analytical program can then focus on the major constituents of concern, resulting in cost savings. Recommended screening tests include: pH; conduc-

---

TABLE 2.2.1   CATEGORIZATION OF PRIORITY POLLUTANTS

Volatile Organics	2-nitrophenol	dimethyl phthalate	endosulfan sulfate
acrolein	4-nitrophenol	2,4-dinitrotoluene	endrin
acrylonitrile	parachlorometacresol	2,6-dinitrotoluene	endrin aldehyde
benzene	1,2,4-trichlorobenzene	1,2-diphenylhyrazine	heptachlor
bis(chloromethyl)ether	phenol	fluoranthene	heptachlor epoxide
bromoform	2,4,6-trichlorophenol	fluorene	PCB-1016
carbon tetrachloride		hexachlorobenzene	PCB-1221
chlorobenzene	**Base and Neutral Organics**	hexachlorobutadiene	PCB-1232
chlorodibromomethane	acenaphthene	hexachlorocyclo-	PCB-1242
pentachlorophenol	acenaphtylene	pentadiene	PCB-1248
2-chloroethyl vinyl ether	anthracene	hexachloroethane	PCB-1254
chloroform	benzidine	indeno(1,2,3-cd)-pyrene	PCB-1260
dichlorobromomethane	benzo(a)anthracene	isophorone	toxaphene
1,2-dichloroethane	benzo(a)pyrene	naphthalene	
1,1-dichloroethane	benzo(ghi)perylene	nitrobenzene	**Metals**
1,1,-dichloroethylene	benzo(k)fluoranthene	N-nitrosodi-n-	antimony
1,2-dichloropropane	3,4-benzo-fluoranthene	propylamine	arsenic
1,2-dichloropropylene	bis(2-chloroethoxy) methane	N-nitrosodimethylamine	beryllium
ethylbenzene	bis(2-chloroethyl)ether	N-nitrosodiphenylamine	cadmium
methyl bromide	bis(2-chloroisopropyl)-	phenanthrene	chromium
methyl chloride	ether	pyrene	copper
methylene chloride	bis(2-ethylhexyl)phthalate	2,3,7,8-tetrachloro-	lead
1,1,2,3-tetrachloroethane	4-bromophenyl phenyl	dibenso-p-dioxin	mercury
tetrachloroethylene	ether		nickel
toluene	butyl benzyl phthalate	**Pesticides and PCBs**	selenium
1,2-trans-dichloroethylene	2-chloro-naphthalene	aldrin	silver
1,1,1-trichloroethane	4-chlorophenyl phenyl	alpha-BHC	thallium
1,1,2-trichloroethane	ether	beta-BHC	zinc
trichloroethylene	chrysene	gamma-BHC	
vinyl chloride	di-n-butyl phthalate	delta-BHC	**Cyanides**
	di-n-octyl phthalate	chlordane	
Acid-Extractable Organics	dibenzo(a,h)anthracene	4,4'-DDD	**Asbestos**
2-chlorophenol	1,2-dichlorobenzene	4,4'-DD chloroethane	
2,4-dichlorophenol	4,4'-DDT	dieldrin	
2,4-dimethylphenol	1,4-dichlorobenzene	alpha-endosulfan	
4,6-dinitro-o-cresol	diethyl phthalate	beta-endosulfan	

Source: Reprinted from U.S. Environmental Protection Agency (EPA), 1980–1988, *National Pollutant Discharge Elimination System,* Code of Federal Regulations, Title 40, Part 122. (Washington, D.C.: U.S. Government Printing Office).

TABLE 2.2.2 RECOMMENDED METHOD FOR ANALYSIS

Analytical Category	Recommended Method for Analysis*
Volatile organics	GC/MS (USEPA Method 624)
Acid-extractable organics	GC/MS (USEPA Method 625)
Base and neutral organics	GC/MS (USEPA Method 625)
TCDD (dioxin)	GC/MS (USEPA Method 608)
Pesticides and PCBs	GC/MS (USEPA Method 625)
Metals	Atomic absorption (flame or graphite)†
Mercury	Cold vapor atomic absorption spectroscopy
Cyanide	EPA colorimetric method
Asbestos	Fibrous asbestos method
Anions (SO_4^{2-}, F^-, Cl^-)	Ion chromatography
Oil and grease	Freon extraction and gravimetric measurement
Purgeable halocarbons	GC (USEPA Method 601)
Purgeable aromatics	GC (USEPA Method 602)
Acrolein and acrylonitrile	GC (USEPA Method 603)
Phenols	GC (USEPA Method 604)
Benzidine	GC (USEPA Method 605)
Pthalate esters	GC (USEPA Method 606)
Nitrosamines	GC (USEPA Method 607)
Pesticides and PCBs	GC (USEPA Method 608)
Nitroaromatics and isophorone	GC (USEPA Method 609)
Polynuclear aromatic hydrocarbons	GC (USEPA Method 610)
Chlorinated hydrocarbons	GC (USEPA Method 611)
TCDD (dioxin screening)	GC (USEPA Method 612)

*GC/MS = gas chromatography/mass spectrometry; GC = gas chromatography.
†Graphite furnace is a more sensitive technique.
Source: Reprinted from U.S. EPA, 1980–1988.

tivity; total organic carbon (TOC); total phenols; organic scan (via GC with flame ionization detector); halogenated (via GC with electron capture detector); volatile organic scan; nitrogen-phosphorous organic scan; and metals (via inductively coupled plasma or atomic emission spectroscopy).

—David H.F. Liu

References

American Public Health Association (APHA). 1980. *Standard methods for the examination of water and wastewater.* 15th ed. APFA. New York, N.Y.

De Vera, E.R., B.P. Simmons, R.D. Stephens, and D.L. Storn, 1980. *Samplers and sampling procedures in hazardous waste streams.* EPA 600–2–80–018, Cincinnati, Oh.

Evans, R.B., and G.E. Schweitzer. 1984. Assessing hazardous waste problems, *Environmental science and technology.* 18(11).

U.S. Environmental Protection Agency (EPA). 1977. *Sampling and analysis procedure for screening of industrial effluent for priority pollutants.* Effluent Guideline Division. Washington, D.C.

———. 1979. *Method for chemical analysis of water and waste.* EPA 600–4–79–020. Washington, D.C.

———. 1979a. *Guidelines establishing procedures for analysis of pollutants.* Code of Federal Regulations, Title 40, Part 136. Washington, D.C.: U.S. Government Printing Office.

———. 1980–1988. *National pollutant discharge elimination system.* Code of Federal Regulations, Title 40, Part 122. Washington, D.C.: U.S. Government Printing Office.

———. 1985. *Protecting health and safety at hazardous waste sites: an overview,* Technology Transfer EPA, 625–9–85–006. Cincinnati, Oh.

———. 1985a. *Characterization of hazardous waste sites—a methods manual; vol II, available sampling methods.* EPA 600–4–84–075. Washington, D.C.

———. 1985b. *Test methods for evaluating solid waste, physical/chemical methods.* 2d ed. SW-846. Washington, D.C.

2.3
COMPATIBILITY

Wasteloads are frequently consolidated before transport from point of generation to point of treatment or disposal. Accurate waste identification and characterization is necessary to:

- Determine whether wastes are hazardous as defined by regulations
- Establish compatibility grouping to prevent mixing incompatible wastes
- Identify waste hazard classes as defined by the Department of Transportation (DOT) to enable waste labeling and shipping in accordance with DOT regulations
- Provide identification to enable transporters or disposal operators to operate as prescribed by regulations.

Most wastes are unwanted products of processes involving known reactants. Thus, the approximate compositions of these wastes are known. Wastes of unknown origin must undergo laboratory analysis to assess their RCRA status, including testing for the hazardous properties of ignitability, reactivity, corrosivity, or toxicity in accordance with methods specified in the regulations (See Section 2.1).

Once a waste is identified, it is assigned to a compatibility group. One extensive reference for assigning groups is a study of hazardous wastes performed for the EPA by Hatayama et al (1980). A waste can usually be placed easily in one of the groups shown in Figure 2.3.1, based on its chemical or physical properties. The compatibility of various wastes is shown in Figure 2.3.1, which indicates the consequences of mixing incompatible wastes. Complete compatibility analysis should be carried out by qualified professionals to ascertain whether any waste can be stored safely in proximity to another waste.

—*William C. Zegel*

Reference

Hatayama et al. 1980. *A method for determining the compatibility of hazardous wastes.* U.S. Environmental Protection Agency (EPA). Office of Research and Development. EPA 600–2–80–076. Cincinnati, Oh.

Reactivity group no.	Reactivity group name	1	2	3	4	5	6	7	8	9	10	11	12	13	14	15	16	17	18	19	20	21	22	23	24	25	26	27	28	29	30	31	32	33	34	101	102	103	104	105	106	107
1	Acids, mineral, nonoxidizing	1																																								
2	Acids, mineral, oxidizing		2																																							
3	Acids, organic		G H	3																																						
4	Alcohols and glycols	H	H F	H P	4																																					
5	Aldehydes	H P	H F	H P		5																																				
6	Amides	H	H GT				6																																			
7	Amines, aliphatic and aromatic	H	H GT	H		H		7																																		
8	Azo compounds, diazo compounds, and hydrazines	H G	H GT	H G	H G	H			8																																	
9	Carbamates	H G	H GT						G H	9																																
10	Caustics	H	H	H		H				H G	10																															
11	Cyanides	GT GF	GT GF	GT GF					G			11																														
12	Dithiocarbamates	H GF F	H GF F	H GF GT		GF GT		U	H G				12																													
13	Esters	H	H F						H G		H			13																												
14	Ethers	H	H F												14																											
15	Flourides, inorganic	GT	GT	GT												15																										
16	Hydrocarbons, aromatic		H F														16																									
17	Halogenated organics	H GT	H F GT					H GT	H G		H GF	H						17																								
18	Isocyanates	H G	H F GT	H G	H P			H P	H G		H P G	H G	U						18																							
19	Ketones	H	H F						H G		H	H								19																						
20	Mercaptans and other organic sulfides	GT GF	H F GT						H G									H	H	H	20																					
21	Metals, alkali and alkaline earth, elemental	GF H F	GF H F	GF H F	GF H F	GF H F	GF H	GF H	GF H	GF H	GF H	GF H	GF GT H	GF H				H E	GF H	GF H	GF H	21																				
22	Metals, other elemental & alloys as powders, vapors, or sponges	GF H F	GF H F	GF					H F GT	U	GF H							H E	GF H		H GF F		22																			
23	Metals, other elemental & alloys as sheets, rods, drops, moldings, etc.	GF H F	GF H F						H F G									H F						23																		
24	Metals and metal compounds, toxic	S	S	S			S	S			S														24																	
25	Nitrides	GF HF	H F E	H GF	GF H E	GF H			U	H G	U	GF H	GF H	GF H				GF H	U	GF H	GF H	E				25																
26	Nitriles	H GT GF	H F GT	H							U											H P			S	GF H	26															
27	Nitro compounds, organic		H F GT			H					H E											H GF E				H GF E		27														
28	Hydrocarbons, aliphatic, unsaturated	H	H F			H																	H E						28													
29	Hydrocarbons, aliphatic, saturated		H F																											29												
30	Peroxides and hydroperoxides, organic	H G	H E		H F	H G		H GT	H F E	H F GT		H E GT	H F GT					H E	H	E	H F GT	H E	P G		H G	H GF E	H P GT		H P		30											
31	Phenols and cresols	H	H F						H G										H P			GF H				GF H					H	31										
32	Organophosphates, phosphothioates, phosphodithioates	H GT	H GT						U		H E											H									U		32									
33	Sulfides, inorganic	GT GF	HF GT	GT		H			E										H												H GT			33								
34	Epoxides	H P	H P	H P	H P	U		H P	H P		H P	H P	U								H P	H P	H P		H P	H P					H P	H P	U	H P	34							
101	Combustible and flammable materials, miscellaneous	H G	H F GT																			H G F				H GF E					H F GT					101						
102	Explosives	H E	H E	H E					H E		H E			H E								H E	H E	H E	E	E					H E	H E		H E	H E	H E	102					
103	Polymerizable compounds	P H	P H	P H					P H		P H	P H	U									P H	P H	P H	P H	P H					P H	P H		P H			H E	103				
104	Oxidizing agents, strong	H GT		H GT	H F	H F	H F GT	H F GT	H E	H F GT		H E GT	H F GT	H F	H F		H F	H GT	H F GT	H F	H F GT	H F E	H F E	H F		H F E	H F GT	H E	H F	H F	H G	H F	H F GT	H F GT	H F G	H F G	H E	H F GT	104			
105	Reducing agents, strong	H GF	H F GT	H GF	H GF F	GF H F	GF H	H GF	H G				H GT	H F				H E	GF H	GF H	GF H						H GF	H E			H E	GF H	GT GF H		H	GF H	H E	H P GF	H F E	105		
106	Water and mixtures containing water	H	H						G										H G			GF H	GF H		S	GF H								GT GF						GF GT	106	
107	Water reactive substances	← EXTREMELY REACTIVE! DO NOT MIX WITH ANY CHEMICAL OR WASTE MATERIAL! EXTREMELY REACTIVE! →																																								107

Reactivity Code	Consequences
H	Heat generation
F	Fire
G	Innocuous and nonflammable gas generation
GT	Toxic gas generation
GF	Flammable gas generation
E	Explosive
P	Violent polymerization
S	Solubilization of toxic substances
U	May be hazardous but unknown

Example:

H F GT	Heat generation, fire, and toxic gas generation

FIG. 2.3.1 Hazardous waste compatibility chart. (Reprinted from Hatayama et al. 1980, *A method for determining the compatibility of hazardous wastes*, U.S. Environmental Protection Agency [EPA] [Office of Research and Development. EPA 600–2–80–076, Cincinnati, Oh].)

3

Risk Assessment and Waste Management

3.1
THE HAZARD RANKING SYSTEM AND THE NATIONAL PRIORITY LIST

The Comprehensive Environmental Response, Compensation, and Liability Act (CERCLA) of 1980, better known as Superfund, became law "to provide for liability, compensation, cleanup and emergency response for hazardous substances released into the environment and the cleanup of inactive hazardous waste disposal sites." CERCLA was intended to give the EPA authority and funds to clean up abandoned waste sites and to respond to emergencies related to hazardous waste.

If a site poses a significant threat, the EPA uses its Hazard Ranking System (HRS) to measure the relative risk. Based upon this ranking system, sites warranting the highest priority for remedial action become part of the National Priority List (NPL).

The HRS ranks the potential threat posed by facilities based upon containment of hazardous substances, route of release, characteristics and amount of substances, and likely targets. HRS methodology provides a quantitative estimate of the relative hazards posed by a site, taking into account the potential for human and environmental exposure to hazardous substances. The HRS score is based on the probability of contamination from three sources—groundwater, surface water, and air—on the site in question. The HRS score assigned to a hazardous site reflects the potential hazards relative to other sites (Hallstedt, Puskar & Levine 1986).

S_M is the potential for harm to humans or the environment from migration of a hazardous substance to groundwater, surface water, or air; it is a composite of scores of each of the three routes

S_{FE} is the potential for harm from flammable or explosive substances

S_{DC} is the potential for harm from direct contact with hazardous substances at the site

The score for each of these hazard modes is obtained from a set of factors characterizing the facility's potential to cause harm as shown in Table 3.1.1. Each factor is assigned a numerical value according to the prescribed criteria. This value is then multiplied by a weight factor, yielding the factor score.

The factor scores are then combined: scores within a factor category are added together, then the total scores for each factor category are multiplied together. S_M is a composite of the scores of three possible migration routes:

$$S_M = \frac{1}{1.73} \sqrt{S_{gw}^2 + S_{sw}^2 + S_a^2} \qquad 3.1(1)$$

Figure 3.1.1 shows a typical worksheet for calculating the score for groundwater. Other worksheets are included in 40 CFR Part 300, Appendix A (1987).

Use of the HRS requires considerable information about the site, its surroundings, the hazardous substances present, and the geology in relation to the aquifers. If the data are missing for more than one factor in connection with the evaluation of a route, then that route score becomes 0, and there is no need to assign scores to factors in a route set at 0.

The factors that most affect an HRS site score are the proximity to a densely populated area or source of drinking water, the quantity of hazardous substances present, and toxicity of those hazardous substances. The HRS methodology has been criticized for the following reasons:

There is a strong bias toward human health effects, with only slight chance of a site in question receiving a high score if it represents only a threat or hazard to the environment.

Because of the human health bias, there is an even stronger bias in favor of highly populated affected areas.

The air emission migration route must be documented by actual release, while groundwater and surface water routes have no such documentation requirement.

The scoring for toxicity and persistence of chemicals may be based on site containment, which is not necessarily related to a known or potential release of toxic chemicals.

TABLE 3.1.1 RATING FACTORS FOR HAZARD RANKING SYSTEM

Hazard Mode	Category	Groundwater Route	Surface Water Route	Air Route
Migration	Route charcteristics	Depth to aquifer of concern	Facility slope and intervening terrain	
		Net precipitation	1-year 24-hour rainfall	
		Permeability of unsaturated zone	Distance to nearest surface water	
		Physical state	Physical state	
	Containment	Containment	Containment	
	Waste characteristics	Toxicity/persistence, Quantity	Toxicity/persistence, Quantity	Reactivity/incompatibility, Toxicity, Quantity
	Targets	Groundwater use	Surface water use	Land use
		Distance to nearest well/population served	Distance to sensitive environment	Population within 4-mile radius
			Population served/distance to water intake downstream	Distance to sensitive environment
Fire and explosion	Containment	Containment		
	Waste characteristics	Direct evidence		
		Ignitability		
		Reactivity		
		Incompatibility		
		Quantity		
	Targets	Distance to nearest population		
		Distance to nearest building		
		Distance to nearest sensitive environment		
		Land use		
		Population within 2-mile radius		
		Number of buildings within 2-mile radius		
Direct contact	Observed incident	Observed incident		
	Accessibility	Accessibility of hazardous substances		
Direct contact	Observed incident	Observed incident		
	Accessiblity	Accessibility of hazardous substances		
	Containment	Containment		
	Toxicity	Toxicity		
	Targets	Population within 1-mile radius		
		Distance to critical habitat		

Source: U.S. Environmental Protection Agency.

A high score for one migration route can be more than offset by low scores for other migration routes.

Averaging the route scores creates a bias against sites with only one hazard, even though that hazard may pose an extreme threat to human health and the environment.

The EPA provides quality assurance and quality control for each HRS score to ensure that site evaluations are performed on a consistent basis. HRS scores range from 0 to 100, with a score of 100 representing the most hazardous sites. Generally, HRS scores of 28.5 or higher will place a site on the NPL. Occasional exceptions have been made in this priority ranking to meet the CERCLA requirement that a site designated as top priority by a state be included on the NPL.

When the EPA places a hazardous waste site on the NPL, it also issues a summary description of the site and its threat to human health and the environment. Some typical examples are in EPA files, and in Wentz's book (1989).

(This discussion follows C.A. Wentz, *Hazardous Waste Management*, McGraw-Hill, pp 392–403, 1989.)

References

Code of Federal Regulations, Title 40, Part 300, Appendix A, 1987.

Hallstedt, G.W., M.A. Puskar, and S.P. Levine, 1986. Application of hazard ranking system to the prioritization of organic compounds identified at hazardous waste remedial action site. *Hazardous waste and hazardous materials*, Vol. 3, No. 2.

Wentz, C.A. 1989. *Hazardous waste management*. McGraw-Hill, Inc.

Facility Name: _____   Date: _____

Surface Water Route Work Sheet					
Rating Factor	Assigned Value (Circle One)	Multi-plier	Score	Max. Score	Ref. (Section)
1 Observed Release	0 45	1		45	4.1
If observed release is given a value of 45, proceed to line 4 . If observed release is given a value of 0, proceed to line 2 .					
2 Route Characteristics					4.2
Facility Slope and Intervening Terrain	0 1 2 3	1		3	
1-yr 24-hr Rainfall	0 1 2 3	1		3	
Distance to Nearest Surface Water	0 1 2 3	2		6	
Physical State	0 1 2 3	1		3	
Total Route Characteristics Score				15	
3 Containment	0 1 2 3	1		3	4.3
4 Waste Characteristics					4.4
Toxicity/Persistence	0 3 6 9 12 15 18	1		18	
Hazardous Waste Quantity	0 1 2 3 4 5 6 7 8	1		8	
Total Waste Characteristics Score				26	
5 Targets					4.5
Surface Water Use	0 1 2 3	3		9	
Distance to a Sensitive Environment	0 1 2 3	2		6	
Population Served/	0 4 6 8 10	1		40	
Distance to Water	12 16 18 20				
Intake Downstream	24 30 32 35 40				
Total Targets Score				55	
6 If line 1 is 45, multiply 1 ∞ 4 ∞ 5					
If line 1 is 0, multiply 2 ∞ 3 ∞ 4 ∞ 5				64,350	
7 Divide line 6 by 64,350 and multiply by 100			S_{SW} =		

FIG. 3.1.1   Surface water route worksheet.

3.2
RISK ASSESSMENT

The term "risk" refers to the probability that an event will have an adverse effect, indirectly or directly, on human health or welfare. Risk is expressed in time or unit activity, e.g., cancer cases per pack of cigarettes smoked. Risk assessment takes into account the cumulative effects of all exposure. For example, in assessing the risk that a person will suffer from air pollution, both indoor and outdoor pollution must be taken into account.

The function of an effective hazardous materials management program is to identify and reduce major risks. This involves both risk assessment and risk management. The flowchart in Figure 3.2.1 shows the factors affecting the hazardous waste risk assessment procedure. This procedure begins with identification of the waste and the laws and regulations pertaining to that waste. When the waste is identified, its toxicity and persistence must be determined

to evaluate the risk of human and the environmental exposure. The risk management process involves selecting a course of action based on the risk assessment.

One way to highlight differences between risk assessment and risk management is to look at differences in the information content of the two processes. Data on tech-

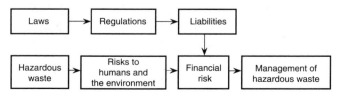

FIG. 3.2.1   Factors affecting the risk assessment of hazardous waste.

nological feasibility, on costs, and on the economic and social consequences of possible regulatory decisions are of critical importance to risk management but not to risk assessment. As statutes require, risk managers consider this information with risk assessment outcomes to evaluate risk management options and make environmental decisions (Figure 3.2.2).

Environmental risk assessment is a multi-disciplinary process. The risk assessment procedure is an iterative loop that the assessor may travel several times. It draws on data, information, and principles from many scientific disciplines, including biology, chemistry, physics, medicine, geology, epidemiology, and statistics. After evaluating individual studies for conformity with standard practices within each discipline, the most relevant information from each is combined and examined to determine the risk. Although studies from single disciplines are used to develop risk assessment, such studies alone are not regarded as risk assessment or used to generate risk assessments.

Review of Basic Chemical Properties

Before exploring the major components of risk assessment, some basic chemical properties and their relationships to biological processes must be reviewed.

Chemical Structure. The chemical structure of a substance is the arrangement of its atoms. This structure determines the chemical's properties, including how a chem-

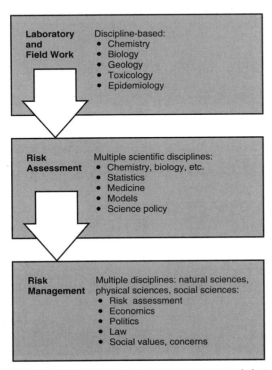

FIG. 3.2.2 Disciplines contributing to environmental decisions.

ical will combine with another substance. Because different structural forms of a chemical may exhibit different degrees of toxicity, the chemical structure of the substance being assessed is critical. For example, the free cyanide ion dissolved in water is highly toxic to many organisms (including humans); the same cyanide combined with iron is much less toxic (blue pigment). Cyanide combined with an organic molecule may have completely different toxic properties.

Solubility. Solubility is a substance's ability to blend uniformly with another. The degree of water and lipid (fat) solubility of a chemical is important in risk assessment. Solubility has significant implications for activities as diverse as cooking or chemical spill cleanup. To estimate the degree of potential water contamination from a chemical spill, it is necessary to know the chemical makeup of the material spilled to judge the extent that chemical contamination will be dispersed by dissolving in water. Likewise, the degree of lipid solubility has important implications, particularly in such processes as bioaccumulation.

Bioaccumulation. The process of chemical absorption and retention within organisms is called bioaccumulation. For example, a fat-soluble organic compound ingested by a microorganism is passed along the food chain when an organism eats the microorganism, then another predator eats the organism. The organic compound, because it is fat-soluble, will concentrate in the fat tissue of each animal in the food chain. The pesticide DDT is an example of a chemical that bioaccumulates in fish, and then in humans and birds eating those fish.

Transformation. Biotransformation and transformation caused by physical factors exemplify how chemical compounds are changed into other compounds. Biotransformation is the change of one compound to another by the metabolic action of a living organism. Sometimes such a transformation results in a less toxic substance, other times in a more toxic substance.

Chemical transformation is prompted by physical agents such as sunlight or water. A pesticide that is converted into a less toxic component by water in a few days following application (e.g., malathion) carries a different long-term risk than a pesticide that withstands natural degradation or is biotransformed into a toxic compound or a metabolite (e.g., DDT). The ability to withstand transformation by natural processes is called *persistence*.

Understanding basic chemical and physical properties helps to determine how toxic a chemical can be in drinking water or in the food chain, and whether the substance can be transported through the air and into the lungs. For example, when assessing the risk of polychlorinated biphenyls (PCBs), it must be recognized that they biodegrade very slowly and that they are strongly fat-soluble, so they readily bioaccumulate. When monitoring their presence, it must also be recognized that they are negligibly soluble in water: concentrations will always be much higher in the fat tissue of a fish, cow, or human than in the blood, which has a higher water content.

RA Paradigms

The risk assessment paradigm published in *Risk Assessment in the Federal Government: Managing the Process* (National Academy of Science [NAS] 1983), provides a useful system for organizing risk science information from these many different sources. In the last decade, the EPA has used the basic NAS paradigm as a foundation for its published risk assessment guidance and as an organizing system for many individual assessments. The paradigm defines four fields of analysis describing the use and flow of scientific information in the risk assessment process (Figure 3.2.3).

The following paragraphs detail those four fields of analysis. Each phase employs different parts of the information database. For example, hazard identification relies primarily on data from biological and medical sciences. Dose-response analysis uses these data in combination with statistical and mathematical modeling techniques, so that the second phase of the risk analysis builds on the first.

HAZARD IDENTIFICATION

The objective of hazard identification is to determine whether available scientific data describes a causal relationship between an environmental agent and demonstrated injury to human health or the environment. In humans, observed injuries may include birth defects, neurological damage, or cancer. Ecological hazards might result in fish kills, habitat destruction, or other environmental effects. If a potential hazard is identified, three other analyses become important for the overall risk assessment.

Chemical toxicities are categorized according to the various health effects resulting from exposure. The health effects, often referred to as *endpoints,* are classified as acute (short-term) and chronic (long-term). *Acute* toxic effects occur over a short period of time (from seconds to days), for example: skin burns from strong acids and poisonings from cyanide. *Chronic* toxic effects last longer and develop over a much longer period of time, and include cancer, birth defects, genetic damage, and degenerative illnesses.

A wide variety of reference materials provide basic toxicity data on specific chemicals, including:

Registry of Toxic Effects of Chemical Substances (RTECS), (U.S. Department of Health and Human Services)
Health Assessment Guidance Manual (U.S. Department of Health and Human Services 1990)
The Handbook of Toxic and Hazardous Chemicals and Carcinogens (Sittig 1985)
Threshold Limit Values (TLVs) for Chemical Substances and Physical Agents and Biological Exposure Indices (BEIs) American Conference of Governmental Industrial Hygienists (ACGIH 1990)
Integrated Risk Information System (IRIS), a database supported by EPA Office of Research and Development, Environmental Criteria and Assessment Office (MS-190), Cincinnati, Oh. 45268, Telephone: 513-569-7916

The U.S. Environmental Protection Agency has classified some 35,000 chemicals as definitely or potentially harmful to human health. However, the risk resulting from ex-

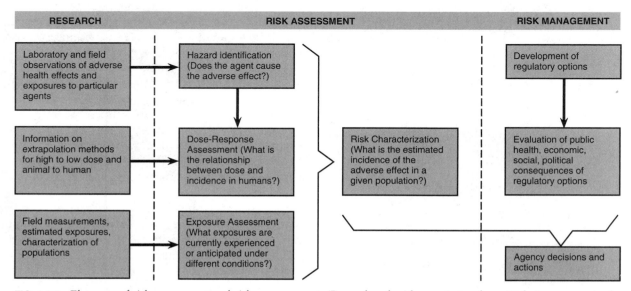

FIG. 3.2.3 Elements of risk assessment and risk management. (Reproduced with permission from *Risk Assessment in the Federal Government: Managing the Process,* 1983, The National Academy of Science [NAS], Washington, D.C.: The National Academy Press.

posure to more than one of these substances at the same time is not known (Enger, Kormelink, Smith & Smith 1989).

The following estimation techniques are commonly used to learn about human toxicity (Nally 1984).

Clinical Studies

The strongest evidence of chemical toxicity to humans comes from observing individuals exposed to the chemical in clinical studies. Scientists can determine direct cause and effect relationships by comparing the control groups (individuals not exposed to the chemical) to the exposed individuals. For obvious moral and ethical reasons, there is a limit to testing toxicity directly on humans. For example, tests for acute toxicity, such as allergic skin reactions, might be permissible, but tests for chronic toxicity, such as cancer, would be unacceptable.

Epidemiological Studies

As clinical studies frequently cannot be performed, scientists gather data on the incidence of disease or other ill effects associated with human exposure to chemicals in real-life settings. The field of *epidemiology* studies the incidence and distribution of disease in a population. This type of information is after the fact and in the case of cancer, comes many years after the exposure. Nevertheless, while epidemiological studies cannot unequivocally demonstrate direct cause and effect, they often can establish convincing and statistically significant associations. Evidence of a positive association carries the most weight in risk assessment.

Many factors limit the number of chemicals examined in epidemiological studies. Often there is no mechanism to verify the magnitude, the duration, or even the route of individual exposure. Control groups for comparing the incidence of disease between exposed and unexposed populations are difficult to identify. In addition, a long latency period between exposure and the onset of disease makes tracking exposure and outcome especially difficult.

On the other hand, epidemiological studies are very useful in revealing patterns of disease or injury distribution, whether these are geographical (i.e., the incidence of stomach cancer in Japan), for a special risk group (i.e., women of child-bearing age), or for an occupation (i.e., the incidence of cancer in asbestos workers). When available, valid epidemiological data are given substantial scientific weight.

Animal Studies

Since evidence from human exposure to a chemical is not usually available, scientists often rely on animal studies to determine the toxicity of a chemical. The objective of animal studies is to determine, under controlled laboratory conditions, the chemical dose that will produce toxic effects in an animal. This information is used to predict what

may occur in humans under normal exposure conditions. Toxic effects that occur in laboratory animals often occur in humans exposed to the same agents. Scientists recognize, however, that animal tests may not be conclusive for humans.

Routes of exposure in animal studies are designed to mimic the routes of possible human exposure. Ideally, a suspected food contaminant would be tested in a feeding study, a suspected skin surface irritant in a dermal irritation study, and a potential air contaminant in an inhalation study. However, it is not always possible to administer a test dose of the chemical to an animal via the expected route of exposure in humans (for instance, if it alters the color or odor of feed) so other methods must be devised.

Test-Tube Studies

Test-tube or in vitro studies involving living cells are particularly useful in testing whether a chemical is a potential carcinogen. Some of these tests are for mutagenicity or the ability to alter genetic material. Mutagenicity is believed to be one way in which carcinogens initiate cancer. These are often referred to as short-term tests because they require only a few hours or days, as opposed to several years required for long-term carcinogenicity studies in laboratory animals. The Ames mutagenicity test, which uses bacteria strains that reproduce only in the presence of a mutagen, is the best-known short-term test.

One of the major drawbacks of these cellular tests is that even with the addition of enzyme mixtures and other useful modifications, they are far simpler than the complex human organism. The human body's sensitive biological systems and remarkable defense mechanisms protect against chemical attack. The cellular tests lack the complexities of whole, integrated organisms, thus, they yield a significant number of false results. Nevertheless, they remain a useful screening process in deciding which chemicals should undergo more meaningful, but far more lengthy and expensive animal testing for carcinogenicity. Cellular tests can also provide insight into a carcinogen's mode of action.

Structure-Activity Relationships

When limited (or no) data are available from the estimation methods above, scientists often turn to structure-activity studies for evidence of chemical toxicity. This technique is based on the principle that chemicals with similar structures may have similar properties. For example, many potential carcinogens are found within categories of structurally similar chemicals.

At present, this method of predicting toxicity is not an exact science; it provides only an indication of potential hazard. However, as the technique develops along with the understanding of biological mechanisms, structure-activity relationships will evolve into a more precise predictive tool.

Animal studies are currently the preferred method for determining chemical toxicity. Although they are less convincing than human studies, animal studies are more convincing than test-tube and structure-activity studies. They are also easier to schedule, an industry has evolved around performing them.

The uncertainties associated with animal toxicity studies are discussed below.

The Testing Scheme

The selection of toxicological tests is crucial to any experimental program. Similarly, decisions regarding: the amount of chemical to be tested; the route of exposure; the test animal species; the composition of the test population (homogeneous or heterogeneous); the effects to be observed; and the duration of the study affect the usefulness and reliability of the resulting data. Although based on scientific judgment, all such decisions introduce elements of subjectivity into the testing scheme. The outcome of the test may be shaped by the specific nature of the test itself. For example, the decision to conduct an inhalation study might preclude discovering toxic effects via a different route of exposure. For this reason, a route of exposure is selected to approximate real-life conditions.

Demonstration of carcinogenicity requires strict observance of analytical protocols. NCI (IRLG 1979) presents criteria for evaluating experimental designs (see Table 3.2.1). Laboratory data not developed in compliance with these protocols are questionable.

Synergism/Antagonism

In vivo animal experiments are controlled studies that allow the isolation of individual factors to determine a specific cause and effect relationship. However, critics point out that such tests, although useful, are not absolute indicators of toxicity. As such tests are specific, synergistic effects from human exposure to more than one chemical are not detected. These tests may also overlook antagonistic effects where one chemical reduces the adverse effect of another.

DOSE-RESPONSE RELATIONSHIP

When toxicological evaluation indicates that a chemical may cause an adverse effect, the next step is to determine the potency of the chemical. The dose-response analysis determines the relationship between the degree of chemical exposure (or dose) and the magnitude of the effect (response) in the exposed organism. Scientists use this analysis to determine the amount of a chemical that causes tumor development in skin irritation, animals, or death in animals.

Dose-response curves are generated from various acute and chronic toxicity tests. Depending on chemical action, the curve may rise with or without a threshold. As Figure 3.2.4 shows, the TD_{50} and TD_{100} points indicate the doses associated with 50% and 100% occurrence of the measured toxic effect; also shown are the No Observable Adverse Effect Level (NOAEL) and Lowest Observable Adverse Effect Level (LOAEL). The NOAEL is assumed to be the basis for the Acceptable Daily Intake (ADI).

Figure 3.2.5 illustrates the threshold and no-threshold dose-response curve. In both cases, the response normally reaches a maximum, after which the dose-response curve becomes flat.

To estimate the effects of low doses, scientists extrapolate from the observed dose-response curve. Extrapolation models extend laboratory results into ranges where observations are not yet available or possible. Most current models are not based exclusively upon known biol-

TABLE 3.2.1   CRITERIA FOR EVALUATING CARCINOGEN EXPERIMENTS ON ANIMALS

Criteria	Recommendations
Experimental design	Two species of rats, and both sexes of each; adequate controls; sufficient animals to resolve any carcinogenic effect; treatment and observation throughout animal lifetimes at range of doses likely to yield maximum cancer rates; detailed pathological examination; statistical analyses of results for significance
Choice of animal model	Genetic homogeneity in test animals, especially between exposed and controls; selection of species with low natural-tumor incidence when testing that type of tumor
Number of animals	Sufficient to allow for normal irrelevant attrition along the way and to demonstrate an effect beyond the level of cancer in the control group
Route of administration	Were tumors found remote from the site of administration? No tumors observed should demonstrate that absorption occurred
Identity of the substance tested	Exposure to chemicals frequently involves mixtures of impurities. What effect did this have on results? What is the significance of pure compound results? Also, consider the carrier used in administration
Dose levels	Sufficient to evoke maximum tumor incidence
Age of treatment	Should be started early
Conduct and duration of bioassays	Refer to NCI's "Guidelines for Carcinogen in Small Rodents"

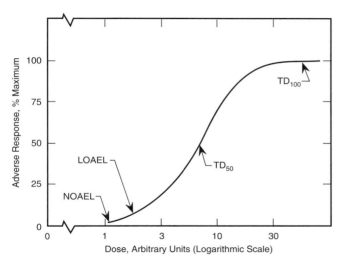

FIG. 3.2.4 Hypothetical dose-response curve. (Adapted from ICAIR Life Systems, Inc., 1985, *Toxicology Handbook*, prepared for EPA Office of Waste Programs Enforcement, Washington, D.C.)

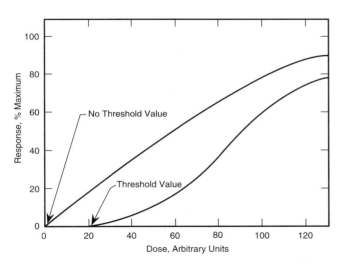

FIG. 3.2.5 Hypothetical dose-response curves. (Adapted from ICAIR Life Systems, Inc., 1985.)

ogy or toxicology, but are largely mathematical constructs based upon assumptions carrying varying degrees of uncertainty. How accurately these extrapolated low-dose responses correspond to true human risk remains a scientific debate.

Animal to Human Extrapolation

In extrapolating from animal data to potential human toxicity, a number of conversion factors are used to account for the differences between humans and animals. Factors must consider individual differences within a species; account for different sensitivities (Table 3.2.2); and note the variations between the two species, such as differences in weight, surface area, metabolism, and absorption.

Extrapolation from animal species to humans has elements of both science and art. It serves as a best estimate, neither invalid nor absolute truth. Although each step of assessment is laden with controversy, animal testing is the generally accepted approach for predicting human toxicity.

EXPOSURE ANALYSIS

Determining toxicity and exposure is necessary in a chemical risk assessment. Exposure to a chemical can occur through direct or indirect routes. Direct exposure is easier to identify, for example, exposure to nicotine and carbon monoxide from smoking cigarettes, or exposure to a pesticide from swimming in a contaminated lake. Indirect exposure can be somewhat more elusive, for example, mercury exposure by eating fish from mercury-contaminated waters. Whether direct or indirect, human exposure to chemicals will be dermal (skin contact), oral (contact by ingestion), and/or inhaled (contact by breathing).

Assessing human exposure to a chemical involves first determining the magnitude, duration, frequency, and route of exposure; and second, estimating the size and nature of the exposed population. Questions include to what concentration of chemical is a person exposed? How often does exposure occur—is it long-term or short-term, continuous or varied? What is the route of exposure—is the chemical in foods, consumer products, or in the workplace? Is the chemical bioactive, or is it purged from the human system without causing any harmful effects? Are special risk groups, such as pregnant women, children, or the elderly, exposed?

Sources of Uncertainty

Exposure measurements and estimates are difficult to obtain, and full of uncertainty. Often, less data exists about human exposure to chemicals than about chemicals' inherent toxicities. When estimating chemical exposure, it is important to be aware that exposure can come from different sources at varying rates—some intermittent, others continuous. Frequently, people assume that exposure comes from only one source and that they only need to monitor levels from that source. However, people may be exposed to different sources at various times and in various quantities. These considerations make estimating exposure very difficult. Below are some sources of uncertainty in estimating chemical exposure.

Monitoring Techniques

When chemical contamination is suspected, it is necessary to identify the baseline, or background concentration, of the chemical before onset of contamination. Subtracting the background concentration from the total concentration detected provides an accurate measure of the exposure resulting from contamination.

TABLE 3.2.2 FACTORS INFLUENCING HUMAN RESPONSE TO TOXIC COMPOUNDS

Factor	Effect
Dose	Larger doses correspond to more immediate effects
Method of administration	Some compounds nontoxic by one route and lethal by another (e.g., phosgene)
Rate of administration	Metabolism and excretion keep pollutant concentrations below toxic levels
Age	Elderly and children more susceptible
Sex	Each sex has hormonally controlled hypersensitivities
Body weight	Inversely proportional to effect
Body fat	Fat bioconcentrates some compounds (large doses can occur in dieters due to stored pollutants)
Psychological status	Stress increases vulnerability
Immunological status	Influences metabolism
Genetic	Different metabolic rates
Presence of other diseases	Similar to immunological status; could be a factor in cancer recurrence
Pollutant pH and ionic states	Interferes or facilitates absorption into the body
Pollutant physical state	Compounds absorbed on particulates may be retained at higher rate
Chemical milieau	Synergisms, antagonisms, cancer "promoters," enhanced absorption
Weather conditions	Temperature, humidity, barometric pressure, and season enhance absorption

It should be noted that scientists cannot measure zero concentration of a chemical—zero concentration is not a scientifically verifiable number. Instead, the terms "nothing detected" or "below the limit of measurement techniques" are used.

Sampling Techniques

In determining the location and number of samples for analysis, samples must accurately represent exposure levels at the place and time of exposure. There may be a difference between soil sampled at the surface or at a depth of six inches. Proper scientific methods must be observed, as an error in sampling will be propagated throughout the entire analysis. Furthermore, enough samples must be taken to allow statistical analysis crucial to ensuring data reliability.

Past Exposure

It is often difficult to determine the past exposure to a chemical. Most epidemiological studies are initiated after symptoms associated with exposure have occurred and after the amount or duration of exposure has changed. Unless detailed records are maintained, as in some workplace environments, the exact amount of exposure must be estimated.

Extrapolation to Lifetime Exposure

When initial exposure is measured (e.g., an industrial worker exposed to about 50 ppm of ethylene oxide for 8 hr per day), an extrapolation is made to determine the exposure over a lifetime of such activity. Information gathered from a small population segment must be extrapolated to the entire population. Such extrapolation often does not account for individual variability in exposures within the population.

RISK CHARACTERIZATION

Although the preceding analyses examine all relevant data to describe hazards, dose-response, or exposure, no conclusions are drawn about the overall risk. The final analysis addresses overall risk by examining the preceding analyses to *characterize* the risk. This process fully describes the expected risk through examining exposure predictions for real-world conditions, in light of the dose-response information from animals, people, and special test systems.

Risk is usually identified as a number. When the risk concern is cancer, the risk number represents the probability of additional cancer cases. For example, an estimate for pollutant X might be expressed as 1×10^{-6} or simply 10^{-6}. This means one additional case of cancer projected in a population of one million people exposed to a certain level of Pollutant X over their lifetimes.

A numerical estimate is only as good as the data it is based on. Scientific uncertainty is a customary and expected factor in all environmental risk assessment. Measurement uncertainty refers to the usual variances accompanying scientific measurement, such as the range (10 ± 1). Sometimes the data gap exists because specific measurements or studies are missing. Sometimes the data gap is more broad, revealing a fundamental lack of understanding about a scientific phenomenon.

The 1983 paradigm and EPA risk assessment guidelines stress the importance of identifying uncertainties and presenting them as part of risk characterization.

The major sources of uncertainty are: (1) difficulty in estimating the amount of chemical exposure to an individual or group; (2) limited understanding of the mechanisms determining chemical absorption and distribution within the body; and (3) reliance on animal experiment data for estimating the effects of chemicals on human organs. All of these areas rely upon scientific judgments even though judgments may vary significantly among experts.

Despite differing views within the scientific community on certain issues, a process has emerged for dealing with these differences. Beginning with peer reviews of each scientific study, this process assures accurate data interpretation by qualified specialists. The next step involves an interdisciplinary review of studies relevant to the risk assessment, where differences of interpretation are fully aired. This structured peer-review process is the best means available to resolve differences among experts.

In summary, despite the limitations of risk assessment, quantifying the best estimate of risk is important in preventing harmful chemical exposure. However, understanding the limits of such estimates and indicating the degree of uncertainty is equally important for sound decision-making.

PUBLIC PERCEPTION OF RISK

The nine criteria in Table 3.2.3 are identified as influencing public perceptions of risk. The characteristics of the criteria on the left contribute to perceptions of low risk, while the criteria on the right contribute to perceptions of higher risk.

Several general observations about perceptions of risk have been made. People tend to judge exposure to involuntary activities or technologies as riskier than voluntary ones. The obvious reason for this perception is that voluntary risk can be avoided whereas involuntary risk cannot. The amount of pesticide residues in food or the concentration of contaminants in drinking water is an involuntary decision for the public. Therefore, the public must turn to the government to regulate these activities and technologies.

Catastrophic events are perceived as riskier than ordinary events. For example, the chance of a plane crash

killing many individuals is perceived as riskier than the chance of an auto accident killing one or two people. Although the severity of a plane crash is higher, the probability of occurrence is much lower, thus the risk may be lower. In addition, delayed effects, such as cancer, are dreaded more than immediate effects such as poisoning.

Determining the acceptability of a risk to society is a social, not scientific, decision. This determination is influenced greatly by public perception of the risk, and is often reflected in legislation. The variation in the perception of risk can be related to the determination of an acceptable level of risk with various value judgments superimposed upon these perceptions. For example, laboratory tests identified saccharin as an animal carcinogen, requiring the FDA to ban it. However, the U.S. Congress determined that using saccharin was an acceptable risk, and prevented a ban due to perceived public benefits. No absolute answer can be provided to the question, "How safe is safe enough?" Determining acceptable levels of risk and making those value judgments is a very difficult and complex task.

To determine the acceptable risk for noncarcinogens, a *safety factor* is applied. Although it is rooted in science, selection of a safety factor is more of a rule of thumb, or an art. This factor is used when determining the safe dose to humans to compensate for uncertainties in the extrapolation process. This safe dose is known as the *acceptable daily intake* (ADI). The ADI amount of a chemical should not cause any adverse effects to the general human population even after long-term, usually lifetime, exposure. An ADI is calculated by dividing the NOAEL by a safety factor.

Risk Management

Risk assessment estimates the magnitude and type of risk from exposure to a potentially hazardous chemical. The government frequently decides to manage the risk. Public decision-makers are called upon to make the judgments: to synthesize the scientific, social, economical, and political factors and determine the acceptable risk for society. They need to reexamine the issues raised in risk assessment and address the following questions:

- Is the chemical economically important or essential?
- Is there a safer alternative?
- Is the risk of chemical exposure voluntary or involuntary?
- Can exposure be reduced?
- What are the benefits associated with use of this chemical?
- Are those individuals or societies subjected to risks the ones receiving the benefits?
- What are the costs of avoidance?
- What are the public perceptions of the risk?
- What level of risk is acceptable?
- Are some risks perceived as unacceptable no matter what the benefits?

TABLE 3.2.3   CRITERIA INFLUENCING PUBLIC PERCEPTION OF RISK

Criteria	Characteristics Perceived as Lower Risk	Characteristics Perceived as Higher Risk
origin	natural	manmade
volition	voluntary	involuntary
effect manifestation	immediate	delayed
severity (number of people affected per incident)	ordinary	catastrophic
controllability	controllable	uncontrollable
benefit	clear	unclear
familiarity of risk	familiar	unfamiliar
exposure	continuous	occasional
necessity	necessary	luxury

Over the years, many laws have been enacted to protect human health, safety, and the environment, providing a basic framework for risk management decisions. Each law reflects state-of-the-art understanding at the time of its enactment, as well as the political concerns and the public perceptions at that time. Regulators must make their decisions within the constraints of the applicable laws. These laws generally do not prescribe risk assessment methodologies. However, many environmental laws do provide very specific risk management directives.

Statutory risk management mandates can be divided into roughly three categories: pure-risk; technology-based standards; and reasonableness of risks balanced with benefits.

PURE-RISK STANDARDS

Pure-risk standards, sometimes termed zero-risk, are mandated or implied by only a few statutory provisions. Following are two examples of such standards:

The "Delaney clause" of the Federal Food, Drug, and Cosmetic Act prohibits the approval of any food additive that has been found to induce cancer in humans or animals.

The provisions of the Clean Air Act pertaining to national ambient air quality standards require standards for listed pollutants that "protect the public health allowing an adequate margin of safety," i.e., that assure protection of public health without regard to technology or cost factors.

TECHNOLOGY-BASED STANDARDS

Technology-based laws, such as parts of the Clean Air Act and the Clean Water Act, impose pollution controls based on the best economically available or practical technology. Such laws tacitly assume that benefits accrue from the use of the medium (water or air) into which toxic or hazardous substances are discharged, and that complete elimination of discharge of some human and industrial wastes into such media currently is not feasible. The basis for imposing these controls is to reduce human exposure, which indirectly benefits health and environment. The goal is to provide an ample margin of safety to protect public health and safety.

NO UNREASONABLE RISK

A number of statutes require balancing risks against benefits in making risk management decisions. Two examples include:

The Federal Insecticide, Fungicide, and Rodenticide Act requires the EPA to register pesticides that will not cause "unreasonable adverse effects on environment." The phrase refers to "any unreasonable risks to man or the environment taking into account the economic, social, and environmental costs and benefits of the use of any pesticide."

The Toxic Substances Control Act mandates that the EPA is to take action if a chemical substance "presents or will present an unreasonable risk of injury to health or the environment." This includes considering the substance's effects on human health and the environment; the magnitude of human and environmental exposure; the benefits and availability of such substances for various uses; and the reasonably ascertainable economic consequences of the rule.

The RCRA embodies both technology-based and pure-risk-based standards. Congress and the EPA have attempted to craft RCRA regulations in pure-risk-based rationales, but the large numbers of mixtures and the variety of generator/source operations have made that approach exceedingly difficult. As a result, the RCRA focuses on the following regulatory mechanisms:

- Identifying wastes that are hazardous to human health and the environment, and capturing them in a cradle-to-grave management system
- Creating physical barriers to isolate the public from contact with identified hazardous wastes
- Minimizing generation of hazardous wastes
- Encouraging reuse, recycling, and treatment of hazardous wastes
- Providing secure disposal for wastes that cannot otherwise be safely managed

—*David H.F. Liu*

References

American Conference of Governmental Industrial Hygienists. 1990. *Threshold limit values for chemical substances and physical agents and biological exposure indices (BEIs).* Cincinnati, Oh.

Enger, E.D., R. Kormelink, B.F. Smith, and Smith, R.J. 1989. *Environmental science: the study of Interrelationships,* Dubuque, Iowa: Wm. C. Brown Publishers.

Interagency Regulatory Liaison Group (IRLG). 1979. Scientific bases for identification of potential carcinogens and estimation of risk. *Journal of The Cancer Institute* 63(1).

Nally, T.L. 1984. *Chemical risk: a primer,* Department of Government Relations and Science Policy, American Chemical Society, Washington, D.C.

National Academy of Science (NAS). 1983. *Risk Assessment in Federal Government: Managing the Process,* Washington, D.C.: National Academy Press.

Patton, D. 1993. The ABC of risk assessment, *EPA Journal,* February/March.

Sittig, M. 1985. *Handbook of toxic and hazardous chemicals and carcinogens,* 2nd ed. Park Ridge, N.J.: Noyes Publications.

U.S. Department of Health and Human Services. 1990. *Health assessment guidance manual.* Published by the Agency for Toxic Substances and Disease Registry. Atlanta, Ga.

U.S. Department of Health and Human Services, *Registry of toxic effects of chemical substances.* Superintendent of Documents. Washington, D.C.: U.S. Government Printing Office.

3.3
WASTE MINIMIZATION AND REDUCTION

The first step in establishing a waste minimization strategy is to conduct a waste audit. The key question at the onset of a waste audit is "why is this waste present?" The environmental engineer must establish the primary cause(s) of waste generation before seeking solutions. Understanding the primary cause is critical to the success of the entire investigation. The audit should be waste-stream oriented, producing specific options for additional information or implementation. Once the causes are understood, solution options can be formulated. An efficient materials and waste trucking system that allows computation of mass balances is useful in establishing priorities. Knowing how much raw material is going into a plant and how much is ending up as waste allows the engineer to decide which plant and which waste to address first.

The first four steps of a waste audit allow the engineer to generate a comprehensive set of waste management options. These should follow the hierarchy of source reduction first, waste exchange second, recycling third, and treatment last.

In the end, production may be abandoned because the product or resulting by-product poses an economic hazard that the corporation is not willing to underwrite. These include cases where extensive testing to meet the TSCA (Toxic Substance Control Act) is required. Other such cases include the withdrawal of pre-manufacturing notice applications for some phthalate esters processes, and the discontinuation of herbicide and pesticide production where dioxin is a by-product.

Source Reduction and Control

INPUT MATERIALS

Source control investigations should focus on changes to input materials, process technology, and the human aspect of production. Input material changes can be classified into three elements: purification, substitution, and dilution.

Purification of input materials prevents inert or impure materials from entering the production process. Such impurities cause waste because the process must be purged to prevent undesirable accumulation. Examples of purified input materials include diionized water in electroplating and oxygen instead of air in oxychlorination reactors for ethylene dichloride production.

Substitution involves replacing a toxic material with a less toxic or more environmentally desirable material. Industrial applications of substitution include: using phosphates instead of dichromates as cooling water corrosion

inhibitors; using alkaline cleaners instead of chlorinated solvents for degreasing; using solvent-based inks instead of water-based inks; and replacing cyanide cadmium plating bath with noncyanide bath.

Dilution is a minor component of input material changes. An example of dilution is the use of a more dilute solution to minimize dragouts in metal parts cleaning.

TECHNOLOGY CHANGES

Technology changes are made to the physical plant. Examples include process changes; equipment, piping or layout changes; changes to operating settings; additional automation; energy conservation; and water conservation.

Process Change

Innovative technology is often used to develop new processes to achieve the same products, while reducing waste. Process redesign includes alteration of existing processes by adding new unit operations or implementation of new technology to replace outmoded operations. For example, a metal manufacturer modified a process to use a two-stage abrasive cleaner and eliminated the need for a chemical cleaning bath.

A classic example of a process change is the staged use of solvent. An electronics firm switched from using three different solvents—mineral spirits for machine parts, perchloroethylene for computer housings, and a fluorocarbon-mineral blend for printed circuit boards—to a single solvent system. Currently, fresh solvent is used for the printed circuit boards, then reused to degrease the computer housings, and finally, to degrease the machine parts. This practice not only reduces solvent consumption and waste, it eliminates potential cross-contamination of solvents, regenerates a single stream for recycling, and simplifies safety and operating procedures (U.S. EPA 1989).

Equipment, Piping, or Layout Changes

Equipment changes can reduce waste generation by reducing equipment–related inefficiencies. The capital required for more efficient equipment is justified by higher productivity, reduced raw material costs, and reduced waste materials costs. Modifications to certain types of equipment can require a detailed evaluation of process characteristics. In this case, equipment vendors should be consulted for information on the applicability of equipment for a process. Many equipment changes can be very simple and inexpensive.

Examples include installing better seals to eliminate leakage or simply putting drip pans under equipment to collect leaking material for reuse. Another minor modification is to increase agitation and alter temperatures to prevent formation of deposits resulting from crystallization, sedimentation, corrosion, or chemical reactions during formulating and blending procedures.

Operational Setting Changes

Changes to operational settings involve adjustments to temperature, pressure, flow rate, and residence time parameters. These changes often represent the easiest and least expensive ways to reduce waste generation. Process equipment is designed to operate most efficiently at optimum parameter settings. Less waste will be generated when equipment operates efficiently and at optimum settings. Trial runs can be used to determine the optimum settings. For example, a plating company can change the flow rate of chromium in the plating bath to the optimum setting and reduce the chromium concentrations used, resulting in less chromium waste requiring treatment.

Additional Control/Automation

Additional controls or automation can result in improved monitoring and adjustment of operating parameters to ensure maximum efficiency. Simple steps involving on-stream set-point controls or advanced statistical process control systems can be used. Automation can reduce human error, preventing spills and costly downtime. The resulting increase in efficiency can increase product yields.

PROCEDURAL CHANGES

Procedural changes are improvements in the ways people affect the production process. All referred to as *good operating practices* or *good housekeeping*, these include operating procedures, loss prevention, waste segregation, and material handling improvement.

Material Loss Prevention

Loss prevention programs are designed to reduce the chances of spilling a product. A hazardous *material* becomes an RCRA hazardous *waste* when it is spilled. A long-term, slow-release spill is often hard to find, and can create a large amount of hazardous waste. A material loss prevention program may include the following directives:

- Use properly designed tanks and vessels for their intended purpose only
- Maintain physical integrity of all tanks and vessels
- Install overflow alarms for all tanks and vessels

- Set up written procedures for all loading, unloading, and transfer operations
- Install sufficient secondary containment areas
- Forbid operators to bypass interlocks or alarms, or to alter setpoints without authorization
- Isolate equipment or process lines that are leaking or out of service
- Install interlock devices to stop flow to leaking sections
- Use seal-less pumps
- Use bellow seal valves and a proper valve layout
- Document all spillage
- Perform overall material balances and estimate the quantity and dollar value of all losses
- Install leak detection systems for underground storage tanks in accordance to RCRA Subtitle I
- Use floating-roof tanks for VOC control
- Use conservation vents on fixed-roof tanks
- Use vapor recovery systems (Metcalf 1989)

Segregating Waste Streams

Disposed hazardous waste often includes two or more different wastes. Segregating materials and wastes can decrease the amount of waste to be disposed. Recyclers and waste exchangers are more receptive to wastes not contaminated by other substances. The following are good operating practices for waste segregation:

- Prevent hazardous wastes from mixing with non-hazardous wastes
- Isolate hazardous wastes by contaminant
- Isolate liquid wastes from solid wastes

Materials Tracking and Inventory Control

These procedures should be used to track waste minimization efforts and target areas for improvement.

- Avoid over-purchasing
- Accept raw materials only after inspection
- Ensure that no containers stay in inventory longer than the specified period
- Review raw material procurement specifications
- Return expired materials to the supplier
- Validate shelf-life expiration dates
- Test outdated materials for effectiveness
- Conduct frequent inventory checks
- Label all containers properly
- Set up manned stations for dispensing chemicals and collecting wastes

Production Scheduling

The following alterations in production scheduling can have a major impact on waste minimization:

- Maximize batch size

- Dedicate equipment to a single product
- Alter batch sequencing to minimize cleaning frequency (light-to-dark batch sequence, for example)
- Schedule production to reduce cleaning frequency

Preventive Maintenance

These programs cut production costs and decrease equipment downtime, in addition to preventing waste release due to equipment failure.

- Use equipment data cards on equipment location, characteristics, and maintenance
- Maintain a master preventive maintenance (PM) schedule
- Maintain equipment history cards
- Maintain equipment breakdown reports
- Keep vendor maintenance manuals handy
- Maintain a manual or computerized repair history file

Cling

The options for minimizing wastes that cling to containers include:

- Use large containers instead of small condensers whenever possible
- Use containers with a one-to-one height-to-diameter ratio to minimize wetted area
- Empty drums and containers thoroughly before cleaning or disposal

PRODUCT CHANGES

Product Substitution

Changing the design, composition, or specifications of end products allows fundamental change in the manufacturing process or in the end use of raw materials. This can lead directly to waste reduction. For example, the manufacture of water-based paints instead of solvent-based paints involves no hazardous toxic solvents. In addition, the use of water-based paints reduces volatile organic emissions to the atmosphere.

Product Reformulation

Product reformulation or composition changes involves reducing the concentration of hazardous substances or changing the composition so that no hazardous substances are present. Reformulating a product to contain less hazardous material reduces the amount of hazardous waste generated throughout the product's lifespan. Using a less hazardous material within a process reduces the overall

amount of hazardous waste produced. For example, a company can use nonhazardous solvents in place of chlorinated solvents.

Dow Chemical Company achieved waste reduction through changes in product packaging. A wettable powder insecticide, widely used in landscape maintenance and horticulture, was originally sold in 2-lb metal cans. The cans had to be decontaminated before disposal, creating a hazardous waste. Dow now packages the product in 4-oz water-soluble packages which dissolve when the product is mixed with water for use (U.S. Congress 1986).

Product Conservation

One of the most successful methods of product conservation is the effective management of inventory with specific shelf-lives. The Holston Army Ammonium Plant reduced waste pesticide disposal from 440 to 0 kg in one year by better management of stocks (Mill 1988).

WASTE EXCHANGE

Waste exchange is a reuse function involving more than one facility. An exchange matches one *industry's* output to the input requirement of another. Waste exchange organizations act as brokers of hazardous materials by purchasing and transporting them as resources to another client. Waste exchanges commonly deal in solvents, oils, concentrated acids and alkalis, and catalysts. Limitations include transport distance, purity of the exchange product, and reliability of supply and demand.

Waste exchanges were first implemented and are now fairly common in Europe; there are few in the U.S. Although more exchanges have recently been set up in this country, they are not widely accepted because of liability concerns. Even when potential users of waste are found, they must be located fairly close to the generator. Waste transportation requires permits and special handling, increasing the cost.

Recycling and Reuse

Recycling techniques allow reuse of waste materials for beneficial purposes. A recycled material is used, reused, or reclaimed [40 CFR §261.1 (c)(7)]. Recycling through use or reuse involves returning waste material to the original process as a substitute for an input material, or to another process as an input material. Recycling through reclamation involves processing a waste for recovery of a valuable material or for regeneration. Recycling can help eliminate waste disposal costs, reduce raw material costs, and provide income from saleable waste.

Recycling is the second option in the pollution prevention hierarchy and should be considered only when all source reduction options have been investigated and implemented. Recycling options are listed in the following order:

- Direct reuse on-site
- Additional recovery on-site
- Recovery off-site
- Sale for reuse off-site (waste exchange)

It is important to note that recycling can increase a generator's risk or liability as a result of the associated material handling and management. Recycling effectiveness depends upon the ability to separate recoverable waste from other process waste.

DIRECT ON-SITE REUSE

Reuse involves finding a beneficial purpose for a recovered waste. Three factors to consider when determining the potential for reuse are:

- The chemical composition of the waste and its effect on the reuse process
- The economic value of the reuse waste and whether this justifies modifying a process to accommodate it
- The availability and consistency of the waste to be reused
- Energy recovery

For example, a newspaper advertising printer purchased a recycling unit to produce black ink from various waste inks. Blending different colors of ink with fresh black ink and black toner, the unit creates black ink. This mixture is filtered to remove flakes of dried ink, and is used in lieu of fresh black ink. The need to ship waste ink for offsite disposal is eliminated. The price of the recycling unit was recovered in nine months, based on savings in fresh ink purchases and costs of waste ink disposal (U.S. EPA 1989).

In another example, an oil skimmer in a holding tank enables annual capture and recycling of 3000 gallons of waste oil from 30,000 gallons of oily waste water disposed to waste landfills. (Metcalf 1989).

ADDITIONAL ON-SITE RECOVERY

Recycling alternatives can be accomplished either on-site or off-site and may depend on a company's staffing or economic constraints. On-site recycling alternatives result in less waste leaving a facility. The disadvantages of on-site recycling lie in the capital outlay for recycling equipment, the need for operator training, and additional operating costs. In some cases, the waste generated does not warrant the installation costs for in-plant recycling systems. However, since on-site alternatives do not involve transportation of waste materials and the resulting liabilities, they are preferred over off-site alternatives.

For instance, sand used in casting processes at foundries contains heavy metal residues such as copper, lead, and zinc. If these concentrations exceed Toxicity Characteristics Leaching Procedure (TCLP) standards, the sand is a hazardous waste. Recent experiments demonstrated that 95% of the copper could be precipitated and recovered (McCoy and Associates 1989). In another example, a photoprocessing company uses an electrolytic deposition cell to recover silver from rinse water used in film processing equipment. By removing the silver from the wastewater, the wastewater can be discharged to the sewer without additional pretreatment.

OFFSITE RECOVERY

If the amount of waste generated on-site is insufficient for a cost-effective recovery system, or if the recovered material cannot be reused on-site, off-site recovery is preferable. Materials commonly reprocessed off-site are oils, solvents, electroplating sludges and process baths, scrap metal, and lead-acid batteries. The cost of off-site recycling depends upon the purity of the waste and the market for the recovered materials.

The photoprocessing company mentioned above also collects used film and sells it to a recycler. The recycler burns the film and collects the silver from residual ash. By removing the silver from the ash, the fly ash becomes non-hazardous (EPA 1989).

SALE FOR REUSE OFF-SITE

See the preceding discussions on waste exchange. The most common reuses of hazardous waste include wastewater used for irrigation and oil field pressurization; sludges used as fertilizers or soil matrix; and sulfuric acid from smelters.

Recycling methods, including numerous physical, chemical and biological technologies will be discussed in Section(s) 4.5 and 4.8.

—*David H.F. Liu*

References

Code of Federal Regulations, Title 40, sec. 261.1.

McCoy and Associates, Inc. 1989. *The hazardous waste consultant.* (March-April).

Metcalf, C., ed. 1989. *Waste reduction assessment and technology transfer (WRATT) training manual.* The University of Tennessee Center for Industrial Services. Knoxville, Tenn.

Mill, M.B. 1988. Hazardous waste minimization in the manufacture of explosives, in *Hazardous waste minimization in the department of defense.* Edited by J.A. Kaminski. Office of the Deputy Assistant Secretary of Defense (Environment). Washington, D.C.

U.S. Congress, Office of Technology Assessment. 1986. *Serious reduction of hazardous waste.* Superintendent of Documents. Washington, D.C.: Government Printing Office.

U.S. Environmental Protection Agency (EPA). 1989. *Waste minimization in metal parts cleaning.* Office of Solid Waste and Emergency Response, Report No. EPA/530-SW-89-049. Washington, D.C.

3.4
HAZARDOUS WASTE TRANSPORTATION

The EPA's cradle-to-grave hazardous waste management system attempts to track hazardous waste from generation to ultimate disposal. The system requires generators to establish a manifest or itemized list form for hazardous waste shipments. This procedure is designed to ensure that wastes are direct to, and actually reach, permitted disposal sites.

Generator Requirements

The *generator* is the first element of the RCRA cradle-to-grave concept, which includes generators, transporters, treatment plants, storage facilities, and disposal sites. Generators of more than 100 kg of hazardous waste or 1 kg of acutely hazardous waste per month must, with a few exceptions, comply with all generator regulations.

Hazardous waste generators must comply with all DOT legislation regulating transport of hazardous materials, as well as other hazardous waste regulations promulgated by both DOT and the EPA. Table 3.4.1 summarizes the requirements, indicates the agency responsible for compliance, and provides a reference to the Code of Federal Regulations.

TABLE 3.4.1   EPA AND DOT HAZARDOUS WASTE TRANSPORTATION REGULATIONS

Required of	Agency	Code of Federal Regulations
Generator/Shipper		
1. Determine if waste is hazardous according to EPA listing criteria	EPA	40 CFR 261 and 262.11
2. Notify EPA and obtain I.D. number; determine that transporter and designated treatment, storage, or disposal facility have I.D. numbers	EPA	40 CFR 262.12
3. Identify and classify waste according to DOT Hazardous Materials Table and determine if waste is prohibited from certain modes of transport	DOT	49 CFR 172.101
4. Comply with all packaging, marking, and labeling requirements	EPA	40 CFR 262.32 (b),
	DOT	49 CFR 173, 49 CFR 172, subpart D, and 49 CFR 172, subpart E
5. Determine whether additional shipping requirements must be met for the mode of transport used.	DOT	49 CFR 174–177
6. Complete a hazardous waste manifest	EPA	40 CFR 262, subpart B
7. Provide appropriate placards to transporter	DOT	49 CFR 172, subpart F
8. Comply with record-keeping and reporting requirements	EPA	40 CFR 262, subpart D
Transporter/Carrier		
1. Notify EPA and obtain I.D. number	EPA	40 CFR 263.11
2. Verify that shipment is properly identified, packaged, marked, and labeled and is not leaking or damaged	DOT	49 CFR 174–177
3. Apply appropriate placards	DOT	49 CFR 172.506
4. Comply with all manifest requirements (e.g., sign the manifest, carry the manifest, and obtain signature from next transporter or owner/operator of designated facility)	DOT	49 CFR 174–177
	EPA	40 CFR 263.20
5. Comply with record-keeping and reporting requirements	EPA	50 CFR 263.22
6. Take appropriate action (including cleanup) in the event of a discharge and comply with the DOT incident reporting requirements	EPA	40 CFR 263.30–31
	DOT	49 CFR 171.15–17

Source: Reprinted from U.S. Environmental Protection Agency.

The regulatory requirements for hazardous waste generators contained in 40 CFR Part 262 include:

- Obtaining an EPA ID number
- Proper handling of hazardous waste before transport
- Establishing a manifest of hazardous waste
- Recordkeeping and reporting

EPA ID NUMBER

The EPA and primacy states monitor and track generator activity by an identification number to each generator. Without this number, the generator is barred from treating, storing, disposing of, transporting, or offering for transportation any hazardous waste. Furthermore, the generator is forbidden from offering the hazardous waste to any transporter, or treatment, storage, or disposal (TSD) facility that does not also have an EPA ID number. Generators obtain ID numbers by notifying the EPA of hazardous waste activity, using the standard EPA notification form.

PRETRANSPORT REGULATIONS

Pretransport regulations are designed to ensure safe transportation of a hazardous waste from origin to ultimate disposal; to minimize the environmental and safety impacts of accidental releases; and to facilitate control of any releases that may occur during transportation. In developing these regulations, the EPA adopted those used by the Department of Transportation (DOT) for transporting hazardous materials (49 CFR 172, 173, 178 and 179). These DOT regulations require:

Proper packaging to prevent hazardous waste leakage under normal or potentially dangerous transport conditions such as when a drum of waste falls from a truck or loading dock;

Labeling, marking, or placarding of the package to identify characteristics and dangers associated with the waste.

These pretransport regulations apply only to generators shipping waste off-site.

Briefly, individual containers are required to display "Hazardous Waste" markings, including the proper DOT shipping name, using the standardized language of 49 CFR Sections 172.101 and .102. The labels on individual containers must display the correct hazard class as prescribed by Subpart E of Part 172. Examples of DOT labels and placards are shown in Figures 3.4.1 and 3.4.2. Placards are important in case of accidents because they are highly visible. Efforts are now in progress for international adoption of hazardous marking, labeling and placarding conventions.

WASTE ACCUMULATION

A generator may accumulate hazardous waste on-site for 90 days or less, provided the following requirements are met:

Proper Storage. The waste must be properly stored in containers or tanks marked "Hazardous Waste" with the date accumulation began.

Emergency Plan. A contingency plan and emergency procedures are developed. Generators must have a written emergency plan.

Personnel Training. Facility personnel must be trained in the proper handling of hazardous waste.

The 90-day period allows more cost effective transportation. Instead of paying to haul several small shipments of waste, the generator can accumulate enough for one big shipment.

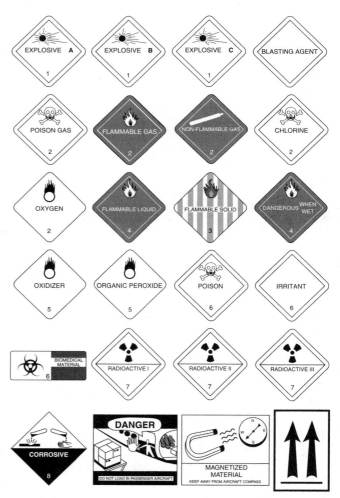

FIG. 3.4.1 DOT labels for hazardous materials packages. Source: Reprinted from U.S. Department of Transportation.

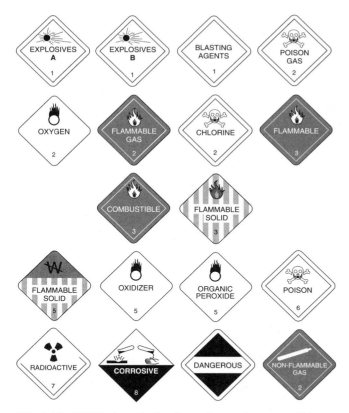

FIG. 3.4.2 DOT placards for hazardous substances. (*Source:* Reprinted from U.S. Department of Transportation.)

If hazardous waste is accumulated on-site for more than 90 days, the generator is considered an operator of a storage facility and becomes subject to Subtitle C requirements including permitting. Under temporary, unforeseen, or uncontrollable circumstances, the 90-day period may be extended for up to 30 days by the EPA Regional Administrator on a case-by-case basis.

Small quantity generators (SQGs), defined as those producing 100–1000 kg of hazardous waste per month, are accorded an exception to this 90-day accumulation period. The Hazardous and Solid Waste Amendments (HSWA) require, and the EPA developed, regulations allowing such generators to accumulate waste for 180 days, or 270 days if waste must be shipped over 200 miles, before SQGs are considered to be operating a storage facility.

THE MANIFEST

The Uniform Hazardous Waste Manifest is the key to the cradle-to-grave management system (Figure 3.4.3). Using the manifest, generators and regulators can track the movement of hazardous waste from the point of generation to the point of ultimate treatment, storage, or disposal (TSD).

The HSWA requires manifests to certify that generators have programs in place to reduce waste volume and toxicity to the degree the generator determines economically practicable. In addition, the treatment, storage, or disposal method chosen by the generator must be the best method currently available to minimize risks to human health and the environment.

Generators must prepare manifests properly since they are responsible for the production and ultimate disposition of hazardous wastes. Some common mistakes found on manifests are (Turner 1992):

Omission of the 24-hr emergency response telephone number. As of December 31, 1990, the DOT required inclusion of a 24-hr telephone number for use if an incident should occur during transportation. Shippers and carriers should look closely at this section to ensure its proper completion.

Omission of the manifest document number. Many generators use this control number to indicate the number of shipments made during a specified period. Others use it to indicate shipments from a specific section of their facility.

Misunderstanding of the generator name and mailing address. The address listed should be the location managing the return manifest form. The 12-digit EPA identification number is site specific in that it is assigned to the physical location where the hazardous waste is generated.

Improper entry of shipping name, hazard class or UN/NA numbers. 49 CFR Sec. 172.202 specifies the proper order for entering a basic description on a shipping document. The technical or chemical group names may be entered in parentheses between the proper shipping name and hazard class.

The manifest is part of a controlled tracking system. Each time waste is transferred from a transporter to a designated facility or to another transporter, the manifest must be signed to acknowledge receipt of the waste. A copy of the manifest is retained by each link in the transportation chain. Once the waste is delivered to the designated facility, the owner or operator of the facility must send a copy of the manifest back to the generator. This system ensures that the generator has documentation that the hazardous waste has reached its destination.

The multiple-copy manifest is initially completed and signed by the hazardous waste generator. The generator retains Part 6 of the manifest, sends Part 5 to the EPA or the appropriate state agency, and provides the remainder to the transporter. The transporter retains Part 4 of the manifest and gives the remaining parts to the TSD facility upon arrival. The TSD facility retains Part 3 and sends Parts 1 and 2 to the generator and the regulatory agency, or agencies, respectively. Throughout

Form Approved. OMB No. 2000-0404. Epires 7-31-86

UNIFORM HAZARDOUS WASTE MANIFEST	1.Generator's US EPA ID No. Manifest Document No.	2. Page 1 of	Information in the shaded areas is not required by Federal law.

3. Generator's Name and Mailing Address

A. State Manifest Document Number

B. State Generator's ID

4. Generator's Phone ( )

5. Transporter 1 Company Name	6. US EPA ID Number	C. State Transporter's ID
		D. Transporter's Phone
7. Transporter 2 Company Name	8. US EPA ID Number	E. State Transporter's ID
		F. Transporter's Phone
9. Designed Facility Name and Site Address	10. US EPA ID Number	G. State Facility's ID
		H. Facility's Phone

11. US DOT Description (Including Proper Shipping Name, Hazard Class, and ID Number)	12. Containers No.	Type	13. Total Quantity	14. Unit Wt/Vol	I. Waste No.
a.					
b.					
c.					
d.					

J. Additional Description for Materials Listed Above

K. Handling Codes for Wastes Listed Above

15. Special Handling Instructions and Additional Information

16. GENERATOR'S CERTIFICATION: I hereby declare that the contents of this consignment are fully and accurately described above by proper shipping name and are classed, packed, marked, and labeled, and are in all respects in proper condition for transport by highway according to applicable international and national government regulations.

Unless I am a small quantity generator who has been exempted by statute or regulation from the duty to make a waste minimization certification under Section 3002(b) of RCRA. I also certify that I have a program in place to reduce the volume and toxicity of waste generated to the degree I have determined to be economically practicable and I have selected the method of treatment, storage, or disposal currently available to me which minimizes the present and future threat to human health and the environment.

| Printed/Typed Name | Signature | Month Day Year |

17. Transporter 1 Acknowledgement of Receipt of Materials

| Printed/Typed Name | Signature | Month Day Year |

18. Transporter 2 Acknowledgement of Receipt of Materials

| Printed/Typed Name | Signature | Month Day Year |

19. Discrepancy Indication Space

20. Facility Owner or Operator: Certification of receipt of hazardous materials covered by this manifest except as noted in item 19.

| Printed/Typed Name | Signature | Month Day Year |

EPA Form 8700-22 (Rev. 4-85) Previous edition is obsolete.

FIG. 3.4.3 Uniform hazardous waste manifest.

this transition, the hazardous waste shipment is generally considered to be in the custody of the last signatory on the manifest.

If 35 days pass from the date when the waste was accepted by the initial transporter and the generator has not received Part 1 of the manifest form from the designated facility, the generator must contact the transporter or the designated facility to determine the whereabouts of the waste. If 45 days pass and the manifest still has not been received, the generator must file an exception report with the EPA regional office. The report must detail the efforts of the generator to locate the waste, and the results of these efforts.

RECORDKEEPING AND REPORTING

Generators are subject to extensive recordkeeping and reporting requirements by 40 CFR Part 262, Subpart D. Generators who transport hazardous wastes off site must submit an annual report to the EPA regional administrator on EPA form 8700-13A. This report covers generator activities during the previous year, and requires detailed accounting of wastes generated and their disposition. Generators must keep copies of each signed manifest for 3 years from the date signed, copies of each exception report, each annual report, copies of analyses, and related determinations made in accord with generator regulations (40 CFR Part 262).

Generators that treat, store, or dispose of their hazardous waste on-site must also notify the EPA of hazardous waste activity, obtain an EPA ID, apply for a permit, and comply with permit conditions. They too must submit an annual report containing a description of the type and quantity of hazardous waste handled during the year, and the method(s) of treatment, storage, or disposal used.

EXPORT AND IMPORT OF HAZARDOUS WASTE

Export of hazardous waste from the U.S. to another country is prohibited unless:

- Notification of intent to export has been provided to the EPA at least 60 days in advance of shipment;
- The receiving country has consented to accept the waste;
- A copy of the EPA "Acknowledgment of Consent" accompanies the shipments; and
- The hazardous waste shipment conforms to the terms of the receiving country's consent (40 CFR §262.52).

Any import of hazardous waste from another country into the U.S. must comply with the requirements of 40 CFR

Part 262, i.e., *the importer becomes the generator*, for RCRA regulatory purposes.

Transporters and Carriers
HAZARDOUS MATERIALS TRANSPORTATION ACT AND OTHER REGULATIONS

Transporters of hazardous waste are the critical link between the generator and the ultimate off-site treatment, storage, or disposal of hazardous waste. Transporter regulations were developed jointly by the EPA and DOT to avoid contradictory requirements. Although the regulations are integrated, they are not contained under the same act. A transporter must comply with regulations under 49 CFR Parts 171–179, The Hazardous Materials Transportation Act, and 40 CFR Part 263 (Subtitle C of RCRA).

A transporter is defined under RCRA as any person or firm engaged in the off-site transportation of hazardous waste within the United States, if such transportation requires a manifest under 40 CFR Part 262. This definition covers transport by air, highway, or water. Transporter regulations do not apply to on-site transportation of hazardous waste by generators with their own TSDs, or TSDs transporting waste within a facility. *However*, generators and TSD owners or operators must avoid transporting waste over public roads that pass through or alongside their facilities (Figure 3.4.4).

Under certain circumstances a transporter may be subject to regulatory requirements other than those contained in 40 CFR Post 263. Once a transporter accepts hazardous waste from a generator or another transporter, the transporter can store it for up to 10 days without being subject to any new regulations. However, if storage time ex-

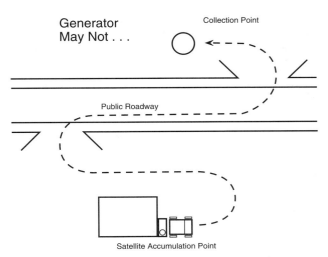

FIG. 3.4.4 Off-site transportation of hazardous waste.

ceeds 10 days, the transporter is considered to be operating a storage facility and must comply with the regulations for such a facility. In addition, transporters who bring hazardous waste into the United States or mix hazardous wastes of different DOT shipping descriptions by placing them in the same container are classified as generators and must comply with the generator regulations.

A transporter is subject to regulations including obtaining an EPA ID number, complying with the manifest system, and dealing with hazardous waste discharges.

The transporter is required to deliver the entire quantity of waste accepted from either the generator or another transporter to the facility designated on the manifest. If the waste cannot be delivered as the manifest directs, the transporter must inform the generator and receive further instructions, such as returning the waste or taking it to another facility. Before handing the waste over to a TSD, the transporter must have the TSD facility operator sign and date the manifest. One copy of the manifest remains at the TSD facility while the other stays with the transporter. The transporter must retain a copy of the manifest for three years from the date the hazardous waste was accepted.

Even if generators and transporters of hazardous waste comply with all appropriate regulations, transporting hazardous waste can still be dangerous. There is always the possibility of an accident. To deal with this possibility, regulations require transporters to take immediate action to protect health and the environment if a release occurs by notifying local authorities and/or closing off the discharge area.

The regulations also give officials special authority to deal with transportation accidents. Specifically, if a federal, state, or local official, with appropriate authority, determines that immediate removal of the waste is necessary to protect human health or the environment, the official can authorize waste removal by a transporter without an EPA ID or a manifest.

MODES OF TRANSPORT

A 1981 report, prepared for the EPA, estimated that 96% of the 264 million tn of hazardous wastes generated each year were disposed at the site where they were generated. By 1989, the National Solid Wastes Management Association (NSWMA 1989) stated that trucks traveling over public highways move over 98% of hazardous wastes that are treated off-site. Another perspective can be gained from statistics for hazardous materials transportation. Rail transportation moves about 8% of the hazardous materials shipped, but 57% of the *ton-miles* of hazardous materials shipped (U.S. Office of Technology Assessment 1986).

The highway transport mode is regarded as the most versatile, and is the most widely used. Tank trucks can access most industrial sites and TSD facilities. Rail shipping requires expensive sidings, and is suitable for very large quantity shipments. Cargo tanks are the main carriers of bulk hazardous materials; however, large quantities of hazardous wastes are shipped in 55-gal drums.

Cargo tanks are the main carriers of bulk hazardous materials over roads. Cargo tanks are usually made of steel or aluminum alloy, or other materials such as titanium, nickel, or stainless steel. They range in capacity from 4,000 to 12,000 gal. Federal road weight laws usually limit motor vehicle weights to 80,000 lb gross. Table 3.4.2 lists DOT cargo tank specifications for bulk shipment of common hazardous materials and example cargos.

As stated above, rail shipments account for about 8% of the hazardous materials transported annually, with about 3,000 loads each day. However, the proportion of hazardous waste shipments is unknown.

TABLE 3.4.2   CARGO TANK TABLE

Cargo Tank Specification Number	Types of Commodities Carried	Examples
MC-306 (MC-300, 301, 302, 303, 305)[a]	Combustible and flammable liquids of low vapor pressure	Fuel oil, gasoline
MC-307 (MC-304)	Flammable liquids, Poison B materials with moderate vapor pressure	Toluene, diisocyanate
MC-312 (MC-310, 311)	Corrosives	Hydrochloric acid, caustic solution
MC-331 (MC-330)	Liquified compressed gases	Chlorine, anhydrous ammonia
MC-338	Refrigerated liquified gases	Oxygen, methane

[a]The numbers in parentheses designate older versions of the specifications; the older versions may continue in service but all newly constructed cargo tanks must meet current specifications. (*Source: Code of Federal Regulations*, Title 49, sections 172.101 and 178.315–178.343).

TABLE 3.4.3 PRESSURE RAIL CARS

Class	Material	Insulation	Test pressure	Relief valve setting	Notes
DOT 105	Steel, aluminum	Required	100	75	No bottom outlet or washout; only one opening in tank; chlorine
			200	150	
			300	225	
			400	300	
			500	375	
			600	450	
DOT 112	Steel	None	200	150	No bottom outlet or washout; anhydrous ammonia
			340	225	
				280	
			400	300	
				330	
			500	375	
DOT 114	Steel	None	340	255	Similar to DOT 105; optional bottom outlet; liquefied petroleum gas
			400	300	

Source: Reprinted from Office of Technology Assessment, 1986, *Transportation of Hazardous Materials* (U.S. Congress, Washington, D.C.).

TABLE 3.4.4 NONPRESSURE RAIL CARS

Class	Material	Insulation	Test pressure	Relief valve setting	Notes
DOT 103	Steel, aluminium, stainless steel, nickel	Optional	60	35	Optional bottom outlet; whiskey
DOT 104	Steel	Required	60	35	Similar to DOT 103
DOT 111	Steel, aluminum	Optional	60	35	Optional bottom outlet and bottom washout
DOT 111A	Steel, aluminum	Optional	100	75	Hydrochloric acid

Source: Reprinted from Office of Technology Assessment, 1986.

The major classifications of rail tank cars are pressure and nonpressure (for transporting both gases and liquids). Both categories have several subclasses, which differ in test pressure, presence or absence of bottom discharge valves, type of pressure relief system, and type of thermal shielding. Ninety percent of tank cars are steel; aluminum is also common.

DOT tank car design specifications are detailed in 49 CFR Part 179. Rail car specification numbers for transporting pressurized hazardous materials are DOT 105, 112, and 114 (Table 3.4.3); for unpressurized shipments the numbers are DOT 103, 104, and 111 (Table 3.4.4). Specifications call for steel jacket plate and thickness ranging from 11 ga (approximately $\frac{1}{8}$ in.) to $\frac{3}{4}$ in and aluminum jacket plate thickness of $\frac{1}{2}$ to $\frac{5}{8}$ in. Capacities for tank cars carrying hazardous materials are limited to 34,500 gal or 263,000 lb gr wt. It is proposed that the gross rail load (GRL) limits on 100-tn trucks be increased to 286,000 lb gr.

Because of regulations and industry initiatives, the tank car of the future may be only three to five years away (Snelgrove 1995). This tank car design will most likely be based on non-accident release (NAR) products. Changes will probably include safety valves or surge devices to replace the safety vent; elimination of bottom loading; improved versions of today's manway design; and the equivalent of pressure heads for non-pressure DOT 111A-specification tank cars.

In recent years, the DOT has significantly revised hazardous materials classifications, hazard communications, and packaging requirements to agree with other national and international United Nations (UN) codes. In the coming years, these regulations will cover every shipping container, from drums and intermediate bulk containers (IBCs), to tractor trailers, and rail tank cars.

—David H.F. Liu

48   HAZARDOUS WASTE

References

Code of Federal Regulations, Title 40, Part 262.
Code of Federal Regulations, Title 49, Parts 172, 173, 178 and 179.
National Solid Waste Management Association (NSWMA). 1989. *Managing hazardous waste: fulfilling the public trust.* Washington, D.C.

Turner, P.L. 1992. Preparing hazwaste transport manifests. *Environmental protection,* December 1992.
U.S. Environmental Protection Agency (EPA) (1986). *RCA Orientation Manual.* Office of Solid Waste. Publication No. EPA 530-SW-86-001. Washington, D.C.: U.S. Government Printing Office.

4
Treatment and Disposal

4.1
TREATMENT, STORAGE, AND DISPOSAL REQUIREMENTS

Treatment, storage, and disposal facilities (TSDs) are the last link in the cradle-to-grave hazardous waste management system. All TSDs handling hazardous waste must obtain operating permits and abide by treatment, storage, and disposal regulations. TSD regulations establish performance standards for owners and operators to minimize the release of hazardous waste into the environment.

The original RCRA establishes two categories of TSDs based on permit status. Section 3005(a) of the act specifies that TSDs must obtain a permit to operate. The first category consists of *interim status* facilities that have not yet obtained permits. Congress recognized that it would take many years for the EPA to issue all permits, therefore, the interim status was established. This allows those who own or operate facilities existing as of November 19, 1980, and who are able to meet certain conditions, to continue operating as if they have a permit until their permit application is issued or denied. The second category consists of facilities with permits.

Under Section 3004(a) of the act, the EPA was required to develop regulations for all TSDs. Although only one set was required, the EPA developed two sets of regulations, one for interim status TSDs and the other for permitted TSDs. While developing TSD regulations, the EPA decided that owners and operators of interim status facilities should meet only a portion of the requirements for permitted facilities.

General Facility Standards

Both interim status and permit standards consist of administrative and nontechnical requirements, and technical and non-specific requirements. The interim status standards, found in 40 CFR Part 265, are primarily good housekeeping practices that owners and operators must follow to properly manage hazardous wastes. The permit standards found in 40 CFR Part 264 are design and operating criteria for facility-specific permits.

As detailed in Section 3.4, all facilities handling hazardous wastes must obtain an EPA ID number. Owners and operators must ensure that wastes are correctly identified and managed according to the regulations. They must also ensure that facilities are secure and operating properly. Personnel must be trained to perform their duties correctly, safely, and in compliance with all applicable laws, regulations, and codes. Owners and operators must:

Conduct waste analyses before starting treatment, storage, or disposal in accord with a written waste analysis plan. The plan must specify tests and test frequencies providing sufficient information on the waste to allow management in accordance with the laws, regulations, and codes.

Install security measures to prevent inadvertant entry of people or animals into active portions of the TSDF. The facility must be surrounded by a barrier with control entry systems or 24-hr surveillance. Signs carrying the warning "Danger—Unauthorized Personnel Keep Out" must be posted at all entrances. Precautions must be taken to avoid fires, explosions, toxic gases, or any other events threatening human health, safety, and the environment.

Conduct inspections according to a written schedule to assess facility compliance status and detect potential problem areas. Observations made during inspections must be recorded in the facility's operating log and kept on file for 3 years. All problems noted must be remedied.

Conduct training to reduce the potential for mistakes that might threaten human health and the environment. In addition, the Occupational Safety and Health Administration (OSHA) now requires each TSD to implement a hazard communication plan, a medical surveillance program, and a health and safety plan. Decontamination procedures must be in place and employees must receive a minimum of 24 hr of health and safety training.

Properly manage ignitable, reactive, or incompatible wastes. Ignitable or reactive wastes must be protected from sources of ignition or reaction, or be treated to eliminate the possibility. Owners and operators must ensure that treatment, storage, or disposal of ignitable, reactive, or incompatible waste does not result in damage to the containment structure, or threaten human health or the environment. Separation of incompatible wastes must be maintained.

Comply with local standards to avoid siting new facilities in locations where floods or seismic events could affect waste management units. Bulk liquid wastes are prohibited from placement in salt domes, salt beds, or underground mines or caves.

PREPAREDNESS AND PREVENTION

Facilities must be designed, constructed, maintained, and operated to prevent fire, explosion, or any unplanned release of hazardous wastes that could threaten human health and the environment. Facilities must be equipped with:

An *internal communication or alarm system* for immediate emergency instructions to facility personnel

Telephone or two-way radio capable of summoning emergency assistance from local police, fire, and emergency response units

Portable fire extinguishers, along with fire, spill control, and contamination equipment

Water at adequate volume and pressure to supply water hoses, foam-producing equipment, automatic sprinklers, or water spray systems

All communication and emergency equipment must be tested regularly to ensure proper emergency operation. All personnel must have immediate access to the internal alarm or emergency communication system. Aisle space must allow unobstructed movement of personnel and equipment during an emergency.

Owners and operators of TSDs must make arrangements to:

Familiarize police, fire, and emergency response teams with the facility, wastes handled and their properties, work stations, and access and evacuation routes

Designate primary and alternate emergency response teams where more than one jurisdiction might respond

Familiarize local hospitals with the properties of hazardous wastes handled at the facility, and the type of injuries or illnesses that could result from events at the facility

CONTINGENCY PLAN AND EMERGENCY PROCEDURE

A contingency plan must be in effect at each TSDF. The plan must minimize hazards from fires, explosions, or any release of hazardous waste constituents. The plan must be implemented immediately whenever there is a fire, explosion or release that would threaten human health or the environment.

The contingency plan must

Describe personnel actions to implement the plan

Describe arrangements with local police, fire, and hospital authorities, as well as contracts with emergency response teams to coordinate emergency services

List names, addresses, and phone numbers of all persons qualified to act as emergency coordinators for the facility

List all emergency equipment, communication, and alarm systems, and the location of each item

Include an evacuation plan for facility personnel

The contingency plan must be maintained at the facility and at all emergency response facilities that might provide services. It must be reviewed and updated whenever any item affecting the plan is changed. A key requirement is designating an emergency coordinator to direct response measures and reduce the adverse impacts of hazardous waste releases.

General Technical Standards for Interim Status Facilities

The objective of the RCRA interim status technical requirements is to minimize the potential for environmental and public health threats resulting from hazardous waste treatment, storage, and disposal at existing facilities waiting for an operating permit. The general standards cover three areas:

- Groundwater monitoring requirements (Subpart F)
- Closure, postclosure requirements (Subpart G)
- Financial requirements (Subpart H)

GROUNDWATER MONITORING

Groundwater monitoring is required for owners or operators of surface impoundments, landfills, land treatment facilities, or waste piles used to manage hazardous wastes. These requirements assess the impact of a facility on the groundwater beneath it. Monitoring must be conducted for the life of the facility, except at land disposal facilities, which must monitor for up to 30 years after closing.

The groundwater monitoring program requires installing a system of four monitoring wells: one up-gradient from the waste management unit and three down-gradient. The down-gradient wells must be placed to intercept any waste from the unit, should a release occur. The up-gradient wells must provide data on groundwater that is not influenced by waste coming from the waste management unit (called background data). If the wells are properly located, data comparisons from up-gradient and down-gradient wells should indicate if contamination is occurring.

After the wells are installed, the owner or operator monitors them for one year to establish background concentrations for selected chemicals. These data form the basis for all future comparisons. There are three sets of parameters for background concentrations: drinking water, groundwater quality, and groundwater contamination.

If a significant increase or decrease in pH is detected for any of the indicator parameters, the owner or operator must implement a groundwater assessment program to determine the nature of the problem. If the assessment shows contamination by hazardous wastes, the owner or operator must continue assessing the extent of groundwater contamination until the problem is ameliorated, or until the facility is closed.

CLOSURE

Closure is the period when wastes are no longer accepted, during which owners or operators of TSD facilities complete treatment, storage, and disposal operations, apply final covers to or cap landfills, and dispose of or decontaminate equipment, structures, and soil.

Following the closure, a 30-yr postclosure period is established for facilities that do not *close clean* as described below. Postclosure care consists of the following at minimum:

- Groundwater monitoring and reporting
- Maintenance and monitoring of waste containment systems
- Continued site security

Clean closure may be accomplished by removing all contaminants from impoundments and waste piles. At a minimum, owners and operators of surface impoundments and waste piles that wish to close clean must conduct soil analyses and groundwater monitoring to confirm that all wastes have been removed from the unit. The EPA or state agency may establish additional clean closure requirements on a case-by-case basis. A successful demonstration of clean closure eliminates the requirement for postclosure care of the site.

FINANCIAL REQUIREMENTS

Financial requirements were established to ensure funds are available to pay for closing a facility, for rendering postclosure care at disposal facilities, and to compensate third parties for bodily injury and property damage caused by accidents related to the facility's operation. There are two kinds of financial requirements:

- Financial assurance for closure and postclosure
- Liability coverage for injury and property damage

To meet financial assurance requirements, owners and operators must first prepare written cost estimates for closing their facilities. If postclosure care is required, a cost estimate for providing this care must also be prepared. These cost estimates must reflect the actual cost of the activities outlined in the closure and postclosure plans, and are adjusted annually for inflation. The cost estimate for closure is based on the point in the facility's operating life when closure would be most expensive. Cost estimates for postclosure monitoring and maintenance are based on projected costs for the entire postclosure period.

The owner or operator must demonstrate to the EPA or state agency an ability to pay the estimated amounts. This is known as financial assurance. The owner/operator may use one or a combination of the following six mechanisms to comply with financial assurance requirements: trust fund, surety bond, letter of credit, closure/postclosure insurance, corporate guarantee for closure, and financial test. All six mechanisms are adjusted annually for inflation or more frequently if cost estimates change.

The Subpart H requirements for these mechanisms are extensive. Readers with particular interest in the details should examine 40 CFR Parts 264 and 265, Subpart H. Liability insurance requirements include coverage of at least $1 million (annual aggregate of at least $2 million) per sudden accidental occurrence, such as fire or explosion. Owners and operators must also maintain coverage of at least $3 million per occurrence (annual aggregate of at least $6 million), exclusive of legal defense costs, for nonsudden occurrences such as groundwater contamination. Liability coverage may be demonstrated using any of the six mechanisms allowed for assurance of closure or postclosure funds.

—David H.F. Liu

4.2
STORAGE

Many early disasters and current Superfund sites grew from uncontrolled accumulation of hazardous wastes. Congress and the EPA sought to impose rigorous controls and accountability on all who accumulate and store hazardous wastes through the RCRA statutes and EPA regulations.

The RCRA defines *storage* as holding hazardous waste for a temporary period, after which the hazardous waste is treated, disposed of, or stored elsewhere. The accumulation of hazardous waste beyond a prescribed period, usually 90 days, is considered storage. The owner or operator of a facility where waste is held for more than 90 days must apply for a permit before starting accumulation, and must comply with regulations pertaining to storage facilities.

Since the primary function of containers and tanks is storage, this overview of 40 CFR Parts 264/265 Subparts I and J includes *permitted* and *interim status* standards for container and tank use. The four general types of land disposal or long-term storage facilities—surface impoundments, waste piles, landfills, and underground injection, are discussed briefly.

Many concerns about storage facilities can be addressed by following proper procedures for storage of materials. Table 4.2.1 lists fundamental storage and handling procedures.

TABLE 4.2.1   PROPER PROCEDURES FOR STORAGE OF HAZARDOUS MATERIALS

Use personal protective equipment
Be familiar with specific hazards of material being handled
Obey all safety rules
Do not smoke while handling materials
Store chemicals according to manufacturer's instructions, away from other chemicals or environmental conditions that could cause reactions
Face labels on containers out
Keep stacks straight and aligned
Check for location accuracy
Do not stack containers too high
Check for loose closures
Place into proper locations as soon as possible
Do not block exits or emergency equipment
Report all spills or leaks immediately

Containers

The 55-gal drum remains a standard container for hazardous waste. The several DOT-specified 55-gal drums are the most frequently used container for collection, storage, shipment, and disposal of liquid hazardous wastes (EPA 1990).

Selecting the proper drum or container requires consulting DOT regulations. The process begins with the 49 CFR Sec. 172.101 Hazardous Materials Transportation Act (MHTA), which is frequently referred to as the heart of MHTA. About 16,000 materials and substances are listed, followed by twelve columns with transport, packaging, and identification requirements.

Drums used in hazardous waste management must be in good condition, clean, free of rust, dents, and creases. In addition, the regulations require:

Containers holding hazardous wastes must always be closed, except when wastes are added or removed
Wastes must be compatible with containers (i.e., corrosive wastes should not be stored in metal containers)
Wastes in leaking or damaged containers must be recontainerized
Containers must be handled properly to prevent ruptures and leaks
Incompatible wastes must be prevented from mixing
Inspections must be conducted to assess container condition

Containers holding ignitable or reactive wastes must be located at least 15 m (50 ft) from the facility's property line.

Permit requirements for containers are similar to the interim status requirements with the following exceptions:

Liquid hazardous waste containers must be placed in a containment system capable of containing leaks and spills. This system must have sufficient capacity to contain 10% of the volume of all containers, or the volume of the largest container, whichever is greater.
When closing a container, all hazardous waste and hazardous waste residues must be removed, unless the container is to be disposed of as hazardous waste.

Tanks

Subpart J regulations apply to stationary tanks storing wastes that are hazardous under Subtitle C of the RCRA. General operating requirements fall into five basic areas:

Tank assessment must be completed to evaluate structural integrity and compatibility with the wastes that the tank system is expected to hold. The assessment covers design standards, corrosion protection, tank tests, waste characteristics, and tankage.

Secondary containment and release detection is required unless the tank does not contain free liquids and is located in a building with impermeable floors. A secondary containment system must be designed, installed, and operated to prevent liquid migration out of the tank system, and to detect and collect any releases that occur. Containment systems include liners, vaults, and double-walled tanks.

Operating and maintenance requirements require the management of tanks to avoid leaks, ruptures, spills, and corrosion. This includes a freeboard or containment structure to prevent and contain escaping wastes. A shut-off or bypass system must be installed to prevent liquid from flowing into a leaking tank.

Response to releases must include immediate removal of the leaking tank contents. The areas surrounding the tank must be visually inspected for leaks and spills. Based on the inspection, further migration of spilled waste must be stopped, and contaminated soils and surface water must be disposed of in accordance with RCRA requirements. All major leaks must be reported to the EPA or state agency.

Closure and postclosure requirements include removing all contaminated soils and other hazardous waste residues from the tank storage area at the time of closure. If decontamination is impossible, the storage area must be closed following the requirements for landfill (EPA 1990).

Surface Impoundments

Surface impoundments are used to reduce waste volume through evaporation, while containing and concentrating residue within liners. Wastes are added directly to lined depressions in the ground known as pits, lagoons, treatment basins, or ponds. Long-term storage more accurately describes the process.

All surface impoundments are required to have at least one liner and be located on an impermeable base. New surface impoundments, replacements, or lateral expansions must include:

- Two or more liners
- A leachate collection system between the liners
- Groundwater monitoring as prescribed in Subpart F

Requirements include preventing liquid from escaping due to overfilling or runoff, and preventing erosion of dikes and dams. Liners must meet permit specifications for materials and thickness. During construction and installation, liners must be inspected for uniformity, damage, and imperfection.

The double-liner system for an impoundment facility is shown in Figure 4.2.1.

Waste Piles

Hazardous waste piles now exist on many industrial sites. Volatile components in such waste piles are available for evaporation and subject to wind and water erosion. They may be leached by percolation of rainfall and runoff. Piles containing minerals or metal values may be leached with weak acid or caustic to recover the value. Unless carefully constructed over an impervious base, leachate escapes to the subsurface, contaminating groundwater or emerging as base flow in streams.

The waste pile has become accepted practice, followed by landfilling, when pile size becomes a problem. Recent determinations of EP or TCLP toxicity will bring many dross and fluff piles within RCRA control. RCRA regulations for waste piles are similar to those for landfills.

Owners or operators of waste piles used for treatment or storage of noncontainerized accumulations of solid, nonflowing hazardous wastes may choose to comply with

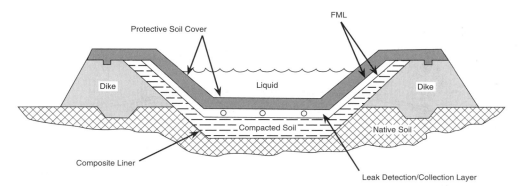

FIG. 4.2.1 Schematic—cross section of a liquid waste impoundment double liner system. (Reprinted, with permission, from W.C. Blackman, 1992, *Basic Hazardous Waste Management* [Boca Raton, Fla.: Lewis Publishers].)

waste pile or landfill requirements. Waste piles used for disposal must comply with landfill requirements. The requirements for managing storage and treatment waste piles include protecting the pile from wind dispersion. The pile must also be placed on an impermeable base compatible with the waste being stored. If hazardous leachate or runoff is generated, control systems must be imposed.

Landfills

Sanitary landfills were developed for municipal refuse disposal to replace open dumps (see Section 10.5). New secure landfills are used to bury non-liquid hazardous wastes in synthetically lined depressions. Secure landfills for hazardous waste disposal are now equipped with double liners, leak detection, leachate monitoring and collection, and groundwater monitoring systems. Synthetic liners are a minimum of 30 mil thickness.

Liner technology has improved greatly and continues to do so. Very large sections of liner fabric now minimize the number of joints. Adjacent sections are welded together to form leak-proof joints with a high degree of integrity. Liners are protected by sand bedding or finer materials free of sharp edges or points which might penetrate the inner fabric. Another layer of bedding protects the inner layer from damage by machinery working the waste. Some states allow one of the liners to be natural clay. The completed liner must demonstrate low permeability and must include a leachate collection system.

Leachate detection and collection systems are equipped with access galleys or other means of leachate removal. Double liner, leakage detection, and leachate collection systems are shown in Figure 4.2.2.

Leachate caps are detailed by the EPA. Figure 4.2.3 is a cross-section of a typical cap design. The objectives of cap design are to protect the cells from erosion, to route

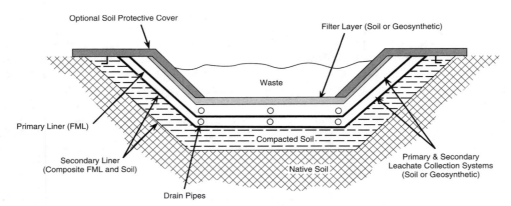

FIG. 4.2.2 Schematic—cross section of a secure landfill double liner system. *Credit:* (Reprinted with permission, from W.C. Blackman, 1992.)

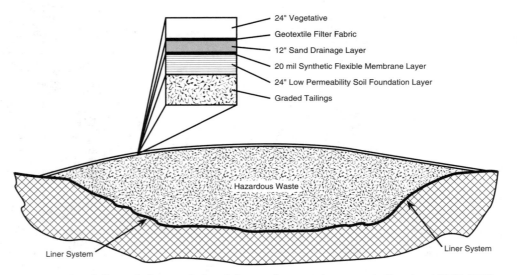

FIG. 4.2.3 Land disposal site cap designed for maximum resistance to infiltration (CH2M-Hill, Denver, CO). (Reprinted, with permission, from W.C. Blackman, 1992.)

potential runoff around and away from the cap, and to prevent buildup of gases generated within the landfill.

Groundwater monitoring schemes are designed to provide up-gradient (background) water quality data, and to detect down-gradient differences in critical water quality parameters. The RCRA requires a minimum of one up-gradient and three down-gradient monitoring wells to detect leakage from landfills (EPA 1981, 1987).

Landfills present two general classes of problems. The first class includes fires, explosions, production of toxic fumes, and related problems from the improper management of ignitable, reactive, or incompatible wastes. Thus, owners and operators are required to analyze wastes to provide enough information for proper management. They must control the mixing of incompatible wastes in landfill cells, and place ignitable and reactive wastes in landfills only when the waste has been rendered unignitable or nonreactive (EPA 1990).

The second class of landfill problems concerns the contamination of surface and groundwater. To deal with problems, interim status regulations require diversion of runoff away from the active face of the landfill; treatment of liquid or semisolid wastes so they do not contain free liquids; proper closure and preclosure care to control erosion and the infiltration of rainfall; and crushing or shredding landfill containers so they cannot collapse later leading to subsidence and breaching of the cover. Groundwater monitoring as described in Subpart F is required, as is collection of rainwater and other runoff from other active faces of the landfill. Segregation of waste such as acids, that would mobilize, make soluble, or dissolve other wastes or waste constituents is required (EPA 1990).

In the HSWA, Congress prohibited disposal of non-containerized liquid hazardous waste, and hazardous waste containing free liquids, in landfills.

Such landfills should be situated away from groundwater sources. These safeguards should be followed because there is no guarantee that engineering solutions will be able to contain the wastes in perpetuity. A well-built facility may allow sufficient leadtime for remedial action before environmental damage occurs.

Secure landfills meeting new RCRA standards may, under temporary variances, be able to accept a few hazardous wastes for which alternative disposal methods have not been developed. Secure landfills may also accept hazardous wastes that are treated to the best demonstrated available technology.

Underground Injection

Underground injection involves using specially designed wells to inject liquid hazardous waste into deep earth strata containing non-potable water. Through this method, a wide variety of waste liquids are pumped underground into deep permeable rocks that are separated from fresh water aquifers by impermeable layers of rock above, below, and lateral to the waste layer. The depth of an injection ranges from 1,000 to 8,000 ft and varies according to the geographical factors of the area. HSWA prohibits the disposal of hazardous waste within $\frac{1}{4}$ mi of an underground source of drinking water.

Figure 4.2.4 is a cross-section of a typical injection well. To prevent plugging of the injection equipment, wastes are usually pretreated to remove solids greater than one micron. The well must be constructed to assure that potable water zones are isolated and protected. At minimum, well casings must be cemented and must extend through all potable water zones.

Deep-well disposal uses limited formation space, is expensive in construction and operation, and the subject of ever-tightening regulations. For hazardous liquid waste to be deep-well injected, the following criteria must apply: the hazardous liquid waste must have a low volume and a high concentration of waste, cannot cause an unfavorable reaction with material in the disposal zone, must be biologically inactive, must be noncorrosive, and must be

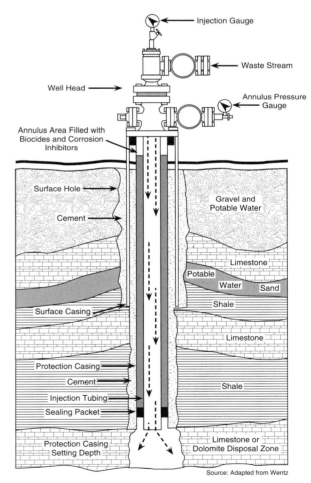

FIG. 4.2.4 Schematic cross section design of a hazardous waste injection well. (Reprinted and adapted, with permission, from C.A. Wentz, 1989, *Hazardous Waste Management* [McGraw-Hill, Inc.].)

difficult to treat by other methods. Thus, the method should be used only for those wastes with no other feasible management options.

Due to faulty construction or deterioration, there is a potential for leakage from some old wells. Detection of a leak and remedial action may not be feasible because of the nature and location of the leakage. Because of the difficulty associated with monitoring subsurface migration of liquid waste, the potential for geographical disturbances to the underground injection system, or the geographical nature of the land, underground injection wells are severely restricted in most states.

—David H.F. Liu

References

U.S. Environmental Protection Agency (EPA). 1981. *Guidance document for subpart F air emission monitoring—land disposal toxic air emissions evaluation guidelines.* Report No. PB87-155578. National Technical Information Service. Springfield, Va.

———. 1987. *Background document on bottom liner performance in double-lined landfills and surface impoundments.* Report PB87-182291. National Technical Information Service. Springfield, Va.

——— 1987. *Background information on proposed liner and leak detection rule.* Report No. PB87-191383. National Technical Information Service. Springfield, Va.

——— 1990. *RCRA orientation manual,* 1990 Edition, Superintendent of Documents. Washington, D.C.: Government Printing Office.

4.3
TREATMENT AND DISPOSAL ALTERNATIVES

Hazardous waste treatment is a rapidly growing, innovative industry. This innovation is driven by the need for effective and economical processes for treating wastes rather than placing waste in landfills without treatment. Among waste management options (Table 4.3.1), the most desirable is source reduction through process modification (Combs 1989). The less desirable options follow.

If waste can be eliminated or reduced significantly, subsequent treatment processes become unnecessary or are reduced in scope. These highly desirable waste minimization alternatives must be carefully considered as reasonable technical and economic solutions to hazardous waste management.

Available Processes

Not all wastes can be eliminated through source reduction or recycling. Most manufacturing waste products require treatment to destroy the wastes or render them harmless to the environment. Technological options for waste handling depend upon waste type, amount, and operating cost. Figure 4.3.1 aligns categories of industrial wastes with the treatment and disposal processes usually applied.

Numerous chemical, physical, and biological treatments are applicable to hazardous wastes. Many such treatment processes are used in by-product recovery and volume reduction processes. All wastes should first be surveyed and characterized to determine which treatment or destruction process should be used.

Hazardous wastes may be organic or inorganic. Water will dissolve many of these substances, while others have limited solubility. Sodium, potassium, and ammonium salts are water soluble, as are mineral acids. Most halogenated inorganics, except fluoride, are soluble; while many carbonates, hydroxides, and phosphates are only slightly soluble. Alcohols are highly soluble, but aromatics and long-chained petroleum-based organics are of low solubility. Solubility is critical in chemical treatment processes.

The following treatment alternatives are detailed in Figure 4.3.1.

Low-concentration effluents and other wastewaters usually require modest capital and operating costs to treat before discharging into municipal sewers.

Strong acids and alkalis can be neutralized to prevent characterization as hazardous wastes under the RCRA corrosivity criteria. Frequently, industrial water may be acid or basic, requiring neutralization before any other treatment. It may be feasible to mix an acidic waste stream with a basic stream to change the pH to a more neutral level of 6 to 8.

TABLE 4.3.1   WASTE MANAGEMENT OPTIONS AND PRIORITIES

- Source reduction (process modification)
- Separation and volume reduction
- Exchange/sale as raw material
- Energy recovery
- Treatment
- Secure ultimate disposal (landfill)

Source: Reprinted with permisstion from G.D. Combs, 1989, *Emerging treatment technologies for hazardous waste,* Section XV, Environmental Systems Company (Little Rock, AR).

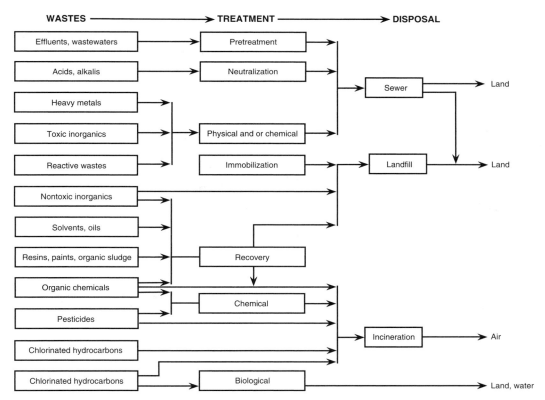

FIG. 4.3.1 Treatment and disposal alternatives for industrial wastes. (Reprinted and adapted with permission from Charles A. Wentz, 1989, *Hazardous Waste Management,* New York, N.Y.: McGraw-Hill, Inc.)

As heavy metals are virtually impossible to destroy, they must be managed by immobilization techniques. After heavy metals have undergone fixation processes and are nonleachable, they can be placed in landfills.

Reactive wastes and toxic inorganics, such as hexavalent chromium and aqueous cyanide-bearing wastes, must be handled carefully prior to the chemical treatments and separation processes that will make them environmentally acceptable. Hexavalent chromium is highly toxic. When it is reduced to trivalent chromium it can be precipitated as chromium hydroxide, which is much less toxic and more acceptable for subsequent recovery or disposal. A common method for treating aqueous cyanide waste is alkaline chlorination.

Should inorganic waste streams contain sufficient amounts of metals or other potentially valuable resources, recovery via physical and chemical processes is highly desirable. Recovery potential must be studied on a case-by-case basis, considering the estimated value of the quantities available, the market acceptance of the recovered materials, and the public perception of recycling and reusing such waste products.

Organic wastes such as solvents, resins, paints, sludges, and chemicals offer considerable recovery potential. Separation techniques such as distillation or extraction can recover valuable hydrocarbon streams for energy or chemical process industry use. However, organic re-covery processes still produce a concentrated but significant volume of hazardous waste that eventually must be destroyed or landfilled.

The destruction of hazardous wastes, such as chlorinated hydrocarbons and pesticides, that cannot be eliminated or recovered involves incineration or biological treatment. Incineration is the third alternative in the EPA's preventive hierarchy, after source reduction and recycling. It is preferred because it eliminates potential problems in landfill disposal or other interim waste management processes.

Biological treatment also offers the potential for complete destruction of biodegradable hazardous wastes. The development of specialized microbes for efficient destruction systems eliminates the need for landfill disposal.

Ultimate disposal of products from hazardous waste management facilities will affect the air, water, and land. There is simply no way to avoid placing the waste by-products of our society and technology into our air, water, and land.

Process Selection

The various waste streams managed in a facility should be surveyed. The waste streams should then be characterized using sampling and analytical techniques to quantify potential threats to human health and the environment. Then the most cost-effective and environmentally safe manner of managing these wastes should be determined.

The hazardous waste activities of other firms provides insight into what needs to be done within an industry to be competitive. Information based on competitive activities is generally accessible and can lead to a shorter learning curve for companies needing to achieve regulatory compliance.

The adaptability of various process technologies to specific hazardous wastes should help to define the limitations of any proposed treatment system. This critique should be made early in the decision-making process to ensure the selection of a technology that is compatible with the waste stream to be controlled (Grisham 1986; Long & Schweitzer 1982).

The selection of treatment systems and ultimate disposal options is usually based on the following considerations.

- Federal, state, and local environmental regulations
- Potential environmental hazards
- Liabilities and risks
- Geography
- Demography

The selection of waste control technologies is based, in part, upon economics (Smith, Lynn & Andrews 1986). Government regulations, adaptability of process technology, public relations and geographic locations are also considerations. The final decision, in the end, can be largely influenced by subjective political reasons.

—*David H.F. Liu*

References

Combs, G.D. 1989. *Emerging treatment technologies for hazardous waste.* Section XV. Environmental Systems Company. Little Rock, Ar.

Grisham, J.W. 1986. *Health aspects of the disposal of waste chemicals.* New York, N.Y.: Pergamon.

Long, F.A., and G.E. Schweitzer. 1982. *Risk assessment at hazardous waste sites.* ACS Symposium Series. Washington, D.C.

Smith, M.A., F.M. Lynn, and R.N.L. Andrews. 1986. Economic impacts of hazardous waste facilities. *Hazardous Waste and Hazardous Materials,* Vol. 3, no. 2.

4.4
WASTE DESTRUCTION TECHNOLOGY

PARTIAL LIST OF SUPPLIERS

Liquid Injection Incinerators
Ensco Environmental Services: TRANE Thermal Co.; Coen Co. Inc.; John Zink Co.; Vent-o-Matic Incinerator Corp.; Lotepro Co.

Rotary Kiln Incinerators
S.D. Myers, Inc.; American Industrial Waste of ENCSO, Inc. (mobile); Exceltech, Inc.; Coen Co.; International Waste Energy Systems; Thermal, Inc.; Lurgi Corp.; Komline Sanderson; Winston Technology, Inc. (mobile); Volland, U.S.A.; Von Roll: DETOXCO Inc.

Fluidized Bed Incinerators
Lurgi Corp.; G.A. Technologies; Waste-Tech Services, Inc.: Dorr-Oliver; Combustion Power; Niro Atomizer

Wet Air Oxidation
Zimpro Inc.; Modar Inc.; Vertox Treatment Systems

Supercritical Water Oxidation
Vertox Corporation; Modar Inc.

Incineration offers advantages over other hazardous waste treatment technologies, and certainly over landfill operations. Incineration is an excellent disposal technology for all substances with high heat release potentials. Liquid and solid hydrocarbons are well adapted to incineration. Incineration of bulk materials greatly reduces the volume of wastes. Any significant reduction in waste volume makes management simpler and less subject to uncertainty.

If wastewater is too dilute to incinerate, yet too toxic to deepwell or biotreat, it is a good candidate for Wet Air Oxidation. Unlike other thermal processes, Wet Oxidation produces no smoke, fly ash or oxides. Spent air from the system passes through an adsorption unit to meet local air quality standards. Operating results show destruction approaching or exceeding 99% for many substances, including cyanides, phenols, sulfides, chlorinated compounds, pesticides, and other organics.

This section focuses on the various types of incineration, wet oxidation, and supercritical water oxidation processes.

Incineration

Incineration is a versatile process. Organic materials are detoxified by destroying the organic molecular structure through oxidation or thermal degradation. Incineration provides the highest degree of destruction and control for a broad range of hazardous substances (Table 4.4.1). Design and operating experience exists and a wide variety of commercial incineration systems are available.

TABLE 4.4.1 SUMMARY OF INCINERATOR DESTRUCTION TEST WORK

Waste	Incineration[a]	Destruction Efficiency of Principal Component (%)
Shell aldrin (20% granules)	MC	99.99
Shell aldrite	MC	99.99
Atrazine (liquid)	MC	99.99
Atrazine (solid)	MC	99.99
Para-arsanilic acid	MS	99.999
Captan (solid)	MC	99.99
Chlordane 5% dust	LI	99.99
Chlordane, 72% emulsifiable concentrate and no. 2 fuel oil	LI	99.999
Chlorinated hydrocarbon, trichloropropane, trichlorethane, and dichloroethane predominating	HT	99.92 99.98
Chloroform	MS	99.999
DDT 5% oil solution	LI	99.99
DDT (solid)	MM	99.970 to 99.98
DDT 10% dust	MC	99.99
20% DDT oil solution	LI	99.98
DDT 25% emulsifiable concentration	LI	99.98
DDT 25% emulsifiable concentrate	MC	99.98 to 99.99
DDT oil 20% emulsified DDT waste oil—1.7% PCB	TO	99.9999
DDT powder	MS	99.998
Dieldrin—15% emulsifiable concentrate	LI	99.999
Dieldrin—15% emulsifiable concentrates and 72% chlordane emulsifiable concentrates (mixed 1:3 ratio)	LI	99.98
Diphenylamine-HCl	MS	99.999
Ethylene manufacturing waste	LI	99.999
GB ($C_4H_{10}O_2PP$)	MS	99.99999969
Herbicide orange	RL	99.999 to 99.985
Hexachlorocyclopentadiene	LI	99.999
Acetic acid, solution or kepone	RKP	99.9999
Toledo sludge and kepone coincineration	RKP	99.9999
Lindane 12% emulsifiable concentrate	LI	99.999
Malathion	MS	99.999 to 99.9998
Malathion 25% wet powder	MC	99.99
Malathion 57% emulsifiable concentrate	MC	99.99
Methyl mathacrylate (MMA)	FB	99.999
0.3% Mirex bait	MC	98.21 to 99.98
Mustard	MS	99.999982
Nitrochlorobenzene	LI	99.99 to 99.999
Nitroethane	MS	99.993
Phenol waste	FB	99.99
Picloram	MC	99.99
Picloram, (tordon 10K pellets)	MC	99.99
PCBs	RK	99.999964
PCB capacitors	RK	99.5 to 99.999

(Continued on next page)

TABLE 4.4.1 *(Continued)*

Waste	Incineration[a]	Destruction Efficiency of Principal Component (%)
PCB	CK	99.9998
Polyvinyl chloride waste	RK	99.99
Toxaphene 20% dust	MC	99.99
Toxaphene 60% emulsifiable concentrate	MC	99.99
Trichlorethane	MS	99.99
2,4-D low-volatile liquid ester	LI	99.99
2,4,5-T (Weedon™)	MM	99.990 to 99.996
2,4,5-T	SH	99.995
2,4,5-T	SH	99.995
2,4,5-T	SH	92
2,4,5-T	SH	99.995
VX ($C_{11}H_{26}O_2PSN$)	MS	99.999989 to 99.9999945
Zineb	MC	99.99

Source: J. Corini, C. Day, and E. Temrowski, 1980 (Sept. 2). Trial Burn Data (unpublished draft) Office of Solid Waste, U.S. Environmental Protection Agency, Washington, D.C.
[a]MC = multiple chamber; MS = molten salt combustion; LI = liquid injection; HT = 2 high-temp. incinerators; MM = municipal multiple-hearth sewage sludge incinerator; TO = thermal oxidizer waste incinerator; RL = 2 identical refractory-lined furnaces; RKP = rotary kiln pyrolyzer; FB = fluidized bed; CK = cement kiln; SH = single-hearth furnace.

Detoxified hazardous wastes include combustible carcinogens, mutagens, teratogens, and pathological wastes. Another advantage of incineration is the reduction of leachable wastes from landfills. Incineration of contaminated soils is increasing. The EPA, for example, employed a mobile incinerator to decontaminate 40 tn of Missouri soil that was contaminated with 4 lb of dioxin compounds.

Different incineration technologies are used to handle various types of hazardous waste. The four most common incinerator designs are liquid injection (sometimes combined with fume incineration), rotary kiln, fixed hearth and fluidized bed incinerators.

The four major subsystems of hazardous waste incineration are: (1) waste preparation and feeding, (2) combustion chamber(s), (3) air pollution control, and (4) residue and ash handling. The normal orientation of these subsystems is shown in Figure 4.4.1, along with typical process component options.

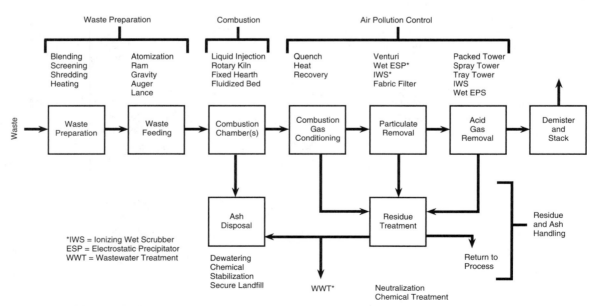

FIG. 4.4.1 General orientation of incineration subsystems and typical process component options. (Reprinted, with permission, from Dempsey and Oppelt 1993).

INCINERATOR SYSTEM DESIGN

Incinerator design plays a key role in ensuring adequate destruction of waste. Important data on waste characteristics needed to design an effective incineration system are listed in Table 4.4.2.

The major incinerator design factors significantly affecting thermal destruction of hazardous waste include:

Temperature

Temperature is probably the most significant factor in ensuring proper destruction of hazardous waste in incinerators. The *threshold temperature* is defined as the operating temperature to initiate thermal destruction of hazardous waste. The threshold temperature ensures waste destruction and allows cost-effective operation.

Residence Time

Incinerator volume determines the residence time for a given flow rate. Residence time, combined with thermal destruction temperature, ensures compliance with destruction and removal efficiency (DRE) regulations. Sufficient residence time must be allowed to achieve DREs, as well as to ensure conversion to desirable incinerator products.

Turbulence

Turbulence is used to attain desirable DREs and to cut operating temperature and residence time requirements. The incinerator configuration affects the degree of turbulence. Pumps, blowers, and baffles should be selected based upon the type of waste to be incinerated and the desired DREs. Heat transfer and fluid flow should be considered in the turbulence requirements.

TABLE 4.4.2   IMPORTANT THERMAL TREATMENT DATA NEEDS

Need	Purpose
Heat Content (HHV and LHV)	Combustiblity
Volatile Matter Content	Furnace Design
Ash Content	Furnace Design, Ash Handling
Ash Characteristics	Furnace Design
Halogen Content	Refractory Design, Flue Gas Ductwork Specification, APC Requirements
Moisture Content	Auxiliary Fuel Requirements
Heavy Metal Content	Air Pollution Control

NOTE: Generally, the data needs for evaluating thermal processes include the data needed for physical treatment for the purpose of feed mechanism design.

Pressure

Thermal destruction systems, which operate at slightly positive elevated pressures, require nonleaking incinerators. Pressurized systems require high-temperature seals for trouble-free operation.

Air Supply

Incomplete combustion products result from insufficient residence time, temperature, or air. The thermal destruction unit must be supplied with amounts of oxygen or air higher than the stoichiometric amount required, to ensure that products of hydrocarbon combustion ultimately result in carbon dioxide and water.

Construction Materials

Most incinerators are constructed with materials selected for continuous trouble-free operation with many hazardous wastes and under many destructive conditions. Materials of construction range from ordinary steel to exotic alloys. The chemical and physical properties of the wastes to be incinerated must be well-defined for selection of materials to ensure a longer operating life and fewer maintenance problems for the incinerator.

Auxiliaries

Numerous additional features must be considered:

Feed systems must be designed to incorporate the hazardous wastes identified by market surveys.
Afterburners may be needed to ensure proper DRE capability.
Downstream treatment is usually necessary to neutralize and remove undesirable destruction products such as mineral acids.
Ash removal may play a key role in the thermal destruction of solid and semi-solid wastes.

Combustion Chambers

Many hazardous wastes are incinerated in industrial boilers and furnaces. However, hazardous waste combustion in boilers is limited by the amount of chlorine in the waste stream, because most industrial boilers do not use scrubbers for hydrogen chloride.

The physical form of the waste and its ash content determine the type of combustion chamber selected. Table 4.4.3 provides selection considerations for the four major combustion chamber designs as a function of different forms of waste (EPA 1981; Dempsey & Oppelt 1993). Incinerator systems derive their names from the types of combustion chambers used.

TABLE 4.4.3 APPLICABILITY OF MAJOR INCINERATOR TYPES TO WASTES OF VARIOUS PHYSICAL FORM

	Liquid Injection	Rotary Kiln	Fixed Hearth	Fluidized Bed
Solids:				
Granular, homogeneous		X	X	X
Irregular, bulky (pallets, etc.)		X	X	
Low melting point (tars, etc)	X	X	X	X
Organic compounds with fusible ash constituents		X	X	X
Unprepared, large, bulky material		X	X	
Gases:				
Organic vapor laden	X	X	X	X
Liquids:				
High organic strength aqueous wastes	X	X	X	X
Organic liquids	X	X	X	X
Solids/liquids:				
Waste contains halogenated aromatic compounds (2,200°F minimum)	X	X	X	
Aqueous organic sludge		X		X

Source: Reprinted with permission from C.R. Dempsey and E.T. Oppelt, 1993, Incineration of hazardous waste: a critical review update, *Air & Waste,* Vol. 43, 1993.

LIQUID INJECTION INCINERATORS

Liquid injection incinerators are applicable for pumpable liquid waste. These units (Figure 4.4.2) are usually simple, refractory-lined cylinders (either horizontally or vertically aligned) equipped with one or more waste burners. Liquid wastes are injected through the burner(s), atomized to fine droplets and burned in suspension. Burners, as well as separate injection nozzles, may be oriented for axial, radial or tangential firing. Improved use of combustion space and higher heat release rates can be achieved by using swirl or vortex burners, or designs involving tangential entry. A forced draft must be supplied to the combustion chamber for the necessary mixing and turbulence.

Good atomization is critical to achieving high destruction efficiency in liquid combustors. Nozzles have been developed to produce mists with mean particle diameters as low as 1 micron (μm), as compared to oil burners, which yield oil droplets in the 10 to 50 μm range. Atomization may be obtained by low pressure air or steam (25 to 100 psig), or mechanical (hydraulic) means using specially designed orifices (25 to 250 psig).

Vertical, downward-oriented liquid injection incinerators are preferred when wastes are high in inorganic salts and fusible ash content; horizontal units may be used with low ash waste. In the past, the typical capacity of liquid injection incinerators was 30 MM Btu/hr heat release. However, units as high as 210 MM Btu/hr are in operation.

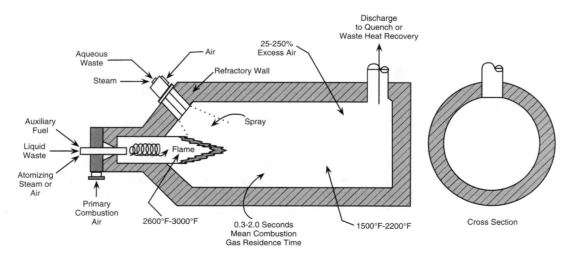

FIG. 4.4.2 Typical liquid injection combustion chamber. (Reprinted, with permission, from Dempsey and Oppelt 1993.)

ADVANTAGES

- Incinerates a wide range of liquid wastes
- Requires no continuous ash removal system, other than for air pollution control
- Capable of a high turndown ratio
- Provides fast temperature response to changes in the waste-fuel flow rate
- Includes virtually no moving parts
- Allows low maintenance costs
- Is a proven technology

DISADVANTAGES

- Must be able to atomize liquids through a burner nozzle except for certain limited applications
- Must provide for complete combustion and prevent flame impingement on the refractory
- Susceptible to plugging. High percent solids can cause problems
- No bulk solids capability

ROTARY KILN INCINERATORS

Rotary kiln incinerators (Figure 4.4.3) are more versatile, as they are used to destroy solid wastes, slurries, containerized wastes, and liquids. Because of this, these units are frequently incorporated into commercial off-site incineration facilities and used for Superfund remediation.

The rotary kiln is a horizontal, cylindrical, refractory-lined shell mounted on a slight slope. Rotation of the shell transports waste through the kiln and mixes the burning solid waste. The waste moves concurrently or countercurrently to the gas flow. The residence of waste solids in the kiln is generally 0.5 to 1.5 hrs. This is controlled by kiln rotation speed (typically 0.5 to 1 rpm), waste feed rate, and in some instances, internal dams to retard waste movement through the kiln. The feed rate is regulated, limiting the waste processed to 20% or less of kiln volume.

Rotary kilns are typically 5–12 ft in diameter and 10–30 ft in length. Rotary kiln incinerators generally have a length-to-diameter ratio (L/D) of 2:8. Smaller L/D ratios result in less particulate carryover. Higher L/D ratios and slower rotational speeds are used when waste materials require longer residence time.

The primary function of the kiln is converting solid wastes to gases through a series of volatilization, destructive distillation, and partial combustion reactions. An afterburner, connected directly to the discharge end of the kiln, completes gas-phase combustion reactions. Gases exiting the kiln are directed to the afterburner chamber.

Some recent systems have a "hot cyclone" installed between the kiln and afterburner to remove solid particles that might create slagging problems in the afterburner. The afterburner may be horizontally or vertically aligned, and functions on the same principles as a liquid injection in-

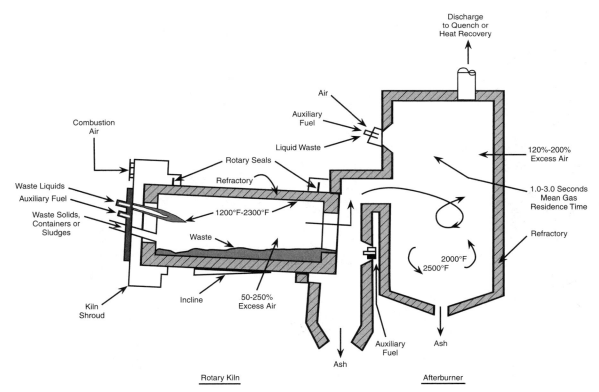

FIG. 4.4.3 Typical rotary kiln/afterburner combustion chamber. (Reprinted, with permission, from Dempsey and Oppelt 1993.)

cinerator. In fact, many facilities also fire liquid hazardous waste through separate waste burners in the afterburner. Afterburners and kilns are usually equipped with auxiliary fuel-firing systems to bring the units up to temperature and to maintain the desired operating temperatures. Some operators fire aqueous waste streams into afterburners as a temperature control measure. Rotary kilns are designed with a heat release capacity of up to 150 MM Btu/hr in the United States; Average units are typically around 60 MM Btu/hr.

ADVANTAGES

- Incinerates a wide variety of liquid and solid wastes
- Receives liquids and solids separately or in combination
- Not hampered by materials passing through a melt phase
- Includes feed capability for drums and bulk containers
- Permits wide flexibility in feed mechanism design
- Provides high turbulence and air exposure of solid wastes
- Continuous ash removal does not interfere with waste burning
- Adapts for use with wet-gas-scrubbing system
- Permits residence time of waste to be controlled by adjusting rotational speed of the kiln

- Allows many wastes to be fed directly into the incinerator without preparations such as preheating or mixing
- Operates at temperatures in excess of 2500°F (1400°C), destroying toxic compounds that are difficult to degrade thermally
- Uses proven technology

DISADVANTAGES

- Requires high capital installation costs, especially for low feed rates
- Necessitates operating care to prevent refractory damage from bulk solids
- Permits airborne particles to be carried out of the kiln before complete combustion
- Frequently requires large excess air intakes due to air leakage into the kiln by the kiln end seals and feed chute. This affects supplementary air efficiency
- Causes high particulate loadings into the air-pollution control system
- Allows relatively low thermal efficiency

FIXED HEARTH INCINERATORS

Fixed hearth incinerators, also called controlled air, starved air, or pyrolytic incinerators, are the third technology for hazardous waste incineration. These units employ a two-stage combustion process, much like rotary kilns.

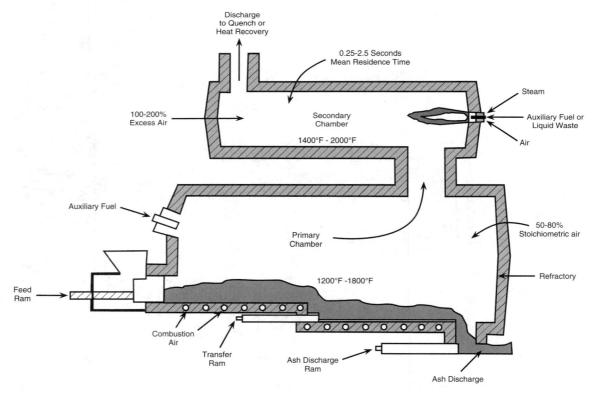

FIG. 4.4.4 Typical fixed hearth combustion chamber. (Reprinted, with permission, from Dempsey and Oppelt 1993.)

As shown in Figure 4.4.4, waste is ram-fed or pumped into the primary chamber, and burned at roughly 50–80% of stoichiometric air requirements. This starved air condition causes the volatile waste to be vaporized by the endothermic heat provided in oxidation of the fixed carbon fraction. The resulting smoke and pyrolytic products consist primarily of methane, ethane, and other hydrocarbons; carbon monoxide and combustion products pass to the secondary chamber. Here additional air is injected to complete combustion, which occurs spontaneously or through the addition of supplementary fuels. Primary chamber combustion reactions and turbulent velocities are maintained at low levels by the starved-air conditions, minimizing particulate entrainment and carryover. With the addition of secondary air, the total excess air for fixed hearth incinerators is 100–200%.

Fixed hearth units tend to be smaller in capacity than liquid injection or rotary kiln incinerators because of the physical limitations in ram-feeding and transporting large amounts of waste material through the combustion chamber. Lower capital costs and reduced particulate control requirements make them more attractive than rotary kilns for smaller on-site installations.

ADVANTAGE
- Represents proven technology

DISADVANTAGES
- Requires more labor
- Operates at a temperature lower than necessary for acceptable waste destruction

FLUIDIZED BED INCINERATORS

Fluidized bed combustion systems have only recently been applied in hazardous waste incineration. Fluidized bed incinerators may be either circulating or bubbling bed designs (Chang et al. 1987). Both types consist of single refractory-lined vessels partially filled with particles of sand, aluminum, calcium carbonate or other such materials. Combustion air is supplied through a distributor plate at the bottom of the combustor (Figure 4.4.5) at a rate sufficient to fluidize (bubbling) or entrain part of the bed material (recirculating bed). In the recirculating bed design, air velocities are higher and the solids are blown overhead, separated in a cyclone, then returned to the combustion chamber (Figure 4.4.6). Operating temperatures are normally in the 1400–1600°F range. Excess air requirements range from 25–150%.

Fluidized bed incinerators are used primarily for liquids, sludges, or shredded solid materials, including soil. To allow good circulation of waste materials and removal of solid residues within the bed, all solids require pre-screening or crushing to a size less than 2 in in dia.

Fluidized bed incinerators offer: high gas-to-solids ratios, high heat transfer efficiencies, high turbulence in both

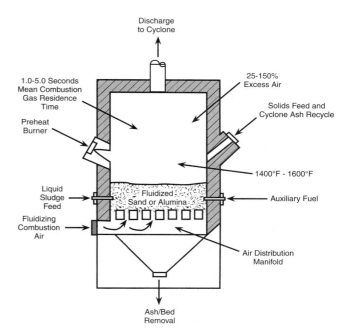

FIG. 4.4.5 Typical fluidized bed combustion chamber. (Reprinted, with permission, from Dempsey and Oppelt 1993).

gas and solid phases, uniform temperatures throughout the bed, and the potential for in-situ gas neutralization by lime, limestone, or carbonate addition. Fluidized beds also have the potential for solid agglomeration in the bed, especially if salts are present in waste feeds.

ADVANTAGES
- Burns solid, liquid, and gaseous wastes
- Simple design has no moving parts
- Compact design due to high heating rate per volume
- Low gas temperatures and excess air requirements minimize nitrogen oxide formation
- Large active surface area enhances combustion efficiency
- Fluctuations in feed rate and composition are easily tolerated due to large heat capacity

DISADVANTAGES
- Residual materials are difficult to remove
- Fluid bed must be prepared and maintained
- Feed selection must prevent bed damage
- Incineration temperatures limited to 1500°F max to avoid fusing bed material
- Little experience on hazardous waste combustion

A wide range of innovative technologies such as high- and low-temperature plasmas, molten salt, molten glass and molten metals baths have merged since the passage of RCRA (Freeman 1990). Many such techniques are now in development.

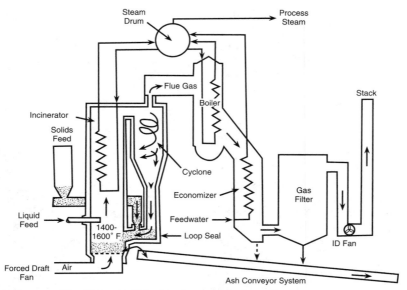

FIG. 4.4.6 Circulating fluid-bed incinerator for hazardous waste.

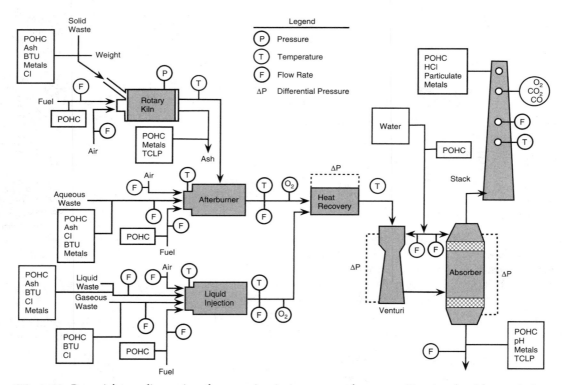

FIG. 4.4.7 Potential sampling points for assessing incinerator performance. (Reprinted, with permission, from Dempsey and Oppelt 1993).

PROCESS PERFORMANCE

Performance measurement is undertaken for any of the following three purposes:

- Establishing initial or periodic compliance with performance standards (e.g., trial burns)
- Routine monitoring of process performance and direct process control (e.g., continuous monitoring)

- Conducting performance measurements for research and equipment development

Figure 4.4.7 illustrates sampling points for assessing incinerator performance. In trial burn activities, sampling activities focus on collecting of waste feed and stack emission samples. Ash and air pollution control system residues are also sampled and analyzed. Sampling of input and output around individual unit components, e.g. scrubbers, may also be conducted in research or equipment evaluation studies.

Trial Burns

Trial burns provide regulatory agencies with the data to issue operating permits. Consequently, trial runs are directed to show that plants achieve the RCRA limits under the desired operating conditions. These RCRA limits are:

A destruction and removal efficiency (DRE) of greater than 99.99% for each of subject principal organic hazardous constituents (POHCs). (Note: the Toxic Substance Control Act [TSCA] requires that incinerators burning PCB and dioxin-containing waste achieve 99.9999% DRE.)

A particulate emission of less than 180 mg per dry standard cu meter (0.08 grains/dry ft) of stack gas (corrected to 7% O_2)

Hydrogen chloride (HCl) emissions less than 4 lb/hr (2.4 kg/hr) or greater than 99% removal efficiency

Trial burns test the plant's operating conditions and ability to meet the three RCRA limits. The EPA recommends three or more runs under any one set of conditions, with varying conditions, or with different waste feed characteristics.

Operating Permits

Permits should allow plants to incinerate the expected types and quantities of waste, at the necessary feed rates and within an acceptable range of operating conditions. Permit conditions must be flexible, with limits that are reasonably achievable. Based on trial burn results, operating permits may specify certain criteria such as:

- Maximum concentration of certain POHCs in waste feed
- Maximum waste feed rate or maximum total heat input rate
- Maximum air feed rate or maximum flue gas velocity
- Minimium combustion temperature
- Maximum carbon monoxide content of stack gas
- Maximum chloride and ash content of waste feed

Sampling and Analysis

The EPA provides guidance on sampling and analysis methods for trial burns designed to measure facility compliance with the RCRA incinerator standards (EPA 1981, 1983, 1989, 1990c; Gorman, et al. 1985).

Table 4.4.4 outlines sampling methods typically involved in RCRA trial burns. Sampling method numbers refer to methods identified in EPA guidance documents and reports (Harris et al. 1984; EPA 1990). The EPA has a computerized data base including a reference directory on the availability and reliability of sampling and analysis methods for designated POHCs.

Assuring Performance

Key control parameters used to trigger fail-safe controls are presented in Table 4.4.5. The parameters are divided into three groups:

Group A parameters are continuously monitored and interlocked to the automatic waste feed cutoff.

Group B parameters are set to ensure that worst-case conditions demonstrated in the trial run are not exceeded during continuous operation. They are not linked with the automatic waste cutoff.

Group C parameters are based on equipment manufacturers' design and operating specifications. They are set independently of trial-run results and are not linked with automatic waste feed cutoff.

No individual real-time monitoring performance indicators appear to correlate with actual organic DRE. No correlation between indicator emissions of CO or HC and DRE has been demonstrated in field-scale incinerator operations, although CO is a conservative indicator of organic emissions. It may be that combinations with other potential real-time indicators, such as surrogate destruction, may be desirable.

Wet Air Oxidation
PROCESS DESCRIPTION

The patented Zimmermann process involves flameless or wet combustion in aqueous solution or dispersions (Zimmermann 1954). Unlike other thermal processes, wet air oxidation does not require dewatering before combustion and creates no air pollution. In aqueous dispersion, a wide range of organic and hazardous industrial wastes can be oxidized to carbon dioxide and water by the addition of air or oxygen. Water, the bulk of the aqueous phase, catalyzes oxidation reactions so they proceed at relatively low temperatures (350–650°F). At the same time, water moderates the oxidation rates by evaporation.

Figure 4.4.8 shows a simplified flow scheme of a continuous air oxidation system. The waste liquor is mixed with air and is preheated by steam during process startup and by hot reactor effluent during operation to 300°–400°F. At this reactor inlet temperature oxidation starts, with the associated heat release further increasing the temperature as the liquid air mixture moves through the reactor. The higher the operating temperature, the greater the destruction of organic pollutants for the same residence time period. The operating temperature cannot

TABLE 4.4.4 SAMPLING METHODS AND ANALYSIS PARAMETERS

Sample	Sampling frequency for each run	Sampling method[a]	Analysis parameter[b]
1. Liquid waste feed	Grab sample every 15 min	S004	V&SV-POHCs, Cl, ash, ult, anal., viscosity, HHV, metals
2. Solid waste feed	Grab sample from each drum	S006, S007	V&SV-POHCs, Cl, ash, HHV, metals
3. Chamber ash	Grab one sample after all runs are completed	S006	V&SV-POHCs, TCLP[d], HHV, TOC, metals
4. Stack gas	Composite	Method 0010 (3h) (MM5)	SV-POHCs
	Composite	Method 5[f]	Particulate, H_2O
	Composite	Method 0011	Formaldehyde
	Composite	Method 0050	HCl, Cl_2
	Composite	Method 0030 (2h)	
	Three pairs of traps	(VOST)	V-POHCs
	Composite in Tedlar gas bag	Method 0040	V-POHCs[c]
	Composite	Method 3 (1-2 h)	CO_2 and O_2 by Orsat
	Continuous	CEM	CO, CO_2, O_2, SO_2
	Composite	Method 0012	Trace metals[e]
5. APCD Effluent (liquid)	Grab sample every $\frac{1}{2}$ h	S004	V&SV-POHCs, Cl, pH, metals
6. APCD Residue (solid)	Grab sample every $\frac{1}{2}$ h	S006	V&SV-POHCs, metals

[a]VOST denotes volatile organic sampling train; MM5 denotes EPA Modified Method 5; SXXX denotes sampling methods found in "Sampling and Analysis Methods for Hazardous Waste Combustion"[83]; CEM denotes Continuous Emission Monitor (usually nondispersive infrared).
[b]V-POHCs denotes volatile principal organic hazardous constituents (POHCs); SV-POHCs denotes semivolatile POHCs; HHV denotes higher heating value; TOC denotes Total Organic Carbon.
[c]Gas bag samples may be analyzed for V-POHCs only if VOST samples are saturated and not quantifiable or if the target POHC is too volatile for VOST.
[d]TCLP = toxicity characteristic leaching procedure[192].
[e]Metals captured by the Multiple Metals Sampling Train.
[f]Method 5 can be combined with Method 0050 or Method 0011.
Source: Reprinted, with permission, from Dempsey and Oppelt, 1993.

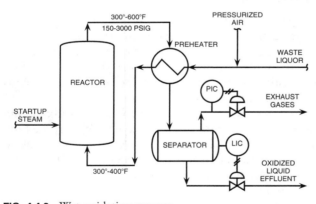

FIG. 4.4.8 Wet oxidation process.

exceed the critical temperature of water (705°F), because the continuous presence of a liquid water phase is essential (Liptak 1974).

A consequence of high operating temperature is the need to run the process at high pressure (300–3000 psig) to keep water from vaporizing. The static pressure energy of the exhaust gases can drive an air compressor or gen-

erate electric power, while the thermal energy of the reactor effluent can be used for steam generation.

Detoxified priority pollutants and products stay in the aqueous phase. Materials such as sulfur compounds, chlorinated hydrocarbons, or heavy metals end up in their highest oxidation state, i.e., sulfates, hydrochloric acid, or salt. Air pollutants are controlled because oxidation takes place in water at low temperatures and no fly ash, dust, sulfur dioxide or nitrogen oxide is formed.

Typically, 80% of the organic substances will be completely oxidized. The system can accommodate some partially halogenated compounds, but highly chlorinated species such as PCBs are too stable for complete destruction without adding a catalyst or very high pressure and temperature (Kiang & Metry 1982).

Control of a wet air oxidation system is relatively simple, as the system is self-regulating. Oxidation occurs in a massive amount of water, which provides an effective heat sink and prevents the reaction from running away. Should a surge of organic material enter the reactor, the air would be depleted, or the heat liberated by additional oxidation would form more steam.

TABLE 4.4.5 CONTROL PARAMETERS

Group	Parameter[a]
Group A Continuously monitored parameters are interlocked with the automatic waste feed cutoff. Interruption of waste feed is automatic when specified limits are exceeded. The parameters are applicable to all facilities.	1. Minimum temperature measured at each combustion chamber exit 2. Maximum CO emissions measured at the stack or other appropriate location 3. Maximum flue gas flowrate or velocity measured at the stack or other appropriate location 4. Maximum pressure in PCC and SCC 5. Maximum feed rate of *each* waste type to *each* combustion chamber[b] 6. The following as applicable to the facility: • Minimum differential pressure across particulate venturi scrubber • Minimum liquid-to-gas ratio (L/G) and pH to wet scrubber • Minimum caustic feed to dry scrubber • Minimum kVA settings to ESP (wet/dry) and kV for ionized wet scrubber (IWS) • Minimum pressure differential across baghouse • Minimum liquid flowrate to IWS
Group B Parameters do *not* require continuous monitoring and are thus *not* interlocked with the waste feed cutoff systems. Operating records are required to ensure that trial burn worst-case conditions are not exceeded.	7. POHC incinerability limits 8. Maximum total halides and ash feed rate to the incinerator system 9. Maximum size of batches or containerized waste[b] 10. Minimum particulate scrubber blowdown or total solids content of the scrubber liquid
Group C Limits on these parameters are set independently of trial burn test conditions. Instead, limits are based on equipment manufacturer's design and operating specifications and are thus considered good operating practices. Selected parameters do *not* require continuous monitoring and are *not* interlocked with the waste feed cutoff.	11. Minimum/maximum nozzle pressure to scrubber 12. Maximum total heat input capacity for each chamber 13. Liquid injection chamber burner settings: • Maximum viscosity of pumped waste • Maximum burner turndown • Minimum atomization fluid pressure • Minimum waste heating value (only applicable when a given waste provides 100% heat input to a given combustion chamber) 14. APCD inlet gas temperature[c]

[a]PCC denotes primary combustion chamber; SCC denotes secondary combustion chamber; APCD denotes air pollution control device; kVA denotes kilovolt-amperes; ESP denotes electrostatic precipitator.
[b]Items 5 and 9 are closely related.
[c]Item 14 can be a group B or C parameter.
Source: Reprinted, with permission, from Dempsey and Oppelt, 1993.

PROCESS CHARACTERISTICS

The wet air oxidation process has three basic reaction mechanisms: hydrolysis, mass transfer, and chemical kinetics. Table 4.4.6 gives brief explanations of the mechanisms and their major influences. The four basic steps encountered in the oxidation of hydrocarbon pollutants are:

$$\text{Hydrocarbon} + \text{oxygen} \longrightarrow \text{alcohol}$$
$$\text{Alcohol} + \text{oxygen} \longrightarrow \text{aldehyde}$$
$$\text{Aldehyde} + \text{oxygen} \longrightarrow \text{acid}$$
$$\text{Acid} + \text{oxygen} \longrightarrow \text{carbon dioxide} + \text{water}$$

Nearly all organic materials in industrial waste break down into several intermediate compounds before complete ox-

TABLE 4.4.6 WET AIR OXIDATION PROCESS REACTION MECHANISMS[a]

Reaction Mechanism	Typical Effects	Strongest Influences
Hydrolysis	Dissolves solids Splits long-chain hydrocarbons	pH Temperature
Mass Transfer	Dissolves, absorbs oxygen	Pressure Presence of liquid-gas interface
Chemical Kinetics	Oxidizes organic chemicals	Temperature Catalysts Oxygen activity

[a]Courtesy of Plant Engineering, Barrington, IL.

idation occurs. The process is efficient in total organic carbon reduction for most compounds, but not for acetates or benzoic acid.

APPLICABILITY/LIMITATIONS

This process is used to treat aqueous waste streams containing less than 5% organics, pesticides, phenolics, organic sulfur, and cyanide wastewaters. At ethylene plants,

Zimpro Passavant provides wet air oxidation units for converting caustic liquors into nonhazardous effluents that can be treated biologically (Zimpro Environmental, Inc. 1993). These liquors are produced during scrubbing of ethylene gases. Table 4.4.7 lists hazardous wastes that are good candidates for the wet air oxidation process.

Skid-mounted units can be situated at disposal sites for pretreatment of hazardous liquid before deep welling, or for carbon regeneration and sludge oxidation.

TABLE 4.4.7   THE EPA HAZARDOUS WASTE LIST—GOOD CANDIDATES FOR WAO

F Classification:

F004, F005	Spent non-halogenated solvents and still bottoms.
F006	Sludges from electroplating operations.
F007, F011, F015	Spent cyanide bath solutions.
F016	Coke oven, blast furnace gas scrubber sludges.

K Classification:

K009-K015	Bottoms, bottom streams, and side cuts from production of acetaldehyde and acrylonitrile.
K017-K020	Heavy ends or still bottoms from epichlorohydrin, ethyl chloride, ethylene dichloride or vinyl chloride operations.
K024-K025	Distillation bottoms from production of phthalic anhydride and nitrobenzene.
K026	Stripping still tails from production of methyl ethyl pyridines.
K027	Residues from toluene diisocyanate production.
K029-K030	Bottoms, ends, stripper wastes from tricholorethylene, perchloroethylene production.
K035	Creosote sludges.
K045	Spent carbon from explosives wastewater.
K052	Leaded petroleum tank bottoms.
K058-K059	Leather tanning, finishing sludges.

P Classification (discarded commercial chemical products):

P024	p-chloroaniline
P029-P030	Copper cyanide, cyanides
P048	2,4-dinitrophenol
P052-P054	Ethylcyanide, ethylenediamine, ethyleneimine.
P063-P064	Hydrocyanic acid, isocyanic acid
P077	p-nitroaniline
P081	Nitroglycerine
P090	Pentachlorophenol
P098	Potassium cyanide
P101	Propionitrile
P106	Sodium cyanide

U Classification:

U007 & U009	Acrylamide, acrylonitrile
U130	Hexachlorocyclopentadiene
U135	Hydrogen sulfide
U152-U153	Methacrylonitrile, methanethiol
U159	Methyl ethyl ketone

Other Hazardous Wastes (SIC Code Numbers):

2865	Vacuum still bottoms from maleic anhydride production.
	Fractionating residues, benzene and chlorobenzene recovery.
	Residues from distillation of 1-chloro-4-nitrobenzene.
	Methanol recovery bottoms, heavy ends, methyl methacrylate production.
2869	Ends, distillation from carbaryl production.
	Ethylene dichloride distillation ends in vinyl chloride production.
	Quench column bottoms, acrylonitrile production.
	Aniline production still bottoms.
3312	Cyanide-bearing wastes from steel finishing.

Source: Reprinted, with permission, from Zimpro Environmental, Inc., 1993, wet air oxidation—solving today's hazardous wastewater problems. *Bulletin WAO-100.*

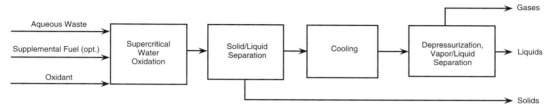

FIG. 4.4.9 SCWO Schematic. (Reprinted from U.S. EPA, 1992).

This technology is not recommended for aromatic halogenated organics, inorganics, or for large volumes of waste. It is not appropriate for solids or viscous liquids.

Status. Available at commercial scale.

Supercritical Water Oxidation

Supercritical water oxidation (SCWO) is an emerging waste treatment technology. There are no full scale SCWO systems in operation, but large bench- and pilot-scale data are available.

PROCESS DESCRIPTION

SCWO is basically a high-temperature, high-pressure process. In SCWO, decomposition occurs in the aqueous phase above the critical point of water (374°C/221 atm or 705°F/3248 psi). A schematic of a generic SCWO process is shown in Figure 4.4.9. The feed is typically an aqueous waste. An oxidant such as air, oxygen, or hydrogen peroxide must be provided unless the waste itself is an oxidant.

Many of the properties of water change drastically near its critical point (374°C/221 atm): the hydrogen bonds disappear and water becomes similar to a moderately polar solvent; oxygen and all hydrocarbons become completely miscible with water; mass transfer occurs almost instantaneously; and solubility of inorganic salts drops to ppm range. Thus, inorganic salt removal must be considered in the design of a SCWO reactor (Thomason, Hong, Swallow & Killilea 1990).

Two process approaches have been evaluated: an above-ground pressure vessel reactor (Modar), and the use of an 8000–1000-ft deep well as a reactor vessel (Vertox). Figure 4.4.10 is a schematic of a subsurface SCWO reactor. Subsurface reactors consist of aqueous liquid waste columns deep enough that the material near the bottom is subject to a pressure of at least 221 atm (Gene Syst, 1990). To achieve this pressure solely through hydrostatic head, a water column depth of approximately 12,000 ft is required. The influent and effluent will flow in opposite directions in concentric vertical tubes. In surface SCWO systems, the pressure is provided by a source other than gravity, and the reactor is on or above the earth's surface.

The supercritical water process is best suited for large volume (200 to 1000 gpm), dilute (in the range of 1–10,000 mg/l COD), aqueous wastes that are volatile and have a sufficiently high heat content to sustain the process. In many applications, high Btu, nonhazardous waste can be mixed with low Btu hazardous waste to provide the heat energy needed to make the process self-sustaining. Emissions or residues include gaseous effluents (nitrogen and carbon dioxide), precipitates of inorganic salts, and liquids containing only soluble inorganic acids and salts. The advantages are rapid oxidation rates, complete oxidation of organics, efficient removal of inorganics, and no off-gas processing required (EPA 1992).

Significant bench- and pilot-scale SCWO performance data are available. Typical destruction efficiencies (DEs) for a number of compounds are summarized in Table 4.4.8. Although several low DEs are included in this table to illustrate that DE is proportional to both temperatures and time, DEs in excess of 99% can be achieved for nearly all pollutants (EPA 1992). Table 4.4.8 shows that using hydrogen peroxide as an oxidant in SCWO systems produces DEs significantly higher than those obtained using of air and oxygen.

APPLICABILITY/LIMITATIONS

Supercritical water oxidation is used to treat a wide variety of pumpable aqueous organic solutions, slurries, and mixed organic and inorganic waste (EPA 1992). Sophisticated equipment and long-term continuous operations have not been demonstrated, thereby limiting its use. Demonstra-

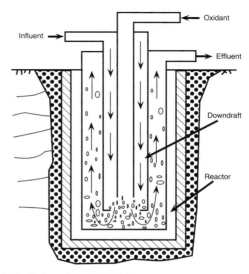

FIG. 4.4.10 Subsurface SCWO Reactor

TABLE 4.4.8 SCWO PERFORMANCE DATA

Pollutant	Temp. (deg. C)	Pressure (atm.)	DE (%)	React TIme (min.)	Oxidant	Feed Conc. (mg/L)
1,1,1-Trichloroethane	495		99.99	4	Oxygen	
1,1,2,2-Tetrachloroethylene	495		99.99	4	Oxygen	
1,2-Ethylene dichloride	495		99.99	4	Oxygen	
2,4-Dichlorophenol	400		33.7	2	Oxygen	2,000
2,4-Dichlorophenol	400		99.440	1	H_2O_2	2,000
2,4-Dichlorophenol	450		63.3	2	Oxygen	2,000
2,4-Dichlorophenol	450		99.950	1	H_2O_2	2,000
2,4-Dichlorophenol	500		78.2	2	Oxygen	2,000
2,4-Dichlorophenol	500		>99.995	1	H_2O_2	2,000
2,4-Dimethylphenol	580	443	>99	10	$H_2O_2 + O_2$	135
2,4-Dinitrotoluene	410	443	83	3	Oxygen	84
2,4-Dinitrotoluene	528	287	>99	3	Oxygen	180
2-Nitrophenol	515	443	90	10	Oxygen	104
2-Nitrophenol	530	430	>99	15	$H_2O_2 + O_2$	104
Acetic acid	400		3.10	5	Oxygen	2,000
Acetic acid	400		61.8	5	H_2O_2	2,000
Acetic acid	450		34.3	5	Oxygen	2,000
Acetic acid	450		92.0	5	H_2O_2	2,000
Acetic acid	500		47.4	5	Oxygen	2,000
Acetic acid	500		90.9	5	H_2O_2	2,000
Activated sludge (COD)	400	272	90.1	2		62,000
Activated sludge (COD)	400	306	94.1	15		62,000
Ammonium perchlorate	500	374	99.85	0.2	None	12,000
Biphenyl	450		99.97	7	Oxygen	
Cyclohexane	445		99.97	7	Oxygen	
DDT	505		99.997	4	Oxygen	
Dextrose	440		99.6	7	Oxygen	
Industrial sludge (TCOD)	425		>99.8	20	Oxygen	
Methyl ethyl ketone	505		99.993	4	Oxygen	
Nitromethane	400	374	84	3	None	10,000
Nitromethane	500	374	>99	0.5	None	10,000
Nitromethane	580	374	>99	0.2	None	10,000
o-Chlorotoluene	495		99.99	4	Oxygen	
o-Xylene	495		99.93	4	Oxygen	
PCB 1234	510		99.99	4	Oxygen	
PCB 1254	510		99.99	4	Oxygen	
Phenol	490	389	92	1	Oxygen	1,650
Phenol	535	416	>99	10	Oxygen	150

Source: (Reprinted from U.S. Environmental Protection Agency, 1992, *Engineering Bulletin: Supercritical water oxidation,* [EPA 540–S–92–006], Office of Research and Development, Cincinnati, Oh. [September]).

tion of use with municipal sewage sludge was completed in 1985.

Possible corrosion problems must be examined when SCWO is considered. High-temperature flames observed during SCWO may present additional equipment problems in both surface and subsurface SCWO systems. There is some concern that these flames will cause hot spots which could weaken SCWO vessels (DOE 1991).

Status. Demonstration of use with municipal sewerage sludge completed in 1986.

—David H.F. Liu (1995)

References

Chang, D.P.Y., et al. 1987. Evaluation of a pilot-scale circulating bed combustor as a potential hazardous waste incinerator. *JAPCA,* Vol. 37, no. 3.

Dempsey, C.R., and E.T. Oppelt. 1993. Incineration of hazardous waste: a critical review update. *Air & Waste,* Vol. 43.

Freeman, H. 1990. *Thermal processes: innovative hazardous waste treatment technology series,* Lancaster, PA: Technomic Publishing Company.

Gene Syst International Inc. 1990. *The gravity pressure vessel* (June).

Gorman, P., et al. 1985. *Practical Guide to Trial Burns for Hazardous Waste Incineration.* U.S. EPA, EPA 600–R2–86–050. NTIS PB 86-190246 (November).

Harris, J.C., D.J. Larsen, and C.E. Rechsteiner. 1984. *Sampling and analysis methods for hazardous waste combustion.* EPA 600–8-84–002. PB 84-155545 (February).

Kiang, Y.H., and A.A. Metry. 1982. *Hazardous waste processing technology.* Ann Arbor, Mi: Ann Arbor Science.

Liptak, B.G. 1974. *Environmental engineers' handbook.* Vol. 3.2.15. Radnor, Pa: Chilton Book Company.

Thomason, Terry B., G.T. Hong, K. Swallow, and W.R. Killilea. 1990. The Modar supercritical water oxidation process. *Innovative hazardous waste treatment technology series.* Vol. 1. Technomic Publishing Company, Inc.

U.S. Department of Energy (DOE). 1991. *Supercritical Oxidation Destroys Toxic Wastes.* NTIS Technical Note (February).

U.S. Environmental Protection Agency (EPA). 1981. *Engineering handbook on hazardous waste incineration.* SW-889. NTIS PB 81-248163 (September).

———. 1983. *Guidance Manual for Hazardous Waste Incinerator Permits.* EPA-SW-966. NTIS PB 84-100577 (March).

———. 1989. *Guidance on setting permit conditions and reporting trial burn results.* EPA 625–6-89–019 (January).

———. 1990. *Handbook: quality assurance/quality control (QA/QC) procedures for hazardous waste incinerators.* EPA 625–6-89–023. NTIS PB 91-145979 (January).

———. 1990c. *Methods manual for compliance with the BIF regulations.* EPA 530-SW-91-010. NTIS PB 90-120-006 (December).

———. 1992. *Engineering bulletin: Supercritical water oxidation.* EPA 540–S–92–006. Office of Research & Development. Cincinnati, Oh. (September).

Zimmerman, F.J., U.S. Patent No. 2,665,249 (Jan. 5, 1954).

Zimpro Environmental Inc. 1993. *Wet air oxidation–solving today's hazardous wastewater problems.* Bulletin #WAO-100.

4.5
WASTE CONCENTRATION TECHNOLOGY

PARTIAL LIST OF SUPPLIERS

Sedimentation: Chemical Waste Management Inc.; Dorr-Oliver Inc.; Eimco Process Equipment Co.; Wyo Ben Inc.; National Hydro Systems Inc.; Sharples Stokes Div., Pennwalt; Water Tech Inc.; AFL Industries

Centrifugation: Clinton Centrifuge Inc.; ALFA Laval Inc.; Tetra Recovery Systems; Dorr-Oliver Inc.; Bird Environmental Systems; Western States Machine; Fletcher; Astro Metallurgical; Barrett Centrifugals; Donaldson Industrial Group; GCI Centrifuges; General Production Services Inc.; IT Corp.; Ingersoll Rand Environmental; Master Chemical Corp. System Equipment; Sartorius Balance Div., Brinkman; Sharples Stokes Div., Pennwalt; Tekmar Co.; Thomas Scientific

Evaporation: Resources Conservation Company (mobile brine concentration systems); Kipin Industries; APV Equipment Inc.; Ambient Technical Div., Ameribrom Inc.; Analytical Bio Chem Labs; Aqua Chem Water Technologies; Capital Control Co., Inc.; Dedert Corp.; HPD Inc.; Industrial Filter & Pump Manufacturing; Kimre Inc.; Fontro Co., Inc.; Lancy International Inc.; Luwa Corp.; Licon Inc.; Rosenmund Inc.; Sasakura International American Corp.; Spraying Systems Co.; Votator Anco Votator Div.; Wallace & Tiernan Div., Pennwalt; Wastesaver Corp.; Weathermeasure Weathertronics; Wheaton Instruments

Air Stripping: OH Materials; Carbon Air Services; Detox Inc.; IT Corporation; Oil Recovery Systems Inc.; Resource Conservation Company; Terra Vac Inc.; Advanced Industrial Technology; Baron Blakeslee Inc.; Beco Engineering Co.; Calgon Carbon Corp.; Chem Pro Corp.; D.R. Technology Inc.; Delta Cooling Towers; Detox Inc.; Hydro Group Inc.; IPC Systems; Kimre Inc.; Munters Corp.; NEPCCO; North East Environmental Products; Oil Recovery System Inc.; Tri-Mer Corp.; Wright R.E. Associates Inc.

Distillation: Exceltech, Inc.; Kipin Industries; Mobil Solvent Reclaimers, Inc.; APV Equipment Inc.; Ace Glass Inc.; Artisan Industries Inc.; Gilmont Instruments Inc.; Glitsch Inc.; Hoyt Corp.; Licon Inc.; Progressive Recovery Inc.; Rosenmund Inc.; Sutcliffe Croftshaw; Tekmar Co.; Thomas Scientific; Vera International Inc.; Vic Manufacturing Co.; Industrial Div.; Wheaton Instruments; York Otto H. Co., Inc.

Soil Flushing: Critical Fluid Systems; IT Corp.

Liquid/Liquid Extraction: Resources Conservation Co.

Filtration: Calgon Carbon Corp.; Carbon Air Services Inc; Chemical Waste Management; Industrial Innovations Inc., Krauss-Maffei; Komline Sanderson; Bird Machine Co.; D.R. Sperry, Inc., Dorr-Oliver

Carbon Adsorption: Calgon Carbon Corp.; Carbon Air Services Inc.; Zimpro Inc.; Chemical Waste Management

Reverse Osmosis: Osmonics, Inc.; Artisan Industries Inc.

Ion Exchange: Calgon Carbon; Dionex; DeVoe-Holbein; Davis Instrument Mfg Co., Inc.; Ecology Protection Systems, Inc.; Envirex Inc.; Industrial Filter & Pump Mfg.; Lancy International Inc.; McCormack Corp.; Osmonic Membrane Sys. Div.; Pace International Corp.; Permutit Co., Inc.; Serfilco LTD.; Techni Chem., Inc.; Thomas Scientific; Treatment Technologies; Water Management Inc.; Western Filter Co.

Chemical Precipitation: Mobile Systems-Rexnord Craig; Ecolochem Inc., Dravo Corp.; Detox Inc.; Envirochem Waste Management Services; Chemical Waste Management Inc.; Andco Environmental Processes Inc.; Ensotech Inc.; Tetra Recovery Systems

Chemical and physical waste treatment processes are used for removal rather than destruction. A more appropriate term for non-destructive processes is *concentration technologies* (Martin & Johnson 1987). Physical treatment processes use physical characteristics to separate or concentrate constituents in a waste stream. Residues then require further treatment and ultimate disposal. Chemical treatment processes alter the chemical structure of wastes, producing residuals that are less hazardous than the original waste.

In this section, physical treatment processes are organized into four groupings: gravity; phase change; dissolution; and size, adsorptivity, or ionic characteristics (Table 4.5.1). Important physical treatment data needs are presented in Table 4.5.2.

TABLE 4.5.1   PHYSICAL TREATMENT PROCESS

Gravity Separation:
- Sedimentation
- Centrifugation
- Flocculation
- Oil/Water Separation
- Dissolved Air Flotation
- Heavy Media Separation

Phase Change:
- Evaporation
- Air Stripping
- Steam Stripping
- Distillation

Dissolution:
- Soil Washing/Flushing
- Chelation
- Liquid/Liquid Extraction
- Supercritical Solvent Extraction

Size/Adsorptivity/Ionic Characteristics:
- Filtration
- Carbon Adsorption
- Reverse Osmosis
- Ion Exchange
- Electrodialysis

The following chemical treatment processes discussed in this section are commonly used for waste treatment applications. These include

- pH adjustment (for neutralization or precipitation)
- Oxidation and reduction
- Hydrolysis and photolysis
- Chemical oxidation (ozonation, electrolytic oxidation, hydrogen peroxide)
- Chemical dehalogenation (alkaline metal dechlorination, alkaline metal/polyethylene glycol, based-catalyzed dechlorination)

Important chemical treatment data needs are presented in Table 4.5.3.

Gravity Separation

SEDIMENTATION

Description

Sedimentation is a settling process in which gravity causes heavier solids to collect at the bottom of a containment vessel, separated from the suspending fluid. Sedimentation

TABLE 4.5.2   PHYSICAL TREATMENT DATA NEEDS

Data Need	Purpose
For Solids	
Absolute Density	Density Separation
Bulk Density	Storage Volume Required
Size Distribution	Size Modification or Separation
Friability	Size Reduction
Solubility	Dissolution
(in H_2O, organic solvents, oils, etc.)	
For Liquids	
Specific Gravity	Density Separation
Viscosity	Pumping & Handling
Water Content (or oil content, etc.)	Separation
Dissolved Solids	Separation
Boiling Pt/Freezing Point	Phase Change Separation, Handling and Storage
For Liquids/Solid Mixtures	
Bulk Density	Storage & Transportation
Total Solids Content	Separation
Solids Size Distribution	Separation
Suspended Solids Content	Separation
Suspended Solids Settling Rate	Separation
Dissolved Solids Content	Separation
Free Water Content	Storage & Transport
Oil and Grease Content	Separation
Viscosity	Pumping and Handling
For Gases	
Density	Separation
Boiling (condensing) Temp.	Phase Change Separation
Solubility (in H_2O, etc.)	Dissolution

TABLE 4.5.3   IMPORTANT CHEMICAL TREATMENT DATA NEEDS

Data Need	Purpose
pH	pH Adjustment Needs, Corrosivity
Turbidity/Opacity	Photolysis
Constituent analysis	Treatment Need
Halogen Content	Dehalogenation

Note: Generally, the data needs for evaluating and comparing chemical treatment technologies include the data needs identified for physical treatment technologies.

TABLE 4.5.4   IMPORTANT SEDIMENTATION DATA NEEDS

Data Need	Purpose
Viscosity of aqueous waste	High viscosity hinders sedimentation
Oil and grease content of waste stream	Not applicable to wastes containing emulsified oils
Specific gravity of suspended solids	Must by greater than 1 for sedimentation to occur

can be accomplished using a batch process or a continuous removal process. Several physical arrangements where the sedimentation process is applied are shown in Figure 4.5.1.

The top diagram illustrates a settling pond. Aqueous waste flows through while suspended solids are permitted to gravitate and settle out. Occasionally the settling particles (sludges) are removed, so this system is considered a semibatch process.

The middle diagram shows a circular clarifier equipped with a solids-removal device. This facilitates continuous clarification, resulting in a lower solid content outlet fluid.

The sedimentation basin is shown in the bottom diagram. It uses a belt-type solids collector mechanism to force solids to the bottom of the basin's sloped edge, where they are removed.

The efficiency of sedimentation treatment depends upon the depth and surface area of the basin, settling time (based

on the holding time), solid particle size, and the flow rate of the fluid.

Applicability/Limitations

Sedimentation is considered a separation process only. Typically, some type of treatment process for aqueous liquids and sludges will follow. Use is restricted to solids that are more dense than water. It is not suitable for wastes consisting of emulsified oils. Important sedimentation data are summarized in Table 4.5.4.

Status.   This is a conventional process.

CENTRIFUGATION

Description

Centrifuge involves physical separation of fluid mixture components based on their relative density. A rapidly rotating fluid mixture within a rigid vessel deposits the more dense solid particles farthest from the axis of rotation, while liquid supernatant lies separated near the axis. Centripetal forces in centrifugation are similar to gravitational forces in sedimentation, except the centripetal forces are thousands of times stronger than gravitational forces, depending upon centrifuge diameter and rotational speed.

Applicability/Limitations

This treatment is limited to dewatering sludges (including metal-bearing sludges), separating oils from water, and clarification of viscous gums and resins. Centrifuges are generally better suited than vacuum filters for dewatering sticky or gelatinous sludges. Disc-type centrifuges (Figure 4.5.2) can be used to separate three component mixtures (e.g. oil, water, and solids). Centrifuges often cannot be used for clarification since they may fail to remove less dense solids and those small enough to remain in suspension. Recovery and removal efficiencies may be improved if paper or cloth filters are used.

Status.   This process is commercially available.

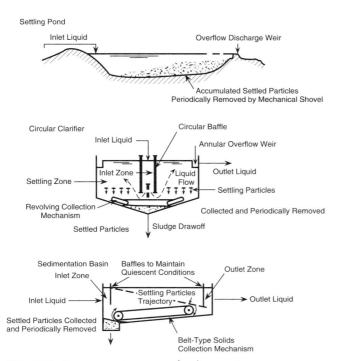

FIG. 4.5.1   Representative types of sedimentation.

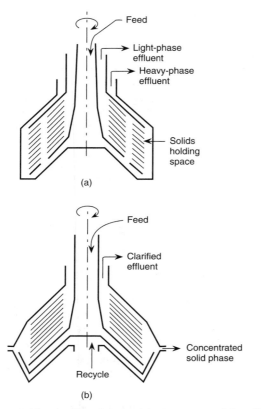

FIG. 4.5.2 Disk-centrifuge bowls. (*a*) separator, solid wall; (*b*) recycle clarifier, nozzle discharge.

FLOCCULATION

Description

Flocculation is used to enhance sedimentation or centrifugation. The waste stream is mixed while a flocculating chemical is added. Flocculants adhere readily to suspended solids and to each other (agglomeration), and the resultant particles are too large to remain in suspension. Flocculation is used primarily for the precipitation of inorganics.

Availability/Limitations

The extent of flocculation depends upon waste stream flow rate, composition, and pH. This process is not recommended for a highly viscous waste stream. Table 4.5.5 presents the important flocculation data needs.

Status. Flocculation is a conventional, demonstrated treatment technique.

OIL/WATER SEPARATION

Description

As in sedimentation, the force of gravity can be used to separate two or more immiscible liquids with sufficiently different densities, such as oil and water. Liquid/liquid sep-

TABLE 4.5.5   IMPORTANT FLOCCULATION DATA NEEDS

Data Need	Purpose
pH of waste	Selection of flocculating agent
Viscosity of waste system	Affects settling of agglomerated solids; high viscosity not suitable
Settling rate of suspended solids	Selection of flocculating agent

aration occurs when the liquid mix settles. Thus, flow rates in continuous processes must be kept low. The waste flows into a chamber, where it is kept quiescent, and permitted to settle. The floating oil is skimmed off the top using an oil skimmer while the water or effluent flows out of the lower portion of the chamber. Acids may be used to break oil/water emulsion and to enhance this process for efficient oil removal.

Availability/Limitations

Effectiveness can be influenced by waste stream flow rate, temperature, and pH. Separation is a pretreatment process if the skimmed oil requires further treatment.

Status. Mobile phase separators are commercially available.

DISSOLVED AIR FLOTATION

Description

Dissolved air flotation involves removing suspended particles or mixed liquids from an aqueous waste stream (Figure 4.5.3). The mixture to be separated is saturated with air or another gas such as nitrogen, then air pressure is reduced above the treatment tank. As air escapes the solution, microbubbles form and are readily adsorbed onto suspended solids or oils, enhancing their flotation characteristics. In the flotation chamber, separate oil or other

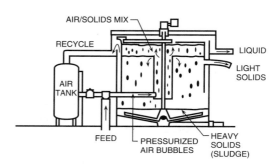

FIG. 4.5.3 Recycle flow dissolved air flotation system. Source: Peabody-Welles, Roscoe, Il.

floats are skimmed off the top while aqueous liquids flow off the bottom.

Applicability/Limitations

This technology is only applicable for waste with densities close to water. Air emission controls may be necessary if hazardous volatile organics are present.

Status. This is a conventional treatment process.

HEAVY MEDIA SEPARATION

Description

Heavy media separation is used to process two solid materials with significantly different absolute densities. Mixed solids are placed in a fluid with a specific gravity adjusted to allow lighter solids to float while heavier solids sink. Usually, the separating fluid or heavy medium is a suspension of magnetite in water. The specific gravity is adjusted by varying the amount of magnetite powder used. Magnetite is easily recovered magnetically from rinsewaters and spills, then reused.

Availability/Limitations

This type of separation is used to separate two insoluble solids with different densities. Limitations include the possibility of dissolving solids and ruining the heavy media; the presence of solids with densities similar to those solids requiring separation; and the inability to cost-effectively separate magnetic materials, because of the need to recover magnetite.

Status. Commonly used in the mining industry to separate ores from tailings.

Phase Change

EVAPORATION

Description

Evaporation is the physical separation of a liquid from a dissolved or suspended solid by applying energy to make the liquid volatile. In hazardous waste treatment, evaporation may be used to isolate the hazardous material in one of the two phases, simplifying subsequent treatment. If the hazardous waste is volatilized, the process is usually called *stripping*.

Availability/Limitations

Evaporation can be applied to any mixture of liquids and volatile solids provided the liquid is volatile enough to evaporate under reasonable heating or vacuum conditions (both the liquid and the solid should be stable under those

conditions). If the liquid is water, evaporation can be carried out in large ponds using solar energy. Aqueous waste can also be evaporated in closed process vessels using steam energy. The resulting water vapor can be condensed for reuse. Energy requirements are minimized by techniques such as vapor recompression or multiple effect evaporators. Evaporation is applied to solvent waste contaminated with nonvolatile impurities such as oil, grease, paint solids or polymeric resins. Mechanically agitated or wipe-thin-film evaporators (Figure 4.5.4) are used. Solvent is evaporated and recovered for reuse. The residue is the bottom stream, typically containing 30 to 50% solids.

Status. This process is commercially available.

AIR STRIPPING

Description

Air stripping is a mass transfer process in which volatile contaminants in water or soils are evaporated into the air. Organics removal from wastewater via air stripping depends upon temperature, pressure, air-to-water ratio, and surface area available for mass transfer. Air-to-water volumetric ratios may range from $10:1$ up to $300:1$. Contaminated off-gas and stripped effluent are the resulting residuals. Volatile hazardous materials must be recaptured for subsequent treatment to preclude air pollution.

Availability/Limitations

This process is used to treat aqueous wastes that are more volatile, less soluble (e.g., chlorinated hydrocarbons such as tetrachloroethylene) and aromatic (e.g., toluene). Limitations include temperature dependency, as stripping efficiency is impacted by changes in ambient temperature. In addition, the presence of suspended solids may reduce efficiency. If the concentration of volatile organic contaminants (VOCs) exceeds about 100 ppm, another separation process, e.g. steam stripping, is usually preferred.

Status. This process is commercially available.

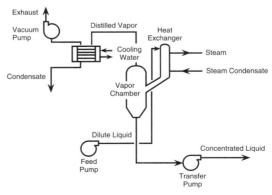

FIG. 4.5.4 Typical single effect evaporator, falling film type.

STEAM STRIPPING

Description

Steam stripping uses steam to evaporate volatile organics from aqueous wastes. Steam stripping is essentially a continuous fractional distillation process carried out in a packed or tray tower. Clean steam, rather than reboiled bottoms, provide direct heat to the column, and gas flows from the bottom to the top of the tower (Figure 4.5.5). The resulting residuals are contaminated steam condensate, recovered solvent and stripped effluent. The organic vapors and the raffinate are sent through a condenser in preparation for further purification treatment. The bottom requires further consideration as well. Possible post-treatment includes incineration, carbon adsorption, or land disposal.

Availability/Limitations

Steam stripping is used to treat aqueous wastes contaminated with chlorinated hydrocarbons, aromatics such as xylenes, ketones such as acetone or MEK, alcohols such as methanol, and high-boiling-point chlorinated aromatics such as pentachlorophenol. Steam stripping will treat less volatile and more soluble wastes than will air stripping and can handle a wide concentration range (e.g., from less than 100 ppm to about 10% organics). Steam stripping requires an air pollution control (APC) mechanism to eliminate toxic emissions.

Status.   Conventional, well documented.

DISTILLATION

Description

Distillation is simply evaporation followed by condensation. The separation of volatile materials is optimized by controlling the evaporation-stage temperature and pressure, and the condenser temperature. Distillation separates miscible organic liquids for solvent reclamation and waste volume reduction. Two types of distillation processes are batch distillation and continuous fractional distillation.

Availability/Limitations

Distillation is used to separate liquid organic wastes, primarily spent solvents, for full or partial recovery and reuse. Both halogenated and nonhalogenated solvents can be recovered via distillation. Liquids to be separated must have different volatilities. Distillation for recovery is limited by the presence of volatile or thermally reactive suspended solids. If constituents in the input waste stream form an *azeotrope* (a specific mixture of liquids exhibiting maximum or minimum boiling point with the individual constituents), the energy cost to break the azeotrope can be prohibitive.

Batch distillation in a heated still pot with condensation of overhead vapors is easily controlled and flexible, but cannot achieve the high product quality typical of continuous fractional distillation. Small, packaged-batch stills treating one drum or less per day are becoming popular for on-site recovery of solvents. Continuous fractional distillation is accomplished in tray columns or packed columns ranging up to 40 ft in diameter and 200 ft high. Each is equipped with a reboiler, a condenser, and an accumulator. Unit capacity is a function of the processed waste, purity requirements, reflux ratios, and heat input. Fractional distillation is not applicable to liquids with high viscosity at high temperature, liquids with high solid concentrations, polyurethanes, or inorganics.

Status.   Commercially available.

Dissolution

SOIL FLUSHING/SOIL WASHING

Soil is comprised of fine-grained (e.g., silt and clay) and coarse-grained (e.g., sand and gravel) particles, organic materials (e.g., decayed plant and animal matter), water, and air. Contaminants bind readily, chemically or physically, to silt, clay, and organic matter. Silt, clay, and organic mat-

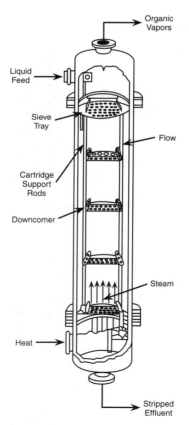

FIG. 4.5.5   Steam stripping column, perforated tray type.

Labels on figure:
- Organic Vapors
- Liquid Feed
- Sieve Tray
- Flow
- Cartridge Support Rods
- Downcomer
- Steam
- Heat
- Stripped Effluent

ter bind physically to sand and gravel. When soil contains large amounts of clay and organic materials, contaminants attach more easily to the soil and are more difficult to remove.

Description

Soil flushing is an in-situ extraction of inorganic and organic compounds from soil, and is accomplished by passing extractant solvents through the soils using an injection and recirculation process. Solvents may include: water, water-surfactant mixtures, acids, bases (for inorganics), chelating agents, oxidizing agents, or reducing agents. Soil washing consists of similar treatments, but the soil is excavated and treated at the surface in a soil washer.

A simplified drawing of the soil washing process is illustrated in Figure 4.5.6. The contaminated soil is removed to a staging area, then sifted to remove debris and large objects such as rocks. The remaining material enters a soil scrubbing unit, is mixed with a washing solution, and agitated. The washing solution may be water, or may contain some additives like detergent to remove contaminants. Then the washwater is drained and the soil is rinsed with clean water. The heavier sand and gravel particles in the processed soil settle out and are tested for contaminants. If clean, these materials can be used on site or taken elsewhere for backfill. If contaminated, these materials may undergo soil washing again.

The contaminated silt and clay in the washwater settle out and are then separated from the washwater. The washwater, which also contains contaminants, undergoes wastewater treatment processes for future recycling use. This wastewater may contain additives that interfere with the wastewater treatment process. If so, the additives must be removed or neutralized by pretreatment methods before wastewater treatment. The silts and clays are then tested for contaminants. If clean, these materials can be used on the site or taken elsewhere for backfill. If contaminated, these materials may undergo soil washing again, or be collected for alternate treatment or off-site disposal in a permitted RCRA landfill.

Availability/Limitations

Soil flushing and washing fluids must have: good extraction coefficients; low volatility and toxicity; capability for safe and easy handling, and, most important, be recoverable and recyclable. This technology is very promising in extracting heavy metals from soil, although problems are likely in dry or organically-rich soils. Surfactants can be used to extract hydrophobic organisms. Soil type and uniformity are important. Certain surfactants, when tested for in-situ extraction, clogged soil pores and precluded further flushing.

Status. The U.S. EPA in Edison, New Jersey, has a mobile soil washer; other systems are under development.

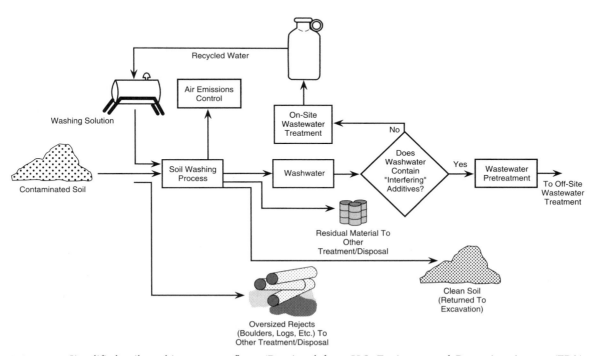

FIG. 4.5.6 Simplified soil washing process flow. (Reprinted from U.S. Environmental Protection Agency (EPA), 1992, *A citizen's guide to glycolate dehalogenation* [EPA 524–F–92–005], Office of Solid Waste and Emergency Response [March]).

CHELATION

Description

A chelating molecule contains atoms that form ligands with metal ions. If the number of such atoms in the molecule is sufficient, and if the molecular shape is such that the final atom is essentially surrounded, then the metal will be unable to form ionic salts which can precipitate out. Thus chelation is used to keep metals in solution and to aid in dissolution for subsequent transport and removal (e.g., soil washing).

Applicability/Limitations

Chelating chemicals are chosen for their affinity to particular metals (e.g., EDTA and calcium). The presence of fats and oils can interfere with the process.

Status.   Chelating chemicals are commercially available.

LIQUID/LIQUID EXTRACTION

Description

Two liquids that are well mixed or mutually soluble may be separated by liquid/liquid extraction. The process requires that a third liquid be added to the original mix. This third liquid must be a solvent for one of the original components, but must be insoluble in and immiscible with the other. The final solvent and solute stream can be separated by distillation or other chemical means, and the extracting solvent captured and reused.

Availability/Limitations

Complete separation is rarely achieved, and some form of post–treatment is required for each separated stream. To effectively recover solvent and solute materials from the process, other treatments such as distillation or stripping are needed.

Status.   This is a demonstrated process.

SUPERCRITICAL EXTRACTION

Description

At a certain temperature and pressure, fluids reach their critical point, beyond which their solvent properties are greatly enhanced. For instance, supercritical water is an excellent non-polar solvent in which most organics are readily soluble. These properties make extraction more rapid and efficient than distillation or conventional solvent extraction methods. Presently, the use of supercritical carbon dioxide to extract hazardous organics is being investigated.

Availability/Limitations

This technology may be useful in extracting hazardous waste from aqueous streams. Specific applicability and limitations are not yet known.

Status.   This process has been demonstrated on a laboratory scale.

Size/Adsorptivity/Ionic Characteristics

FILTRATION

Description

Filtration is the separation and removal of suspended solids from a liquid by passing the liquid through a porous medium. The porous medium may be a fibrous fabric (paper or cloth), a screen, or a bed of granular material. The filter medium may be precoated with a filtration aid such as ground cellulose or diatomaceous earth. Fluid flow through the filter medium may be accomplished by gravity, by inducing a partial vacuum on one side of the medium, or by exerting mechanical pressure on a dewaterable sludge enclosed by filter medium.

Availability/Limitations

Filtration is used to dewater sludges and slurries as pretreatment for other processes. It is also a polishing step for treated waste, reducing suspended solids and associated contaminants to low levels. Pretreatment by filtration is appropriate for membrane separation, ion exchange, and carbon adsorption to prevent plugging or overloading these processes. Filtration of settled waste is often required to remove undissolved heavy metals present as suspended solids. Filtration does not reduce waste toxicity, although powdered activated carbon may be used as an adsorbent and filter aid. Filtration should not be used with sticky or gelatinous sludges, due to the likelihood of filter media plugging.

Status.   This process is commercially available.

CARBON ADSORPTION

Description

Most organic and inorganic compounds will readily attach to carbon atoms. The strength of that attachment—and the energy for subsequent desorption—depends on the bond formed, which in turn depends on the specific compound being adsorbed. Carbon used for adsorption is treated to produce a high surface-to-volume ratio (900 : 1,300 sq.m/g), exposing a practical maximum number of carbon atoms for active adsorbtion. This treated carbon is said to be *activated* for adsorption. When activated carbon has adsorbed so much contaminant that its adsorptive capacity is severely depleted, it is said to be *spent*. Spent carbon can be regenerated, but for strongly adsorbed contaminants, the cost of such regeneration is higher than simple replacement with new carbon.

Availability/Limitations

This process is used to treat single-phase aqueous organic wastes with high molecular weight and boiling point, and low solubility and polarity; chlorinated hydrocarbons such as tetrachloroethylene; and aromatics such as phenol. It is also used to capture volatile organics in gaseous mixtures. Limitations are economic, relating to how rapidly the carbon becomes spent. As an informal guide, concentrations should be less than 10,000 ppm; suspended solids less than 50 ppm; and dissolved inorganics, oil, and grease less than 10 ppm.

Status. Conventional, demonstrated.

REVERSE OSMOSIS

Description

In normal osmotic processes, solvent flows across a semi-permeable membrane from a dilute solution to a more concentrated solution until equilibrium is reached. Applying high pressure to the concentrated side causes the process to reverse. Solvent flows from the concentrated solution, leaving an even higher concentration of solute. The semi-permeable membrane can be flat or tubular, and acts like a filter due to the pressure driving force. The waste stream flows through the membrane, while the solvent is pulled through the membrane's pores. The remaining solutes, such as organic or inorganic components, do not pass through, but become more and more concentrated on the influent side of the membrane.

Availability/Limitations

For efficient reverse osmosis, the semi-permeable membrane's chemical and physical properties must be compatible with the waste stream's chemical and physical characteristics. Some membranes will be dissolved by some wastes. Suspended solids and some organics will clog the membrane material. Low-solubility salts may precipitate onto the membrane surface.

Status. Commercial units are available.

ION EXCHANGE

Description

Although some ion exchange media occur naturally, this process normally uses specially formulated resins with an

exchangeable ion bonded to the resin with a weak ionic bond. Ion exchange depends upon the electrochemical potential of the ion to be recovered versus that of the exchange ion; it also depends upon the concentration of the ions in the solution. After a critical relative concentration of recoverable ion to exchanged ion in the solution is exceeded, the exchanged resin is said to be spent. Spent resin is usually recharged by exposure to a concentrated solution of the original exchange ion, causing a reverse exchange. This results in regenerated resin and a concentrated solution of the removed ion, which can be further processed for recovery and reuse. This process is used to remove toxic metal ions from solution to recover concentrated metal for recycling. The residuals include spent resins and spent regenerants such as acid, caustic, or brine.

Availability/Limitations

This technology is used to treat metal wastes including cations (e.g., Ni^{2+}, Cd^{2+}, Hg^{2+}) and anions (e.g., CrO_4^{2-}, SeO_4^{2-}, $HAsO_4^{2-}$). Limitations are selectivity and competition, pH, and suspended solids. Concentrated waste streams with greater than 25,000 mg/L contaminants can be more cost-effectively separated by other means. Solid concentrations greater than 50 mg/L should be avoided to prevent resin blinding.

Status. This is a commercially available process.

ELECTRODIALYSIS

Description

Electrodialysis concentrates or separates ionic species contained in a water solution. In electrodialysis, a water solution is passed through alternately placed cation-permeable and anion-permeable membranes (Figure 4.5.7). An

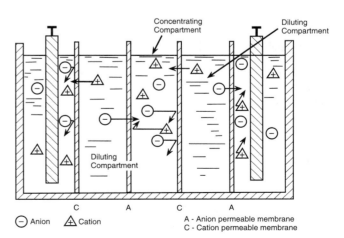

FIG. 4.5.7 Electrodialysis. An electric current concentrates the dissolved ions in compartments adjacent to those between the electrodes.

electrical potential is applied across the membrane to provide the motive force for ion migration. The ion-selective membranes are thin sheets of ion exchange resins reinforced by a synthetic fiber backing.

Availability/Limitations

The process is well established for purifying brackish water, and was recently demonstrated for recovery of metal salts from plating rinse.

Status. Units are being marketed to reclaim metals of value from rinse streams. Such units can be skid mounted and require only piping and electrical connections.

Chemical Treatment Processes
NEUTRALIZATION

Description

When an ionic salt is dissolved in water, several water molecules break into their ionic constituents of H^+ and OH^-. Neutralization is the process of changing the constituents in an ionic solution until the number of hydrogen ions (H^+) is balanced by the hydroxyl (OH^-) ions. Imbalance is measured in terms of the hydrogen ion (H^+) concentration, and is described as the solution's pH. Neutrality, on the pH scale, is 7; an excess of H^+ ions (acidity) is listed at between 0 and 7; and an excess of hydroxy or OH^- ions (alkalinity) is indicated as between 7 and 14. Neutralization is used to treat waste acids and alkalis (bases) to eliminate or reduce reactivity and corrosivity. Neutralization is an inexpensive treatment, especially if waste alkalis can be used to treat waste acid and vice/versa. Residuals include neutral effluents containing dissolved salts, and any precipitated salts.

Applicability/Limitations

This process has extremely wide application to aqueous and nonaqueous liquids, slurries, and sludges. Some applications include pickle liquors, plating wastes, mine drainage, and oil emulsion breaking. The treated stream undergoes essentially no change in physical form, except precipitation or gas evolution.

The process should be performed in a well-mixed system to ensure completeness (Figure 4.5.8). Compatibil-ity of the waste and treatment chemicals should be ensured to prevent formation of more toxic or hazardous compounds than were originally present.

Status. This is a common industrial process.

CHEMICAL PRECIPITATION

Description

Like neutralization, chemical precipitation is a pH adjustment process. To achieve precipitation, an acid or base is added to a solution to adjust the pH to a point where the constituents to be removed reach their lowest solubility. Chemical precipitation facilitates the removal of dissolved metals from aqueous wastes. Metals may be precipitated from solutions by the following methods.

Alkaline agents, such as lime or caustic soda, are added to waste streams to raise the pH. The solubility of metals decreases as pH increases, and the metal ions precipitate out of the solution as hydroxide (Figure 4.5.9).

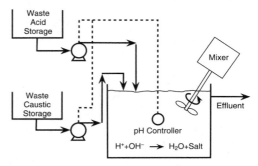

FIG. 4.5.8 Simultaneous neutralization of acid and caustic waste.

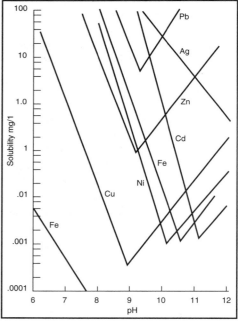

Experimentally determined solubilities of metal hydroxides.

FIG. 4.5.9 Solubilities of metal hydroxides at various pH's. (Reprinted, with permission from Graver Water.)

Soluble sulfides, such as hydrogen or sodium sulfide, and insoluble sulfides, such as ferrous sulfide, are used for precipitation of heavy metals. Sodium bisulfide is commonly used for precipitating chromium out of solution.

Sulfates, including zinc sulfate or ferrous sulfate, are used for precipitation of cyanide complexes.

Carbonates, especially calcium carbonate, are used directly for precipitation of metals. In addition, hydroxides can be converted into carbonates with carbon dioxide, and easily filtered out.

Hydroxide precipitation with lime is most common; however, sodium sulfide is sometimes used to achieve lower effluent metal concentrations. Solid separation is effected by standard flocculation/coagulation techniques. The residuals are metal sludge and treated effluent with an elevated pH and, in the case of sulfide precipitation, excess sulfide.

The metal's valence state is important in the process of precipitation. For example, ferrous iron is considerably more soluble than ferric iron, making oxidizing agent treatment to convert ferrous iron to ferric iron an essential part of the iron-removal process. Another example is hexavalent chromium, Cr^{+6}, which is more soluble than the less hazardous trivalent form. Chromates must be reduced before removal of trivalent chromium in a precipitation process. Also, the engineer must consider the possibility of complex ion formation when dealing with waste water containing ammonia, fluoride, cyanide, or heavy metals. For example, an iron complex may be the ferrocyanide ion, which is soluble, and remains in solution unless the complex is broken by chemical treatment.

Applicability/Limitations

This technology is used to treat aqueous wastes containing metals. Limitations include the fact that metals have different optimum pH levels for precipitation. Chelating and complexing agents can interfere with the process. Organics are not removed except through adsorptive carryover. The resulting sludge may be hazardous by definition, but often may be taken off the list by special petition.

Precipitation has many useful applications to hazardous waste treatment, but laboratory jar tests should be made to verify the treatment. The jar test is used to select the appropriate chemical; determine dosage rates; assess mixing, flocculation and settling characteristics; and estimate sludge production and handling requirements.

Status.   Commercially available.

OXIDATION AND REDUCTION

Description

Oxidation and reduction must take place in any such reaction. In any oxidation reaction, the oxidation state of one compound is raised, while the oxidation state of another compound is reduced. Oxidation and reduction change the chemical form of a hazardous material: rendering it less toxic; changing its solubility, stability, or separability; or otherwise changing it for handling or disposal purposes. In the reaction, the compound supplying oxygen, chlorine or another negative ion, is called the oxidizing agent while the compound supplying the positive ion and accepting the oxygen is called the reducing agent. The reaction can be enhanced by catalysis, electrolysis or irradiation.

Reduction lowers the oxidation state of a compound. Reducing agents include: iron, aluminum, zinc, and sodium compounds. For efficient reduction, waste pH should be adjusted to an appropriate level. After this is accomplished, the reducing agent is added and the resulting solution is mixed until the reaction is complete. This treatment can be applied to chemicals such as hexavalent chromium, mercury, and lead. Other treatment processes may be used in conjunction with chemical reduction.

Cyanide-bearing wastewater generated by the metal-finishing industry, is typically oxidized with alkaline chlorine or hypochlorite solutions. In this process, the cyanide is initially oxidized to a less toxic cyanate and then to carbon dioxide and and nitrogen in the following reactions:

$$NaCN + Cl_2 + 2\,NaOH \longrightarrow NaCNO + 2\,NaCl + H_2O \tag{4.5(1)}$$

$$2\,NaCNO + 3\,Cl_2 + 4\,NaOH \longrightarrow \\ 2\,CO_2 + N_2 + 6\,NaCl + 2\,H_2O \tag{4.5(2)}$$

In the first step, the pH is maintained at above 10, then the reaction proceeds in a matter of minutes. In this step great care must be taken to maintain relatively high pH values, because at lower pHs there is a potential for the evolution of highly toxic hydrogen cyanide gas. The second reaction step proceeds most rapidly around a pH of 8, but not as rapidly as the first step. Higher pH values may be selected for the second step to reduce chemical consumption in the following precipitation steps. However, cyanide complexes of metals, particularly iron and to some extent nickel, cannot be decomposed easily by the cyanide oxidation method.

Cyanide oxidation can also be accomplished with hydrogen peroxide, ozone, and electrolysis.

Applicability/Limitations

The process is nonspecific. Solids must be in solution. Reaction can be explosive. Waste composition must be well known to prevent the inadvertent production of a more toxic or more hazardous end product.

Status.   This is a common industrial process.

HYDROLYSIS

Description

Hydrolysis is the breaking of a bond in a non-water-soluble molecule so that it will go into ionic solution with water.

$$XY + H_2O \longrightarrow HY + XOH \qquad 4.5(3)$$

Hydrolysis can be achieved by: adding chemicals, e.g., acid hydrolysis; irradiation, e.g., photolysis; or biological means, e.g., enzymatic bond cleavage. The cloven molecule can then be further treated by other means to reduce toxicity.

Applicability/Limitations

Chemical hydrolysis applies to a wide range of otherwise refractory organics. Hydrolysis is used to detoxify waste streams of carbamates, organophosphorous compounds and other pesticides. Acid hydrolysis as an in-situ treatment must be performed carefully due to potential mobilization of heavy metals. In addition, depending on the waste stream, products may be unpredictable and the mass of toxic discharge may be greater than the waste originally input for treatment.

Status. Common industrial process.

CHEMICAL OXIDATION

Oxidation destroys hazardous contaminants by chemically converting them to nonhazardous or less toxic compounds that are stable, less mobile, or inert. Common oxidizing agents are ozone, hydrogen peroxide, hypochlorites, chlorine, and chlorine dioxide. Current research shows that combining these reagents, or combining ultraviolet (UV) light and oxidizing agent(s) makes the process more effective.

The effectiveness of chemical oxidation on general contaminant groups is shown in Table 4.5.6 (U.S. EPA 1991). Chemical oxidation depends on the chemistry of the oxidizing agents and the chemical contaminants. Table 4.5.7 lists selected organic compounds by relative oxidation ability. The oxidation process is nonselective; any oxidizable material reacts. Chemical oxidation is also a part of the treatment process for cyanide-bearing wastes and metals such as arsenic, iron, and manganese. Metal oxides formed in the oxidation process precipitate more readily out of the solution.

Some compounds require a combination of oxidizing agents or the use of UV light with an oxidizing agent.

TABLE 4.5.6 EFFECTIVENESS OF CHEMICAL OXIDATION ON GENERAL CONTAMINANT GROUPS FOR LIQUIDS, SOILS, AND SLUDGES[a]

Contaminant Groups	Liquids	Soils, Sludges
Organic		
Halogenated volatiles	■	▼
Halogenated semivolatiles	■	▼
Nonhalogenated volatiles	■	▼
Nonhalogenated semivolatiles	■	▼
PCBs	■	☐
Pesticides	■	▼
Dioxins/Furans	▼	☐
Organic cyanides	■	■
Organic corrosives	▼	▼
Inorganic		
Volatile metals	■	▼
Nonvolatile metals	■	▼
Asbestos	☐	☐
Radioactive materials	☐	☐
Inorganic corrosives	☐	☐
Inorganic cyanides	■	■
Reactive		
Oxidizers	☐	☐
Reducers	■	▼

■ Demonstrated Effectiveness: Successful treatability test at some scale completed
▼ Potential Effectiveness: Expert opinion that technology will work
☐ No Expected Effectiveness: Expert opinion that technology will not work
[a] Enhancement of the chemical oxidation process is required for the less easily oxidizable compounds for some contaminant groups.

Source: Reprinted, from U.S. Environmental Protection Agency (EPA), 1991, *Engineering Bulletin: chemical oxidation treatment,* (EPA 540–2–91–025, Office of Research and Development, Cincinnati, Oh. [September]).

TABLE 4.5.7 SELECTED ORGANIC COMPOUNDS BY RELATIVE ABILITY TO BE OXIDIZED

Ability to be Oxidized	Examples
High	phenols, aldehydes, amines, some sulfur compounds
Medium	alcohols, ketones, organic acids, esters, alkyl-substituted aromatics, nitro-substituted aromatics, carbohydrates
Low	halogenated hydrocarbons, saturated aliphatics, benzene

Source: Reprinted, from U.S. EPA, 1991.

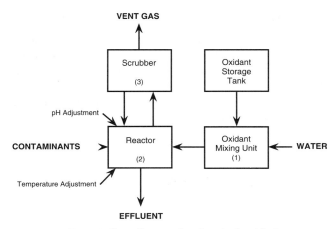

FIG. 4.5.10 Process flow diagram for chemical oxidation system. (Reprinted from U.S. EPA, 1991.)

Polychlorinated biphenyls (PCBs) do not react with ozone alone, but have been destroyed by combined UV and ozone treatment. Enhanced chemical oxidation has been used at several Superfund sites (U.S. EPA 1990a).

Description

Chemical oxidation increases the oxidation state of a contaminant and decreases the oxidation state of the reactant. The electrons lost by the contaminant are gained by the oxidizing agent. The following equation is an example of oxidation reaction:

$$NaCN + H_2O_2 \longrightarrow NaCNO + H_2O \qquad 4.5(4)$$

Figure 4.5.10 details the process flow for a chemical oxidation system. The main component is the process reactor. Oxidant is fed into the mixing unit (1), then the reactor (2). Reaction products and excess oxidant are scrubbed before venting to the ambient air. Reactor pH and temperature are controlled to ensure completion at the reaction. The reaction can be enhanced by adding UV light.

Systems that combine ozone with hydrogen peroxide or UV radiation are catalytic ozonation processes. They accelerate ozone decomposition, increasing hydroxyl radical concentration, and promoting oxidation of the compounds. Specifically, hydrogen peroxide, hydrogen ion, and UV radiation have been found to initiate ozone decomposition and accelerate oxidation of refractory organics via free radical reaction. Reaction times can be 100 to 1000 times faster in the presence of UV light. Minimal emissions result from the UV-enhanced system.

Applicability/Limitations

This process is nonspecific. Solids must be in solution. It may be exothermic or explosive or require addition of heat. Waste composition must be well-known to prevent producing a more toxic or hazardous end product. Oxidation by hydrogen peroxide is not applicable for in situ treatment. However, it may be used for surface treatment of contaminated groundwater sludge. Oxidation is not cost-effective for highly concentrated waste because of the large amount of oxidizing agent required.

Ozone can be used to pretreat wastes to break down refractory organics or to oxidize untreated organics after biological or other treatment processes. Ozone is currently used to destroy cyanide and phenolic compounds. Rapid oxidation offers advantages over the slower alkaline chlorination method. Limitations include the physical form of the waste (i.e., sludges and solids are not readily treated) and non-selective competition with other species. Ozonation systems have higher capital costs because ozone generators must be used.

The cost of generating UV lights and the problems of scaling or coating on the lamps are two of the major drawbacks to UV-enhanced chemical oxidation systems. They do not perform well in turbid waters or slurries because reduced light transmission lowers the effectiveness.

Status. Commercially available.

ELECTROLYTIC OXIDATION

Description

In this process, cathodes and anodes are immersed in a tank containing waste to be oxidized, and a direct current is imposed on the system. This process is particularly applicable to cyanide-bearing wastes. Reaction products are ammonia, urea, and carbon dioxide. During decomposition, metals are plated out on the cathodes.

Applicability/Limitations

Electrolytic oxidation is used to treat high concentrations of up to 10% cyanide and to separate metals to allow their potential recovery. Limitations include the physical form of the feed (solids must be dissolved), non-selective competition with other species and long process times. Electrolytic recovery of single metal species can be 90% or higher.

Status.   Commercially available.

ALKALINE METAL DECHLORINATION

Description

This process of chemical dechlorination displaces chlorine from chlorinated organic compounds contained in oils and liquid wastes. Typically, wastes are filtered before entering the reactor system and encountering the dechlorinating reagent. The great affinity of alkali metals for chlorine (or any halide) is the chemical basis of this process.

Successive treatment includes additional centrifugation and filtration. By-products include chloride salts, polymers, and heavy metals. Several chemical dechlorination processes are based on a method developed by the Goodyear Tire and Rubber Company in 1980. The original method uses sodium naphthalene and tetrahydrofuran to strip chlorine atoms from PCBs, polymerizing the biphenols into an inert condensible sludge. The reactor is blanketed with nitrogen because the reagents are sensitive to air and water, and an excess of reagent to chlorine is required. The Goodyear Company has not commercially developed this technology; however, several companies have modified the method by substituting their own proprietary reagents for the naphthalene. The equipment is mobile and can be transported on semi-trailers.

Applicability/Limitations

Such processes are used to treat PCBs, other chlorinated hydrocarbons, acids, thiols, and dioxins. Moisture content

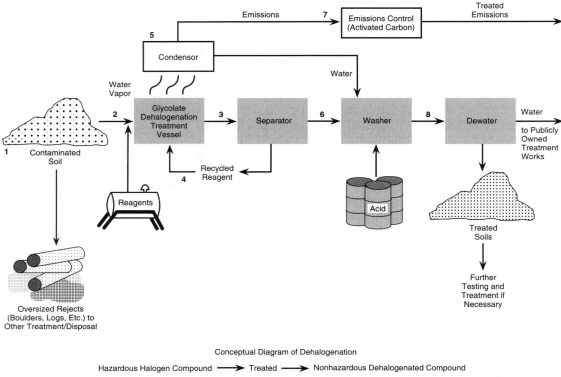

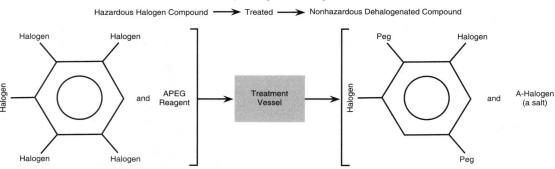

FIG. 4.5.11 Glycolate dehalogenation process flow. (Reprinted, from U.S. EPA, 1992 [March].)

adversely affects the rate of reaction, therefore dewatering should be a pretreatment step. Waste stream concentrations are also important.

Status. Commercially available.

ALKALINE METAL/POLYETHYLENE GLYCOL (APEG)

Description

In 1978, EPA-sponsored research led to the development of the first in a series of APEG reagents, which effectively dechlorinate PCBs and oils. These reagents were alkali metal/polyethylene glycols which react rapidly to dehalogenate halo-organic compounds of all types (Figure 4.5.11).

In the APEG reagents, alkali metal is held in solution by large polyethylene anions. PCBs and halogenated molecules are soluble in APEG reagents. These qualities combine in a single-phase system where the anions readily displace the halogen atoms. Halogenated aromatics react with PEGs resulting in the substitution of halogenated aromatics for chlorine atoms to form a PEG ether. The PEG ether decomposes to a phenol.

The effectiveness of APEG on general contaminant groups for various matrices is shown in Table 4.5.8 (U.S. EPA, 1990b).

A variation of APEG, referred to as ATEG, uses potassium hydroxide or sodium hydroxide/tetraethylene glycol, and is more effective on halogenated aliphatic compounds.

Figure 4.5.11 is a schematic of the APEG treatment process. Waste preparation includes excavating and/or moving the soil to the process where it is normally screened (1) removing debris and large objects and producing particles small enough to allow treatment in the reactor without binding the mixer blades.

Typically, reagent components are mixed with contaminated soil in the reactor (2). Treatment proceeds inefficiently without mixing. The mixture is heated to between 100°C and 180°C. The reaction proceeds for 1–5 hrs. depending upon the type, quantity, and concentration of the contaminants. The treated material goes from the reactor to a separator (3), where the reagent is removed and can be recycled (4).

During the reaction, water is evaporated in the reactor, condensed (5), and collected for further treatment or recycled through the washing process, if required. Carbon filters (7) are used to trap any volatile organics that are

TABLE 4.5.8　EFFECTIVENESS OF APEG TREATMENT ON GENERAL CONTAMINANT GROUPS FOR VARIOUS MATRICES

	Effectiveness			
Contaminant Groups	Sediments	Oils	Soil	Sludge
Organic				
Halogenated volatiles	▼	▼	▼	▼
Halogenated semivolatiles	▼	▼	▼	▼
Nonhalogenated volatiles	☐	☐	☐	☐
Nonhalogenated semivolatiles	☐	☐	☐	☐
PCBs	■	■	■	■
Pesticides (halogenated)	▼	■	■	▼
Dioxins/Furans	■	■	■	■
Organic cyanides	☐	☐	☐	☐
Organic corrosives	☐	☐	☐	☐
Inorganic				
Volatile metals	☐	☐	☐	☐
Nonvolatile metals	☐	☐	☐	☐
Asbestos	☐	☐	☐	☐
Radioactive materials	☐	☐	☐	☐
Inorganic corrosives	☐	☐	☐	☐
Inorganic cyanides	☐	☐	☐	☐
Reactive				
Oxidizers	☐	☐	☐	☐
Reducers	☐	☐	☐	☐

■ Demonstrated Effectiveness: Successful treatability test at some scale completed
▼ Potential Effectiveness: Expert opinion that technology will work
☐ No Expected Effectiveness: Expert opinion that technology will not work

Source: Reprinted, from U.S. Environmental Protection Agency (EPA), 1990, *Engineering Bulletin: Chemical dehalogenation treatment: APEG treatment* (EPA 540–2–90–015, Office of Research and Development, Cincinnati, Oh. [September]).

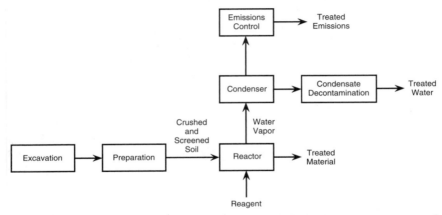

FIG. 4.5.12 BCD process flow schematic. (Reprinted, from U.S. Environmental Protection Agency, 1992, *BCD detoxification of chlorinated wastes*. Office of Research and Development, Cincinnati, Oh. [September].)

not condensed. In the washer (6), soil is neutralized by the addition of acid. It is then dewatered (8) before disposal.

Applicability/Limitations

Dehalogenation is effective in removing halogens from hazardous organic compounds such as dioxins, furans, PCBs, and chlorinated pesticides; rendering them non-toxic. APEG will dehalogenate aliphatic compounds if the mixture reacts longer and at temperatures significantly higher than for aromatics. This technology usually costs less than incineration.

Treatability tests should be conducted before the final selection of the APEG technology. Operating factors such as quantity of reagents, temperature, and treatment time should be defined. Treated soil may contain residual reagents and treatment by-products that should be removed by washing the soil with water. The soil should also be neutralized by lowering the pH before final disposal.

Specific safety aspects must be considered. Treatment of certain chlorinated aliphatics in high concentrations with APEG may produce potentially explosive compounds (e.g., chloroacetylenes) or cause a fire hazard.

Status.   This process has been field tested.

BASED-CATALYZED DECOMPOSITION (BCD)

Description

Based-catalyzed decomposition is another technology for removing chlorine molecules from organic substances.

$$\text{Chlorinated Products} \xrightarrow[\text{Catalyst, base, } \Delta]{\text{Hydrogen donors}}$$

$$\begin{array}{c}\text{Dechlorinated products} \\ + \\ \text{sodium chloride}\end{array} \qquad 4.5(5)$$

The BCD process (Figure 4.5.12) embodies the following steps: mixing the chemicals with the contaminated matrix (such as excavated soil or sediment or liquids, containing these toxic compounds), and heating the mixture at 320–340°C for 1–3 hr. The off-gases are treated before releasing to the atmosphere. The treated receptor remains are nonhazardous, and can be either disposed of according to standard methods, or further processed to separate components for reuse.

Applicability/Limitations

Laboratory and bench-scale tests demonstrated this technology's ability to reduce PCBs from 4,000 ppm to less than 1 ppm. The BCD process requires only 1–5% reagent by weight. The reagent is also much less expensive than the APEG reagent. BCD also is regarded as effective for pentachlorophenol (PCP), PCBs, pesticides (halogenated), herbicides (halogenated), dioxins and furans. Again, BCD is not intended as an in situ treatment. Treatability studies should be conducted before the final selection.

Status.   This process has been field tested.

—David H.F. Liu

References

Martin, E.J., and J.H. Johnson, Jr. 1987. *Hazardous waste management engineering*. Chapter 3. New York, N.Y.: Van Nostrand Reinhold.

U.S. Environmental Protection Agency (EPA). 1990. *Technology evaluation report: SITE program demonstration of the Ultrox R international ultraviolet radiation/oxidation technology*. EPA 540–5–89–012 (January).

———. 1990b. *Engineering Bulletin: Chemical dehalogenation treatment: APEG treatment*. EPA 540–2–90–015. Office of Research and Development. Cincinnati, Oh. (September).

———. 1991. *Engineering Bulletin: Chemical oxidation treatment*. EPA 540–2–91–025. Office of Research and Development. Cincinnati, Oh. (October).

4.6 SOLIDIFICATION AND STABILIZATION TECHNOLOGIES

Solidification techniques encapsulate hazardous waste into a solid material of high structural integrity. Encapsulation involves either fine waste particles, microencapsulation, or a large block or container of wastes, macroencapsulation (Conner 1990). Stabilization techniques treat hazardous waste by converting it into a less soluble, mobile, or toxic form. Solidification/stabilization (S/S) processes utilize one or both of these techniques.

The goal of S/S processes is the safe ultimate disposal of hazardous waste. Four primary reasons for treating the waste are to:

- Improve handling characteristics for transport on-site or to an off-site TSD facility
- Limit the mobility or solubility of pollutants contained in the waste
- Reduce the exposed area allowing transfer or loss of contained pollutants
- Detoxify contained pollutants

Applications

Table 4.6.1 summarizes the effectiveness of S/S on general contaminant groups for soils and sludges. The fixing and binding agents for S/S immobilize many heavy metals and solidify a wide variety of wastes including spent pickle liquor, contaminated soils, incinerator ash, wastewater treatment filter cake, waste sludge, and many radionuclides (EPA 1990). In general, S/S is considered as an established full-scale technology for nonvolatile heavy metals, although the long-term performance of S/S in Superfund applications has yet to be demonstrated (EPA 1991).

Technology Description

S/S processes can be divided into the following broad categories: inorganic processes (cement and pozzolanic) and organic processes (thermoplastic and thermosetting).

CEMENT-BASED PROCESSES

These processes generally use Portland cement and sludge along with certain other additives (some proprietary) including fly ash or other aggregate to form a monolithic, rock-like mass. Type I Portland cement, the cement normally used in construction, is generally used for waste fixation. Type II is used in the presence of moderate sulfate concentrations (150–1500 mg/kg), and Type V, for high sulfate concentrations (greater than 1500 mg/kg).

These processes are successful on many sludges generated by the precipitation of heavy metals. The high pH of the cement mixture keeps the metals as insoluble hydrox-

TABLE 4.6.1 EFFECTIVENESS OF S/S ON GENERAL CONTAMINANT GROUPS FOR SOIL AND SLUDGES

Contaminant Groups	Effectiveness Soil/Sludge
Organic	
Halogenated volatiles	□
Nonhalogenated volatiles	□
Halogenated semivolatiles	■
Nonhalogenated semivolatiles and nonvolatiles	■
PCBs	▼
Pesticides	▼
Dioxins/Furans	▼
Organic cyanides	▼
Organic corrosives	▼
Inorganic	
Volatile metals	■
Nonvolatile metals	■
Asbestos	■
Radioactive materials	■
Inorganic corrosives	■
Inorganic cyanides	■
Reactive	
Oxidizers	■
Reducers	■

KEY: ■ Demonstrated Effectiveness: Successful treatability test at some scale completed.

▼ Potential Effectiveness: Expert opinion that technology will work.

□ No Expected Effectiveness: Expert opinion that technology will/does not work.

Source: Reprinted, from U.S. Environmental Protection Agency, 1993, *Engineering bulletin: solidification/stabilization of organics and inorganics,* (EPA 540–5–92–015. Office of Research and Development, Cincinnati, Oh.

ides or carbonate salts. Metal ions may also be taken up into the cement matrix.

Additives such as clay, vermiculite, and soluble silicate improve the physical characteristics and decrease leaching losses from the resulting solidified sludge. Many additives are proprietary.

POZZOLANIC PROCESSES

These lime-based stabilization processes depend on the reaction of lime with a fine-grained siliceous material and water to produce a hardened material. The most common pozzolanic materials used in waste treatment are fly ash, ground blast-furnace slag, and cement-kiln dust. As all these materials are waste products to be disposed of, the fixation process can reduce contamination from several wastes. Other additives, generally proprietary, are often added to the sludge to enhance material strength or to help limit migration of problem contaminants from the sludge mass.

Lime and Portland cement are the setting agents, but gypsum, calcium carbonate, and other compounds may also be used. Lime-based and cement-based processes are better suited for stabilizing inorganic wastes rather than organic wastes. Decomposition of organic material in sludge after curing can result in increased permeability and decreased strength of the material.

Certain processes fall in the category of cement-pozzolanic processes. In this case both cement and lime-siliceous materials are combined to give the best and most economical containment for waste.

THERMOPLASTIC PROCESSES

Bitumen stabilization techniques (including bitumen, paraffin and polyethylene) were developed for use in radioactive waste disposal and later adapted for handling industrial wastes. In a bitumen process, the waste is dried and then mixed with bitumen, paraffin or polyethylene (usually at temperatures greater than 100°C). The mixture solidifies as it cools, then is placed in a container, such as a steel drum or a thermoplastic coating, before disposal.

A variation of the bitumen process uses an asphalt emulsion that is miscible with the wet sludge. This process can be conducted at a lower temperature than a bitumen process. The emulsion-waste mixture must be dried before disposal.

The type of waste sludges that can be fixed with bitumen techniques is limited. Organic chemicals that act as solvents with bitumen cannot be stabilized. High concentrations of strong oxidizing salts such as nitrates, chlorates, or perchlorates will react with bitumen and cause slow deterioration.

ORGANIC POLYMER PROCESSES

The major organic polymer process (including urea-formaldehyde, unsaturated polyesters) currently in use is the urea-formaldehyde process. In the process, a monomer is added to the waste or sludge and thoroughly mixed. Next, a catalyst is added to the mixture and mixing continues until the catalyst is dispersed. The mixture is transferred to another container and allowed to harden. The polymerized material does not chemically combine with the waste. Instead, a spongy mass forms, trapping the solid particles while allowing some liquid to escape. The polymer mass can be dried before disposal.

Table 4.6.2 compares the advantages and disadvantages of the above S/S processes. Table 4.6.3 illustrates the compatibility of selected waste categories with S/S processes.

Figures 4.6.1 and 4.6.2 depict generic elements of typical ex situ and in situ S/S processes for soils and sludges. Ex situ processing involves: (1) excavation to remove the contaminated waste from the subsurface; (2) classification to remove oversize debris; (3) mixing; and (4) off-gas treatment. In situ processing has only two steps: (1) mixing; and (2) off-gas treatment. Both processes require a system for delivering water, waste, and S/S agents in proper proportions; and a mixing device (e.g., rotary drum paddle or auger). Ex situ processing requires a system for delivering treated waste to molds, surface trenches, or subsurface injection. The need for off-gas treatment using vapor collection and treatment modules is specific to the S/S process.

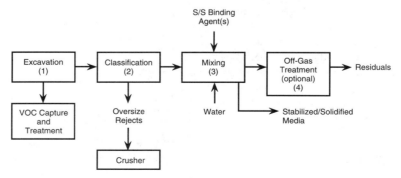

FIG. 4.6.1 Generic elements of a typical ex situ S/S process. (Reprinted, from U.S. EPA, 1993.)

TABLE 4.6.2 COMPARISON FOR SOLIDIFICATION AND STABILIZATION PROCESSES

Process	Description	Advantages	Disadvantages
Cement	Slurry of wastes and water is mixed with portland cement to form a solid	Low costs; readily available mixing equipment; relatively simple process; suitable for use with metals	Solids are suspended, not chemically bound; therefore are subject to leaching; doubles waste volume; requires secondary containment; incompatible with many wastes (organics, some sodium salts, silts, clays, and coal or lignite)
Pozzolanic	Waste is reacted with lime and a fine-grained siliceous material (fly ash, ground blast furnace slag, cement kiln dust) to form a solid	Low cost; readily available mixing equipment; suitable for power-plant wastes (FGD sludges, etc.) as well as a wide range of industrial wastes, including metals, waste oil, and solvents	Increases waste volume; may be subject to leaching; requires secondary containment
Thermoplastic	Waste is dried, heated, and dispensed through a heated plastic matrix of asphalt bitumen, paraffin or polyethylene	Less increase in volume than with cement- or lime-based processes; reduced leaching relative to cement- or lime-based processes; suitable for radioactive wastes and some industrial wastes	Wastes must be dried before use; high equipment costs; high energy costs; requires trained operators; incompatible with oxidizers, some solents and greases, some salt, and chelating/complexing agents; requires secondary containment
Organic polymers	Waste is mixed with a prepolymer and a catalyst that causes solidification through formation of a spongelike polymer matrix; urea-formaldehyde or vinyl ester-styrene polymers are used	Suitable for insoluble solids; very successful in limited limited applications	Pollutants are not chemically bound, subject to leaching; strongly acidic leach water may be produced; requires special equipment and operators; some of the catalysts used are corrosive; harmful vapors may be produced; incompatible with oxidizers and some organics; some resins are biodegradable and decompose with time

Source: Reprinted, from U.S. Environmental Protection Agency, 1980, *Guide to the disposal of chemically stabilized and solidified waste,* (EPA SW–872, Cincinnati, Oh. [September]).

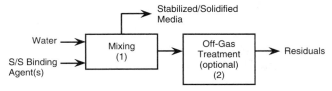

FIG. 4.6.2 Generic elements of a typical in situ S/S process. (Reprinted, from U.S. EPA, 1993.)

Also, hazardous residuals from some pretreatment technologies must be disposed of using appropriate procedures.

Technology Limitations

Tables 4.6.4 and 4.6.5 summarize factors that interfere with stabilization and solidification processes.

TABLE 4.6.3 COMPATIBILITY OF WASTE CATEGORIES WITH SOLIDIFICATION/STABILIZATION TECHNIQUES

Waste Component	Treatment Type						
	Cement-Based	Lime-Based	Thermoplastic	Organic Polymer (UF)[a]	Surface Encapsulation	Self-Cementing	Classification and Synthetic Mineral Formation
Organics							
1. Organic solvents and oils	Many impede setting, may escape as vapor	Many impede setting, may escape as vapor	Organics may vaporize on heating	May retard set of polymers	Must first be absorbed on solid matrix	Fire danger on heating	Wastes decompose at high temperatures
2. Solid organics (e.g., plastics, resins, tars)	Good–often increases durability	Good–often increases durability	Possible use as binding agent	May retard set of polymers	Compatible–many encapsulation materials are plastic	Fire danger on heating	Wastes decompose at high temperatures
Inorganics							
1. Acid wastes	Cement will neutralize acids	Compatible	Can be neutralized before incorporation	Compatible	Can be neutralized before incorporation	May be neutralized to form sulfate salts	Can be neutralized and incorporated
2. Oxidizers	Compatible	Compatible	May cause matrix breakdown, fire	May cause matrix breakdown	May cause deterioration of encapsulating materials	Compatible if sulfates are present	High temperatures may cause undesirable reactions
3. Sulfates	May retard setting and cause spalling unless special cement is used	Compatible	May dehydrate and rehydrate, causing splitting	Compatible	Compatible	Compatible	Compatible in many cases
4. Halides	Easily leached from cement, may retard setting	May retard setting, most are easily leached	May dehydrate	Compatible	Compatible	Compatible if sulfates are also present	Compatible in many cases
5. Heavy metals	Compatible	Compatible	Compatible	Acid pH solubilizes metal hydroxides	Compatible	Compatible if sulfates are present	Compatible in many cases
6. Radioactive materials	Compatible	Compatible	Compatible	Compatible	Compatible	Compatible if sulfates are present	Compatible

Source: Reprinted, from U.S. EPA, 1980.
[a]Urea-formaldehyde resin.

Physical mechanisms that interfere with the S/S process include incomplete mixing due to high moisture or organic chemical content. This results in only partial wetting or coating of particles with stabilizing and binding agents, and the aggregation of untreated waste into lumps (EPA 1986). Chemical mechanisms that interfere with S/S of cement-based systems include chemical adsorption, complexation, precipitation, and nucleation (Conner 1990).

Environmental conditions must be considered in determining whether and when to implement S/S process technology. Extremes of heat, cold, and precipitation can adversely affect S/S applications. For example, the viscosity of one or more of the materials may increase rapidly with

TABLE 4.6.4 SUMMARY OF FACTORS THAT MAY INTERFERE WITH STABILIZATION PROCESSES

Characteristics Affecting Processing Feasibility	Potential Interference
VOCs	Volatiles not effectively immobilized; driven off by heat of reaction. Sludges and soils containing volatile organics can be treated using a heated extruder evaporator or other means to evaporate free water and VOCs prior to mixing with stabilizing agents.
Use of acidic sorbent with metal hydroxide wastes	Solubilizes metal.
Use of acidic sorbent with cyanide wastes	Releases hydrogen cyanide.
Use of acidic sorbent with waste containing ammonium compounds	Releases ammonia gas.
Use of acidic sorbent with sulfide wastes	Releases hydrogen sulfide.
Use of alkaline sorbent (containing carbonates such as calcite or dolomite) with acid waste	May create pyrophoric waste.
Use of siliceous sorbent (soil, fly ash) with hydrofluoric acid waste	May produce soluble fluorosilicates.
Presence of anions in acidic solutions that form soluble calcium salts (e.g., calcium chloride acetate, and bicarbonate)	Cation exchange reactions—leach calcium from S/S product increases permeability of concrete, increases rate of exchange reactions.
Presence of halides	Easily leached from cement and lime.

Source: Reprinted, from United States Environmental Protection Agency, 1991, *Technical resources document on solidification/stabilization and its application to waste material (Draft),* (Contract No. 68–CO–003, Office of Research and Development, Cincinnati, Oh.).

falling temperature, or the cure rate may be unacceptably slowed.

Depending on the waste and binding agents involved, S/S processes can produce hot gases, including vapors that are potentially toxic, irritating, noxious to workers or communities downwind from the processes. In addition, if volatile substances with low flash points are involved, and the fuel-air ratio is favorable, there is a potential for fire and explosion.

Taking S/S processes from bench-scale to full-scale operation involves inherent uncertainties. Variables such as ingredient flow-rate control, material balance, mixing, materials handling and storage, and weather may all affect a field operation. These potential engineering difficulties emphasize the need for field demonstration before full-scale implementation.

Performance Testing

Treated wastes are subjected to physical tests to (1) determine particle size and distribution, porosity, permeability, and dry and wet density; (2) evaluate bulk-handling properties; (3) predict the reaction of a material to applied stresses in embankments and landfills; and (4) evaluate durability under freeze/thaw and wet/dry weathering cycles.

Chemical leach testing determines the chemical stability of treated wastes when in contact with aqueous solutions encountered in landfills. The procedures demonstrate the immobilization of contaminants by the S/S processes. Many techniques for leach testing are available. The major variables encountered in different leaching procedures are: the nature of the leaching solution; waste to leaching solution ratios; number of elutions of leaching solutions used; contact time of waste and leaching solution; surface area of waste exposed; and agitation technique used. Treated wastes must meet certain maximum leachate concentrations when subject to the Toxicity Characteristic Leaching Procedure (TCLP) determination (see Section 2.1).

—*David H.F. Liu*

References

Connor, J.R. 1990. *Chemical fixation and solidification of hazardous waste.* New York, N.Y.: Van Nostrand Reinhold.

U.S. Environmental Protection Agency (EPA). 1987. *Handbook—remedial action at waste disposal sites (rev.).* EPA 625–6–85–006. Washington, D.C. (January).

———. 1990. *Stabilization/solidification of CERCLA and RCRA wastes: physical test, chemical testing procedures, technology, and field activities.* EPA 625–6–89–022. Cincinnati, Oh. (May).

———. 1991. *Technical resources document on solidification/stabilization and its application to waste material (draft).* Contract No. 68–C0–0003. Office of Research and Development. Cincinnati, Oh.

TABLE 4.6.5 SUMMARY OF FACTORS THAT MAY INTERFERE WITH SOLIDIFICATION PROCESSES

Characteristics Affecting Processing Feasibility	Potential Interference
Organic compounds	Organics may interfere with bonding of waste materials with inorganic binders.
Semivolatile organics or poly-aromatic hydrocarbons (PAHs)	Organics may interfere with bonding of waste materials.
Oil and grease	Weaken bonds between waste particles and cement by coating the particles. Decrease in unconfined compressive strength with increased concentrations of oil and grease.
Fine particle size	Insoluble material passing through a No. 200 mesh sieve can delay setting and curing. Small particles can also coat larger particles, weakening bonds between particles and cement or other reagents. Particle size $>\frac{1}{4}$ inch in diameter not suitable.
Halides	May retard setting, easily leached for cement and pozzolan S/S. May dehydrate thermoplastic solidification.
Soluble salts of manganese, tin, zinc, copper, and lead	Reduced physical strength of final product caused by large variations in setting time and reduced dimensional stability of the cured matrix, thereby increasing leachability potential.
Cyanides	Cyanides interfere with bonding of waste materials.
Sodium arsenate, borates, phosphates, iodates, sulfides, and carbohydrates	Retard setting and curing and weaken strength of final product.
Sulfates	Retard setting and cause swelling and spalling in cement S/S. With thermoplastic solidification may dehydrate and rehydrate, causing splitting.
Phenols	Marked decreases in compressive strength for high phenol levels.
Presence of coal or lignite	Coals and lignites can cause problems with setting, curing, and strength of the end product.
Sodium borate, calcium sulfate, potassium dichromate, and carbohydrates	Interferes with pozzolanic reactions that depend on formation of calcium silicate and aluminate hydrates.
Nonpolar organics (oil, grease, aromatic hydrocarbons, PCBs)	May impede setting of cement, pozzolan, or organic-polymer S/S. May decrease long-term durability and allow escape of volatiles during mixing. With thermoplastic S/S, organics may vaporize from heat.
Polar organics (alcohols, phenols, organic acids, glycols)	With cement or pozzolan S/S, high concentrations of phenol may retard setting and may decrease short-term durability; all may decrease long-term durability. With thermoplastic S/S, organics may vaporize. Alcohols may retard setting of pozzolans.
Solid organics (plastics, tars, resins)	Ineffective with urea formaldehyde polymers; may retard setting of other polymers.
Oxidizers (sodium hypochlorite, potassium permanganate, nitric acid, or potassium dichromate)	May cause matrix breakdown or fire with thermoplastic or organic polymer S/S.
Metals (lead, chromium, cadmium, arsenic, mercury)	May increase setting time of cements if concentration is high.
Nitrates, cyanides	Increase setting time, decrease durability for cement-based S/S.
Soluble salts of magnesium, tin, zinc, copper and lead	May cause swelling and cracking within inorganic matrix exposing more surface area to leaching.
Environmental/waste conditions that lower the pH of matrix	Eventual matrix deterioration.
Flocculants (e.g., ferric chloride)	Interference with setting of cements and pozzolans.
Soluble sulfates >0.01% in soil or T50 mg/L in water	Endangerment of cement products due to sulfur attack.
Soluble sulfates >0.5% in soil or 2000 mg/L in water	Serious effects on cement products from sulfur attacks.
Oil, grease, lead, copper, zinc, and phenol	Deleterious to strength and durability of cement, lime/fly ash, fly ash/cement binders.
Aliphatic and aromatic hydrocarbons	Increase set time for cement.
Chlorinated organics	May increase set time and decrease durability of cement if concentration is high.
Metal salts and complexes	Increase set time and decrease durability for cement or clay/cement.
Inorganic acids	Decrease durability for cement (Portland Type I) or clay/cement.
Inorganic bases	Decrease durability for clay/cement; KOH and NaOH decrease durability for Portland cement Type III and IV.

Source: Reprinted from U.S. EPA, 1991.

4.7
BIOLOGICAL TREATMENT

Activated Sludge: Polybac Corp.; Detox Inc.; Ground Decontaminaton Systems

Rotating Biological Contactors: Polybac Corp.; Detox Inc.; Ground Decontaminaton Systems

Bioreclamation: FMC

Biological degradation of hazardous organic substances is a viable approach to waste management. Common processes are those originally utilized in treating municipal wastewaters, based on aerobic or anaerobic bacteria. In-situ treatment of contaminated soils can be performed biologically. Cultures used in biological degradation processes can be native (indigenous) microbes, selectively adapted microbes, or genetically altered microorganisms.

Table 4.7.1 shows that every class of anthropogenic compound can be degraded by some microorganism. Anthropogenic compounds such as halogenated organics are relatively resistant to biodegradation. One reason for this is the naturally present organisms often cannot produce the enzymes necessary to transform the original compound to a point where resultant intermediates can enter common metabolic pathways and be completely mineralized.

Several of the most persistent chlorinated compounds, such as TCE, appear to be biodegradable only through co-metabolism. Co-metabolism involves using another substance as a source of carbon and energy to sustain microbial growth. The contaminant is metabolized gratuitously due to a lack of enzyme specificity. To stimulate co-metabolism in bioremediation, a co-substrate is added to the

TABLE 4.7.1 EXAMPLES OF ANTHROPOGENIC COMPOUNDS AND MICROORGANISMS THAT CAN DEGRADE THEM

Compound	Organism
Aliphatic (nonhalogenated)	
Acrylonitrile	Mixed culture of yeast mold, protozoan bacteria
Aliphatic (halogenated)	
Trichloroethane, trichloroethylene, methyl chloride, methylene chloride	Marine bacteria, soil bacteria, sewage sludge
Aromatic compounds (nonhalogenated)	
Benzene, 2,6-dinitrotoluene, creosol, phenol	*Pseudomonas* sp, sewage sludge
Aromatic compounds (halogenated)	
1,2-; 2,3-; 1,4-Dichlorobenzene, hexachlorobenzene, trichlorobenzene	Sewage sludge
Pentachlorophenol	Soil microbes
Polycyclic aromatics (nonhalogenated)	
Benzo(a)pyrene, naphthalene	*Cunninghamells elogans*
Benzo(a)anthracene	*Pseudomonas*
Polycyclic aromatics (halogenated)	
PCBs	*Pseudomonas, Flavobacterium*
4-Chlorobiphenyl	Fungi
Pesticides	
Toxaphene	*Corynebacterium pyrogenes*
Dieldrin	Anacystic nidulans
DDT	Sewage sludge, soil bacteria
Kepone	Treatment lagoon sludge
Nitrosamines	
Dimethylnitrosamine	*Rhodopseudomonas*
Phthalate esters	Micrococcus 12B

Source: Reprinted, with permission from Table 1 of H. Kobayashi and B.E. Rittmann, 1982, Microbial removal of hazardous organic compounds, *Environmental Science and Technology,* (vol. 16, p. 173A).

contaminated site, to induce growth of microorganisms whose enzymes can degrade both the co-substrate and the original pollutant. Even inherently toxic inducers, such as phenol or toluene, are sometimes added to stimulate bacterial production of enzymes to degrade polyaromatic hydrocarbons and chlorinated aliphatics.

Table 4.7.2 lists important treatment data needs for biological treatments.

This section describes biological processes applicable to hazardous waste.

Aerobic Biological Treatment

DESCRIPTION

Hydrocarbons are catabolized or broken down into simpler substances by microorganisms using aerobic respiration, anaerobic respiration, and fermentation. In general, aerobic degradation processes are more often used for biodegradation because the degradation process is more rapid and more complete, and problematic products such as methane and hydrogen sulfide are not produced. However, anaerobic degradation is important for dehalogenation.

In aerobic respiration, organic molecules are oxidized to carbon dioxide (CO_2), water, and other end products using molecular oxygen as the terminal electron acceptor. Oxygen is also incorporated into the intermediate products of microbial catabolism through oxidase enzyme action, making these products more susceptible to further biodegradation. Microorganisms metabolize hydrocarbons by anaerobic respiration in the absence of molecular oxygen using inorganic substrates as terminal electron acceptors. Naturally occurring aerobic bacteria can decompose natural and synthetic organic materials to harmless or stable forms by mineralizing them to CO_2 and water. Some anthropogenic compounds appear refractory to biodegradation by naturally occurring microbial populations because of environmental influences, lack of solubility, and the absence of required enzymes, nutrients or other factors. However, properly selected or engineered micro-

bial populations, maintained under environmental conditions conducive to their metabolic activity are an important means of biological transformation or degrading these otherwise refractory wastes.

All microorganisms require adequate levels of inorganic and organic nutrients, growth factors (vitamins, magnesium, copper, manganese, sulfur, potassium, etc.), water, oxygen, carbon dioxide and sufficient biological space for survival and growth. One or more of these factors is usually in limited supply. In addition, various microbial competitors adversely affect each other in struggling for these limited resources. Other factors influencing microbial degradation rates include microbial inhibition by chemicals in the waste to be treated, the number and physiological state of the organisms as a function of available nutrients, the seasonal state of microbial development, predators, pH, and temperature. Interaction between these and other potential factors can cause wide variations in degradation kinetics.

For these and other reasons, aerobic biodegradation is usually carried out in processes where many of the requisite conditions can be controlled. Such processes include conventional activated-sludge processes, with modifications such as sequencing batch reactors, and aerobic-attached growth biological processes such as rotating biological contactors and trickling filters. Recently developed genetically engineered bacteria are reported to be effective for biological treatment of specific, relatively uniform, hazardous wastes.

APPLICABILITY/LIMITATIONS

Used to treat aqueous waste contaminated with low levels (e.g., BOD less than 10,000 mg/L) of nonhalogenated organic and certain halogenated organics. Treatment requires consistent, stable operating conditions.

Status. Conventional, broadly used technology

Activated Sludge

DESCRIPTION

Activated sludge treatment breaks down organic contaminants in aqueous waste streams through aerobic microorganisms' activity. These microorganisms metabolize biodegradable organics. This treatment includes conventional activated sludge processes and modifications such as sequencing batch reactors. The aeration process includes pumping the waste to an aeration tank where biological treatment occurs. Following this, the stream is sent to a clarifier where the liquid effluent (the treated aqueous waste) is separated from the sludge biomass (Figure 4.7.1). Aerobic processes can significantly reduce a wide range of organic, toxic and hazardous compounds. However, only dilute aqueous wastes (less than 1%) are normally treatable.

TABLE 4.7.2   IMPORTANT BIOLOGICAL TREATMENT DATA NEEDS

Data Need	Purpose
Gross Organic Component (BOD, TOC)	Treatability
Priority Pollutant Analysis	Toxicity to Process Microbes
Dissolved Oxygen	Aerobic Reaction Rates/ Interference with Anaerobic System
Nutrient Analysis (NH_3, NO_3, PO_4, etc.)	Nutrient Requirements
pH	pH Adjustment
ORP	Chemical Competition

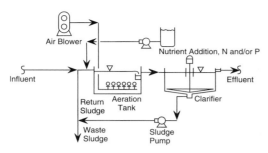

FIG. 4.7.1 Activated sludge process.

APPLICABILITY/LIMITATIONS

The treatment requires consistent, stable operating conditions. Activated sludge processes are not suitable for removing highly chlorinated organics, aliphatics, amines, and aromatic compounds from waste streams. Some heavy metals and organic chemicals are harmful to the organisms. When using conventional open aeration tanks and clarifiers, volatile hazardous materials may escape.

Status.   Conventional, well developed

Rotating Biological Contactors

DESCRIPTION

Rotating biological contactors (RBCs) aerobically treat aqueous waste streams, especially those containing alcohols, phenols, phthalates, cyanides, and ammonia. Primary treatment (e.g., clarifiers or screens) to remove materials that could settle in RBC tanks or plug the discs, is often essential for good operation. Influents containing high concentrations of floatables (e.g., grease) require treatment with a primary clarifier or an alternate removal system (EPA 1984; EPA 1992).

A typical RBC unit consists of 12-ft-dia plastic discs mounted along a 25-ft horizontal shaft. The disc surface is normally 100,000 sq ft for a standard unit and 150,000 sq ft for a high density unit. Figure 4.7.2 details a typical RBC system.

As the discs rotate through leachate at 1.5 rpm, a microbial slime forms on the discs. These microorganisms degrade the organic and nitrogenous contaminants present in the waste stream. During rotation, about 40% of the discs' surface area is in contact with the aqueous waste, while the remaining area is exposed to the atmosphere. The rotation of the media through the atmosphere causes oxygenation of the attached organisms. When operated properly, the shearing motion of the discs through the aqueous waste causes excess biomass to shear off at a steady rate. Suspended biological solids are carried through the successive stages before entering the secondary clarifier.

The RBC treatment process involves a number of steps as indicated in Figure 4.7.3. Typically, aqueous waste is transferred from a storage or equalization tank (1) to a mixing tank (2) where chemicals are added for metal precipitation, nutrient adjustment, and pH control. The waste stream then enters a clarifier (3) where solids are separated from the liquid. The clarifier effluent enters the RBC (4) where the organics and/or ammonia are converted to innocuous products. The treated waste is then pumped into a second clarifier (5) for removal of biological solids. After secondary clarification, the effluent enters a storage tank (6) where, depending upon remaining contamination, the waste may be stored pending additional treatment or discharged to a sewer system or surface stream. Throughout this process, offgases should be collected for treatment (7).

In addition to maximizing the system's efficiency, staging can improve the system's ability to handle shock loads by absorbing the impact in stages. Staging, which employs a number of RBCs in series, enhances biochemical kinetics and establishes selective biological cultures acclimated to successively decreasing organic loading. As the waste stream passes from stage to stage, progressively increasing levels of treatment occur.

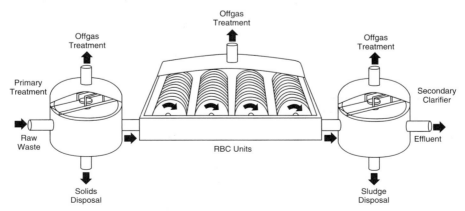

FIG. 4.7.2 Typical RBC plant schematic. (Reprinted, from U.S. Environmental Protection Agency (EPA), 1992, *Engineering Bulletin: rotating biological contactors*, [EPA 540–5–92–007, October].)

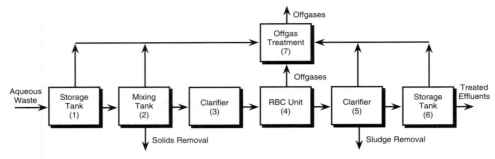

FIG. 4.7.3 Block diagram of the RBC treatment process. (Reprinted, from U.S. EPA, 1992.)

Factors effecting the removal efficiency of RBC systems include the type and concentration of organics present, hydraulic residence time, rotational speed, media surface area exposed and submerged, and pre- and post-treatment activities.

APPLICABILITY/LIMITATIONS

Rotating biological contacts are not sufficient to remove highly chlorinated organics, aliphatics, amines or aromatic compounds. Some heavy metals and organic chemicals are harmful to the organisms.

Table 4.7.3 lists the important data needed for screening RBCs.

Status.   Conventional process

Bioreclamation

DESCRIPTION

Bioreclamation uses aerobic microbial degradation in treating contaminated areas. It is used for in-situ treatment using injection/extraction wells or excavation processes. Extracted water, leachates or wastes are oxygenated, nutrients and bacteria are added, and the liquids are reinjected into the ground. Bacteria can then degrade wastes still in the soil. This treatment has successfully reduced the contamination levels of biodegradable nonhalogenated organics in soils and groundwater.

TABLE 4.7.3   IMPORTANT DATA NEEDS FOR SCREENING RBCs

Data Need	Purpose
Gross organic components (BOD, TOC)	Waste strength, treatment duration
Priority pollutant analyses (organics, metals, pesticides, CN, phenols)	Suitability for treatment, toxic impact assessment
Influent temperature	Feasibility in climate

APPLICABILITY/LIMITATIONS

For in-situ treatment, limitations include site geology and hydrogeology restricting waste pumping and extraction, along with reinjection and recirculation. Ideal soil conditions are neutral pH, high permeability and a moisture content of 50–75%. Biological treatment systems are used to treat soils contaminated with pentachlorophenol, creosote, oils, gasoline, and pesticides.

Table 4.7.4 lists important bioreclamation data needs.

Status.   Demonstrated process

Anaerobic Digestion

DESCRIPTION

All anaerobic biological treatment processes reduce organic matter, in an oxygen-free environment, to methane and carbon dioxide. This is accomplished using bacteria cultures, including facultative and obligate anaerobes. Anaerobic bacterial systems include:

- Hydrolytic bacteria (catabolized saccharide, proteins, lipids)

TABLE 4.7.4   IMPORTANT BIORECLAMATION DATA NEEDS

Data Need	Purpose
Gross organic components (BOD, TOC)	Waste strength, treatment duration
Priority analysis	Identify refractory and biodegradable compounds, toxic impact
Microbiology cell enumerations	Determine existence of dominant bacteria
Temperature	Feasibility in climate
Dissolved oxygen	Rate of reaction
pH	Bacteria preference
Nutrient analysis NH$_3$, NO$_3$, PO$_4$, etc.	Nutrient requirements

- Hydrogen-producing acetogenic bacteria (catabolized products of hydrolytic bacteria, e.g., fatty acids and neutral end products)
- Homolactic bacteria (catabolized multicarbon compounds to acetic acid)
- Methanogenic bacteria (metabolized acetic acid and higher fatty acids to methane and carbon dioxide)

Figure 4.7.4 is a schematic diagram of biological reaction in an anaerobic system.

Strict anaerobics require totally oxygen-free environments and an oxidation reduction potential of less than −0.2 V. Microorganisms in this group are known as methanogenic consortia and are found in anaerobic sedi-ments or sewage sludge digesters. These organisms play an important role in reductive dehalogenation reactions, ni-trosamine degradation, reduction of epoxides to olefins, reduction of nitro-groups, and ring fission of aromatic structures.

Available anaerobic treatment concepts are based on approaches such as the classic well-mixed system, the two-stage system and the fixed bed system.

In a well-mixed digester system, a single vessel is used to contain the wastes being treated and all bacteria must function in that common environment. Such systems typically require long retention times, and the balance between acetogenic and methanogenic populations is easily upset.

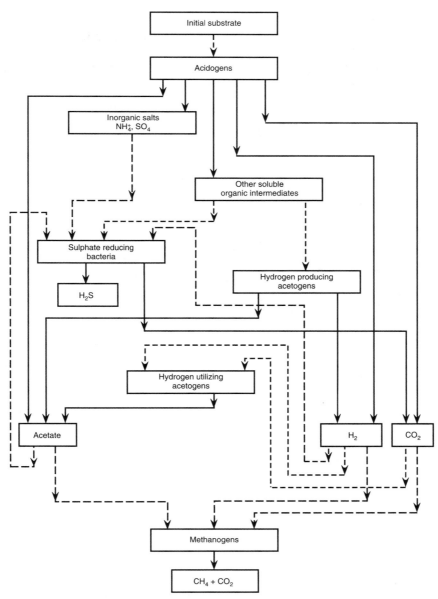

FIG. 4.7.4 Schematic diagram of biological reactors in anaerobic systems.

In the two-stage approach, two vessels are used to maintain separate environments, one optimized for acetogenic bacteria (pH 5.0), and the other optimized for methanogenic bacteria (pH 7.0). Retention times are significantly lower and upsets are uncommon in this approach.

The fixed bed approach (for single- or 2-staged systems) utilizes an inert solid media to which the bacteria attach. Low solids wastes are pumped through columns of the bacteria-rich media. Use of such supported cultures allows reduced retention times since bacterial loss through washout is minimized. Organic degradation efficiencies can be quite high.

A number of proprietary engineered processes based on these types of systems are being actively marketed. Each has distinct features, but all utilize the fundamental anaerobic conversion to methane and carbon dioxide (Figure 4.7.5).

APPLICABILITY/LIMITATIONS

This process is used to treat aqueous waste with low to moderate levels of organics. Anaerobic digestion can handle certain halogenated organics better than aerobic treatment. Stable, consistent operating conditions must be

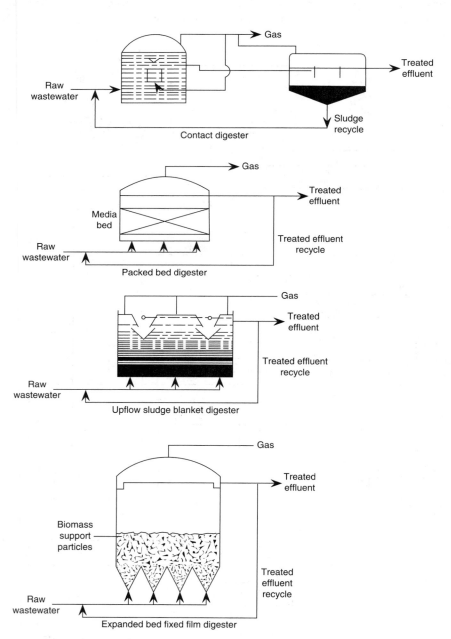

FIG. 4.7.5 Schematic diagrams of anaerobic digesters currently in use.

maintained. Anaerobic degradation can take place in native soils, but in a controlled treatment process, an air-tight reactor is required. Since methane and carbon dioxide gases are formed, it is common to vent the gases or burn them in flare systems. However, volatile hazardous materials could readily escape via such gas venting and flare systems. Thus controlled off-gas burning could be required. Depending upon the nature of the waste to be treated, the off-gas could be used as a source of energy.

Status.   Available and widely used in POTWs.

White-Rot Fungus

DESCRIPTION

Lignin-degrading white-rot fungus (phanerochaete chrysosporium) degrades a broad spectrum of organopollutants, including chlorinated, lignin-derived by-products of the Kraft pulping process. White-rot fungus degrades aliphatic, aromatic, and heterocyclic compounds. Specifically, white-rot fungus has been shown to degrade indane, benzo(a)pyrene, DDT, TCDD, and PCBs to innocuous end products. The studies performed, to date, suggest that the white-rot fungus may prove to be an extremely useful microorganism in the biological treatment of hazardous organic waste.

Note: Certain plants, such as specific strains of *Brassica* (mustards), accumulate heavy metals when growing in metal contaminated soils, forming the basis for a process called phytoremediation. These plants can accumulate up to 40% of their biomass as heavy metals, including lead [Atlas 1995].

APPLICABILITY/LIMITATIONS

This technology is in the development phase and has been applied only in laboratory environs.

Status.   Demonstrated on laboratory scale

—David H.F. Liu

References

Atlas, R.M. 1995. Bioremediation. *C&E News* (3 April).
U.S. Environmental Protection Agency (EPA). 1984. *Design information on rotating biological contactors.* EPA 600–2–84–106 (June).
———. 1992. *Engineering Bulletin: rotating biological contactors.* EPA 540–S–92–007 (October).

4.8
BIOTREATMENT BY SEQUENCING BATCH REACTORS

FEATURE SUMMARY

Type of Process: Biological treatment of liquid hazardous wastewaters.

Type of Reactor: Sequencing Batch Reactor (SBR), a fill-and-draw, activated sludge-type system where aeration and settling occur in the same tank.

Type of Aeration and Mixing Systems: Jet-aeration systems are common and allow mixing either with or without aeration; however, other aeration and mixing systems are used.

Type of Decanters: Most decanters, including some which are patented, float or otherwise maintain inlet orifices slightly below the water surface to avoid removal of both settled and floating solids.

Type of Tanks: Steel tanks, appropriately coated for corrosion control, are most common; however, concrete and other materials may be used. Concrete tanks are favored for municipal treatment of domestic wastewaters.

Partial List of Suppliers of SBR Equipment: Aqua-Aerobics Systems; Austgen-Biojet; Bioclear Technology; Envirodyne Systems; Fluidyne; Jet Tech; Mass Transfer; Purestream; Transenviro

The Sequencing Batch Reactor (SBR) is a periodically operated, activated sludge-type, dispersed-growth, biological wastewater treatment system. Both biological reactions and solids separation are accomplished in a single reactor, but at different times during a cyclic operation. In comparison, continuous flow activated sludge systems use two reactors, one dedicated to biological reactions and the other dedicated to solids separation. Once constructed, these two-tank systems offer little flexibility because changing reactor size is difficult. However, the SBR is flexible, as the time dedicated to each function can be adjusted. For example, reducing the time dedicated to solids separation provides additional time for the completion of biological reactions. Other advantages of the SBR system are described below, along with a description of the SBR operating cycle. The SBR originally was developed to treat domestic wastewater and is now used to effectively treat industrial and other organic wastewaters containing hazardous substances.

Process Description

Influent wastewater is added to a partially filled reactor. The partially filled reactor contains biomass acclimated to the wastewater constituents during preceding cycles. Once the reactor is full, it behaves like a conventional activated sludge aeration basin, but without inflow and outflow. After biological reactions are completed, and aeration and mixing is discontinued, the biomass settles and the treated supernatant is removed. Excess biomass is wasted at any convenient time during the cycle. Frequent wasting results in holding the mass ratio of influent substrate to biomass nearly constant from cycle to cycle. In contrast, continuous flow systems hold the mass ratio of influent substrate to biomass constant by adjusting return sludge flow rates continually as influent flow rates and characteristics and settling tank underflow concentrations vary.

No specific SBR reactor shape is required. The width-to-length ratio is unimportant, although this is a concern with conventional continuous flow systems. Deep reactors improve oxygen transfer efficiency and occupy less land area. The SBR shown in Figure 4.8.1 uses an egg-shaped reactor that offers most of the advantages of a spherical reactor, and provides a deeper reactor. Along with improved oxygen transfer efficiency, deep reactors allow a higher fraction of treated effluent removal during decanting. Similar to a spherical reactor, the egg-shaped reactor has a minimum surface area to volume ratio resulting in lower heat loss, less material needed in reactor construction, and less energy required for mixing.

The small reactor top is easily enclosed to contain volatile organics, or to direct exhaust gases for removal in an absorber. During filling, floating materials are forced together towards the top center for easy removal. The con-

verging bottom improves thickening of the settled solids, reduces sediment accumulation, and allows easy solids removal from a single center point.

The egg-shaped reactor is constructed as a single piece, eliminating the need for special reinforcement at static tension points or seams. The bottom section of the egg is buried in the ground, supporting the reactor without a special foundation.

Modes of Operation

The illustration of an SBR in Figure 4.8.1 includes five discrete periods of time, with more than one operating strategy possible during any time period. The five discrete periods are defined as:

1. *Idle:* waiting period
2. *Fill:* influent is added
3. *React:* biological reactions are completed
4. *Settle:* solids separate from treated effluent
5. *Draw:* treated effluent is removed

Each discrete period is detailed in the following sections.

IDLE

Idle is considered the beginning of the cycle, although there exists no true beginning after the initial start-up. Idle occurs between draw, the removal of treated effluent, and fill, the beginning of influent addition. Idle time may be short or long depending on flow rate variation and operating strategy. When the influent flow rates are constant and predictable, and when flow equalization is provided upstream, idle is nearly eliminated. Idle time is long during periods of low influent flow, and short during periods of high influent flow. During operations with variable idle times, the SBR also provides flow equalization. Idle can also be used to accomplish other functions, such as sludge wasting, and mixing to condition the biomass to a low substrate concentration.

STATIC, MIXED AND AERATED FILL

Developing an operating strategy for the fill period is a complex problem for designers of hazardous wastewater SBR systems. Domestic wastewater treatment systems seldom require laboratory treatability studies, because these systems follow conservative design approaches and municipal wastewater flow rates and characteristic variations are predictable. Laboratory treatability studies are almost always needed to design SBRs and to select the appropriate fill policy for hazardous wastes. The following describes alternative fill policies that must be developed during treatability studies.

SBR influent may require pretreatment. The decision to provide screening and degritting is made on the same basis used by designers of conventional continuous flow

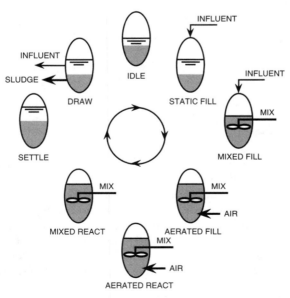

FIG. 4.8.1 Illustration of one cycle in a single SBR.

plants. Upstream flow equalization allows for rapid fill rates (i.e., higher than influent flow rates). This results in reduced cycle times and reduced reactor size. The use of rapid fill periods also results in the accumulation of high substrate concentrations. Upstream flow equalization is not necessary because idle is a normal SBR function. However, upstream flow equalization may be selected, allowing wastewaters with highly variable characteristics, or from more than one process to be blended for more uniformity. In addition, high concentrations resulting from spill events can be caught and kept from interfering with the biological process. Flow equalization basins are often included in SBR systems because they are inexpensive, provide added system flexibility, and reduce or eliminate idle time.

Static Fill

Static fill introduces influent wastewater into the SBR with little or no mixing and contact with the settled biomass, resulting in a high substrate concentration when mixing first begins. High substrate concentrations result in high reaction rates. In addition, such concentrates control sludge bulking because they favor organisms that form more dense floc particles over filament forming organisms. Bulking sludge is a common problem in continuous flow systems where substrate concentrations are always low. Finally, static fill favors organisms that produce internal storage products during high substrate conditions, a requirement for biological phosphorous removal. However, static fill time should be limited if an influent biodegradable constituent is present at concentrations toxic to the organisms.

Mixed Fill

Mixed fill begins the biological reactions by bringing influent organics into contact with the biomass. Mixing without aeration reduces the energy needed for aeration, because some organics are biologically degraded using residual oxygen or alternative electron acceptors, such as nitrate-nitrogen. If nitrate-nitrogen is the electron acceptor, a desirable denitrification reaction occurs. The period when alternative electron acceptors are present and oxygen is absent is called *anoxic*. Anaerobic conditions develop after all electron acceptors are consumed, and fermentation reactions may occur. Thus, in a single reactor, aerobic, anoxic and anaerobic treatment conditions exist by correctly varying mixing and aeration policies during fill.

Aerated Fill

Aerated fill begins the aerobic reactions that are later completed during react. Aerated fill reduces cycle time because the aerobic reactions occur during the fill period. In some cases, a biodegradable influent constituent may be present in concentrations that are toxic to the organisms. When that condition exists, aeration during fill begins early to limit concentration of that constituent. For example, if the wastewater constituent is toxic at 10 mg/L, and present in the influent at a concentration of 30 mg/L, aeration should begin prior to the reactor becoming three-quarters full, if the SBR liquid volume at the end of draw was one-half the volume at the end of fill. This assumes toxic constituent degradation is at a rate greater than or equal to its rate of addition. If the rate of degradation is low, then aeration should begin earlier, or a higher volume of treated effluent should be held in the SBR for more dilution.

REACT

Aeration is often provided during the fill react period to complete the aerobic reactions. However, aerated react alternated with mixed react will provide alternating periods of aerobic and anoxic, or even anaerobic, conditions. This is a normal procedure for nitrification and denitrification. During periods of aeration, nitrate concentration increases as organic nitrogen and ammonia are converted to nitrites and nitrates. The mixed react results in anoxic conditions needed for denitrification and the conversion of nitrates to nitrogen gas. Anaerobic conditions are necessary if some waste constituents are degraded only anaerobically, or partially degraded anaerobically followed by a complete degradation under aerobic conditions.

For mixed wastes, the easily degraded constituents are removed first, and the more difficult to degrade constituents are removed later during extended periods of aeration. Long periods of aerated react, after removal of soluble substrates, may be necessary to condition the biomass, to remove internal storage products, or to aerobically digest the biomass. Aerated react may also be stopped soon after the soluble substrate is removed. This saves energy and maximizes sludge production, which is desirable when separate anaerobic sludge digestion is used to stabilize these waste solids and to produce methane, an energy-rich and useful by-product.

SETTLE

Settle is normally provided under quiescent conditions in the SBR; however, gentle mixing during the early stages of settle may result in both a clearer effluent and a more concentrated settled biomass. Unlike continuous flow systems, settle occurs without inflow or outflow, and the accompanying currents that interfere with settling.

DRAW

The use of a floating decanter, or a decanter that moves downward during draw, offers several advantages. Draw is initiated earlier because the effluent is removed from near the surface while the biomass continues to settle at lower depths. The effluent is removed from a selected depth below the surface by maintaining outlet orifices or slots at a fixed depth below the variable water surface. This avoids removal of floating materials and results in effluent removal from high above the settled biomass. Floating decanters allow maximum flexibility, because fill and draw volumes can be varied from time to time, or even from cycle to cycle. However, lower-cost fixed-level decanters can be used if the settle period is extended, to assure that the biomass has settled below the decanter orifices. Fixed-level decanters can be made somewhat more flexible, if they are designed to allow operators to occasionally lower or raise the location of the decanter.

Rapid draw rates allow use of smaller reactors, but cause high surges of flow in downstream units and in receiving waters. Effluent flow equalization tanks or reduced draw rates will reduce peak flow discharges.

Figure 4.8.2 illustrates the hydraulic conditions over two days in a three-tank SBR system under design flow conditions. The illustration shows influent flow into Tank 1 beginning at 6:00 A.M. The treatment strategy provides a static fill (Fs) for 1.67 hr followed by an aerated fill for 1.0 hr. The influent is diverted to Tank 2 at 8:40 A.M., and to Tank 3 at 11:20 A.M. In Tank 1, a 2.33 hr react is followed by a 1.0 hr settle and a 1.0 hr draw. Tank 1 idle occurs from 1:00 P.M. to 2:00 P.M. At 2:00 P.M. the influent is again diverted to Tank 1, and the cycle is repeated. As shown in Figure 4.8.2, each tank cycle is the same, with an 8-hr cycle divided as follows: fill 2.67 hr (i.e., static fill, 1.67 hr and aerated fill, 1.00 hr), react 2.33 hr, settle 1.0 hr, draw 1.0 hr, and idle 1.0 hr. Total aeration time is 3.33 hr in both aerated fill and react.

The shaded areas of the illustration show that influent flows continuously into one of the three tanks. Under design conditions, flow occurs at a constant rate and every cycle is identical. Effluent is not continuous, and is illustrated by the cross-hatched areas. In this design flow example, each cycle fill time is 2.67 hr and draw time is 1.0

hr, resulting in an effluent flow rate equal to 2.67 times the influent flow rate.

Under actual flow conditions, diurnal flow rates vary. Typically, flow rates increase throughout the morning, peak in the early afternoon, and decrease later in the day with minimum flow in the early morning hours. Figure 4.8.3 illustrates a three-tank SBR system with a typical diurnal flow variation. Note the short fill and idle periods during high flow rate times, and the long fill and idle periods during low flow rate times. During peak flow, no idle period exists (e.g., about 2:00 P.M. of Day 2 in Tank 1).

Laboratory Treatability Studies

Most SBR systems for treatment of industrial wastewaters, especially those containing hazardous wastes, must be designed based on treatability studies. This section outlines the procedures and equipment needed to perform these studies. Figure 4.8.4 illustrates a laboratory SBR system.

The reactors are 1-L to 4-L *reaction kettles,* typically covered to control volatile organics. The small reactors require less influent wastewater and effluent disposal, and the larger reactors allow collection of larger sample volumes. A gas collection tube, containing activated carbon or other organic absorbent, prevents the escape of volatile organics and measures the extent of organic volatilization. Two gas collection tubes are connected in series to prevent volatile organics in laboratory air from contributing to those released from the reactor. The reactor can be mixed with or without aeration using a magnetic stirrer and a star-head stir bar. Air is supplied through a diffuser from a laboratory air supply. A gas collection tube containing an organic absorbent may be used to prevent organics in the supply air from entering the SBR (not illustrated). To minimize evaporation during SBR aeration, a water humidifier is included in the air stream. Peristaltic tubing pumps add influent and remove effluent from the

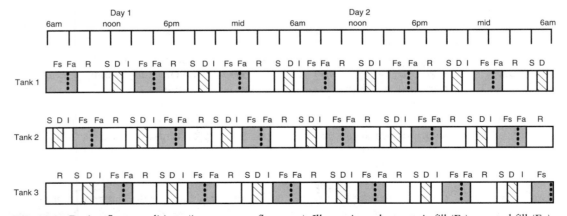

FIG. 4.8.2 Design flow conditions (i.e., constant flow rate). Illustrations show static fill (Fs), aerated fill (Fa), react (R), settle (S), draw (D), and idle (I) for a three-tank system. Fill periods are shaded to demonstrate that inflow continuously occurs into one of the three tanks, and draw periods are cross-hatched to show the periodic nature of effluent flow.

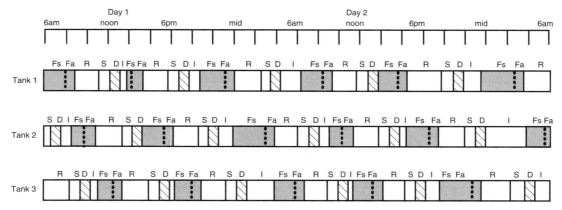

FIG. 4.8.3 Typical flow conditions (i.e., diurnal flow rate variation). Illustrations show static fill (Fs), aerated fill (Fa), react (R), settle (S), draw (D), and idle (I) for a three-tank system. Fill periods are shaded to demonstrate that inflow continuously occurs into one of the three tanks, and draw periods are cross-hatched to show the periodic nature of effluent flow.

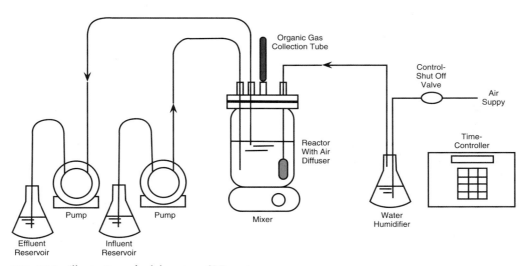

FIG. 4.8.4 Illustration of a laboratory SBR system.

SBR. Liquid level switches can be used to control the fill and draw volume, or, as an alternative, suction tubes can be placed at an elevation to prevent pumping excessive volumes. The suction tube alternative will result in too little addition or removal if minor changes occur in the pumping rates, but this is minimized by frequent monitoring and occasional manual over-ride at the end of a pumping cycle, or simply recorded to reflect a slightly different loading rate. System control is provided with a microprocessor-based timer and controller that turns pumps on and off, opens and closes the air shut-off valve, and turns the mixer on and off at appropriate times.

Laboratory studies include a start-up period to develop a biomass enriched for organisms acclimated to the wastewater constituents. Start-up time is minimized by including these organisms in the initial seed biomass. Aeration of some wastewaters alone will result in an accumulation of a suitable biomass. At other times, activated sludge collected from nearby municipal or industrial wastewater treatment plants may be needed. Finally, for difficult to degrade wastes, organisms are taken from: soil samples collected from waste spill sites; sediments collected from nearby receiving waters; or from residues in contact with waste constituents for long periods of time. During start-up, fill and draw periods are controlled manually after substrate reduction is observed. Once an enriched and acclimated biomass is developed, automatic SBR operation is used to determine the appropriate operating strategy, as described above under the explanation of the different phases of the SBR cycle.

—Lloyd H. Ketchum, Jr.

5

Storage and Leak Detection

5.1
UNDERGROUND STORAGE TANKS

The terms *underground storage tanks, USTs, and UST systems* include underground storage tank vessels, and the connected underground piping. Thus, above–ground tanks with extensive underground piping may also be regulated under RCRA Subtitle I. Usually associated with gasoline service stations, these tanks are also used to store materials that are classified as hazardous due to flammability or combustibility. Leaking underground storage tanks can cause fires or explosions or contaminate groundwater, threatening public health and the environment.

Problems and Causes

Large numbers of older USTs are "bare" steel. Tanks that are over 10 years old and unprotected from corrosion are likely to develop leaks. A leaking UST, if undetected or ignored, can cause large amounts of subsurface petroleum product loss. In a recent survey of motor fuel storage tanks, the EPA found that 35% of the estimated 796,000 tanks leak. Abandoned tanks were found at 14% of the surveyed establishments, but the EPA did not conduct leakage tests on these abandoned tanks.

Underground storage tanks release contaminants into the subsurface environment because of one or more of the following factors:

GALVANIC CORROSION

The most common failure of underground tank systems is galvanic corrosion of the tank or piping. Corrosion can be traced to failure of corrosion protection systems due to pinholes in the coating or taping, depletion of sacrificial anodes, corrosion from the inside due to the stored product, and various other reasons. Many corrosion–related leaks are found in systems that have no corrosion protection at all. However, tanks with corrosion protection can also corrode if the protective coating is damaged during installation; the sacrificial anodes are not replaced when required; the current is switched off in impressed current systems; or the protection system is not designed properly for the soil condition and stored liquids.

FAULTY INSTALLATION

Installation failures include inadequate backfill, allowing movement of the tank, and separation of pipe joints. These tanks receive a substantial portion of structural support from backfill and bedding. Mishandling can cause structural failures of RFP tanks, or damage to steel tank coating and cathodic protection.

PIPING FAILURES

The underground piping that connects tanks to each other, to delivery pumps, and to fill drops is even more frequently made of unprotected steel. An EPA study found that piping failure accounts for a substantial portion of large spills at UST facilities. The study concluded that piping failure is caused equally by corrosion and poor workmanship. Threaded metal areas, made electrically active by the threading, have strong tendency to corrode if not coated and cathodically protected. Improper layout of pipe runs, incomplete tightening of joints, inadequate cover pad construction, and construction can accidents lead to failure of delivery piping. Figure 5.1.1 diagrams a typical service station tank and piping layout.

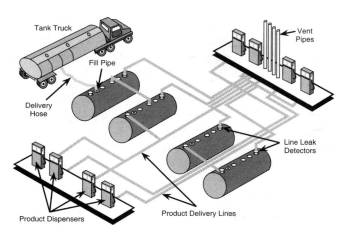

FIG. 5.1.1 Typical service station tank and piping layout (EPA).

SPILLS AND OVERFILLS

Spills and overfills, usually the result of human error, aggravate release problems at UST facilities. In addition to direct contamination effects, repeated spills of petroleum or hazardous waste can intensity the corrosiveness of the soils. These mistakes can be corrected by following the correct tank filling practices required by 40 CFR 280. Other causes include: delivery source failure, shutoff valve failure, and tank level indicator failure.

COMPATIBILITY OF UST AND CONTENTS

Compatibility between tanks and contents means that fuel components will not change the physical or mechanical properties of the tank. Compatibility for liners requires that fuel components not cause blistering, underfilm corrosion or internal stress or cracking. There are concerns that some FRP tanks or liners may not be compatible with some methanol-blended (or possible ethanol-blended) fuels. A fuel tank should be designed to handle fuel compositions of the future.

Owners and operators of businesses with FRP-constructed or lined tanks should consult appropriate standards of the American Petroleum Institute.

UST Regulations

EPA regulations establishing controls for new and existing underground storage tanks became effective in December 1988. Following is a list of important requirements:

DESIGN, CONSTRUCTION, AND INSTALLATION

Tanks and piping may be constructed of fiberglass, coated steel (asphalt or paint), or metal without additional corrosion protection. If constructed of coated steel, tanks and piping must also contain corrosion-protection devices such as cathodic protection systems. Such corrosion-protection devices must be regularly inspected. If constructed of metal without corrosion protection, records must be maintained showing that a corrosion expert has determined that the site is not corrosive enough to cause leaks during the tank's operating life. Tanks must be installed properly and precautions must be taken to prevent damage.

Information on the design (including corrosion protection), construction, and installation of tanks and piping is available on notification forms filed in the designated state office, usually the state UST office. Reports of inspections, monitoring, and testing of corrosion protection devices are on file at the UST site or must be made available to the implementing agency upon request.

Recent developments in preventing leakage include using asphalt coated steel tanks, double-walled fiberglass tanks, double-walled steel tanks, epoxy-coated steel tanks, fiberglass-coated steel tanks, fiberglass-coated double-walled tanks, synthetic underground containment liners, and tanks placed in subgrade vaults (see Figure 5.1.2). Some suppliers offer double-walled pipes for added safety. Lined trenches offer secondary containment for single-walled pipes.

SPILLS AND OVERFILLS CONTROL

Except for systems filled by transfer of no more than 25 gallons at one time, UST systems must use one or more overfill prevention devices. These devices include sensors to detect tank capacity level, automatic flow shutoff valves, and spill catchment basins.

Owners and operators must report spills and overfills to the implementing agency within a reasonable time period.

The standards apply to new UST systems and some existing tanks.

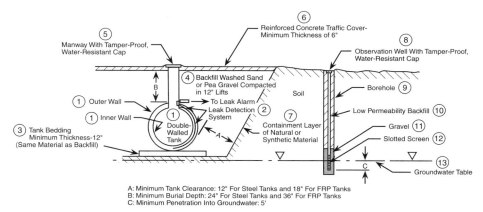

FIG. 5.1.2 Double wall tank and leak detection system. (Reprinted from New York Department of Environmental Conservation, 1982, *Siting manual for storing hazardous substances: a practical guide for local officials.* New York, N.Y.)

REPAIRS

Tank repairs must be conducted in accordance with a code of practice developed by a nationally recognized association or an independent testing laboratory. Following repairs, tightness tests may be required.

Owners must maintain repair records on-site or must make them available to the implementing agency upon request.

These repair standards apply to new and existing UST systems.

LEAK DETECTION

Release detection requirements differ between petroleum USTs and hazardous waste USTs. Petroleum UST systems may choose from among five primary release detection methods, for example: 1) automatic tank gauging that tests for product loss and conducts inventory control; 2) testing or monitoring for vapor within the soil gas of the tank area; and 3) testing or monitoring for liquids in the groundwater. Hazardous substance tanks must use secondary containment systems such as double-walled tanks or external liners, unless the owner has obtained a variance.

Owners must maintain records pertaining to: system leak detection methods; recent test results to detect possible leaks; and maintenance of release detection equipment; and must make these available to the agency on request. Information on leak detection method(s) used by UST systems is also on the notification form on file in each state's UST office.

These leak detection standards apply to new UST systems.

TABLE 5.1.1   MINIMUM REQUIREMENTS FOR COMPLIANCE WITH UST REGULATIONS (EPA)

Leak Detection	
NEW TANKS *2 Choices*	• Monthly Monitoring* • Monthly Inventory Control and Annual Tank Tightness Every 5 Years (You can only use this choice for 10 years after installation.)
EXISTING TANKS *3 Choices*	• Monthly Monitoring* • Monthly Inventory Control and Annual Tank Tightness Testing (This choice can be used until December 1998.) • Monthly Inventory Control and Tank Tightness Testing Every 5 Years (This choice can only be used for 10 years after adding corrosion protection and spill/overfill prevention or until December 1998, whichever date is later.)
NEW & EXISTING PRESSURIZED PIPING *Choice of one from each set*	• Automatic Flow Restrictor • Annual Line Testing • Automatic Shutoff Device -and- • Monthly Monitoring* • Continuous Alarm System (except automatic tank gauging)
NEW & EXISTING SUCTION PIPING *3 Choices*	• Monthly Monitoring* (except automatic tank gauging) • Line Testing Every 3 Years • No Requirements—if slope and check valve conditions are met
Corrosion Protection	
NEW TANKS *3 Choices*	• Coated and Cathodically Protected Steel • Fiberglass • Steel Tank clad with Fiberglass
EXISTING TANKS *4 Choices*	• Same Options as for New Tanks • Add Cathodic Protection System • Interior Lining • Interior Lining and Cathodic Protection
NEW PIPING *2 Choices*	• Coated and Cathodically Protected Steel • Fiberglass
EXISTING PIPING *2 Choices*	• Same Options as for New Piping • Cathodically Protected Steel
Spill/Overfill Prevention	
ALL TANKS	Catchment Basins -and- • Automatic Shutoff Devices -or- • Overfill Alarms -or- • Ball Float Valves

*Monthly Monitoring includes: Automatic Tank Gauging, Ground-water Monitoring, Vapor Monitoring, Other Approved Methods, Interstitial Monitoring

OUT-OF-SERVICE SYSTEMS AND CLOSURE

During temporary closures, owners must continue all usual system operation, maintenance, and leak detection procedures, and must comply with release reporting and cleanup regulations if a leak is suspected or confirmed. Release detection, however, is not required if the UST system is empty.

If a UST system is taken out of service for more than 12 months, it must be permanently closed unless it meets certain performance standards and upgrading requirements. Before a tank is permanently closed, the owner or operator must test for system leaks: if a leak is found, the owner or operator must comply with corrective-action regulations. Once the tank is permanently out of service, it must be emptied, cleaned, and either removed from the ground or filled with an inert solid material.

Closure procedures are available at each tank site or must be made available to the implementing agency upon request.

These closure standards apply to all new and existing UST systems.

Financial Responsibility

These regulations specify the amount and scope of coverage required for corrective action and for compensating third parties for bodily injury and property damage from leaking tanks. The minimum coverage required varies depending on the number of tanks owned, from $1,000,000 for up to 100 USTs to $2,000,000 for more than 100 USTs.

Various mechanisms may be used to fulfill coverage requirements, such as self-insurance, indemnity contracts, insurance, standby trust funds, or state funds. Quick action may ensure that available funds are directed toward coverage. Particular attention should be paid to self-insured tank owners. However, even large companies may go bankrupt, leaving the contracting engineer unprotected.

Information on financial responsibility for new tanks must be filed with the EPA regional office. Owners must also maintain evidence of financial responsibility at tank sites or places of business, or make such evidence available upon request of the implementing agency.

These financial responsibility standards apply to most new and existing UST systems.

The technical requirements for UST rules are summarized in Table 5.1.1.

—*David H.F. Liu*

5.2
LEAK DETECTION AND REMEDIATION

Contamination caused by leaky USTs often may not be detected until it is widespread, and difficult and expensive to correct. Regular tests and inspections of tanks and piping are necessary to ensure that leaks are detected early and prevented promptly. The extent of releases and their migration are characterized to plan corrective actions.

If a tank is leaking more than 0.05 gph, it is a leaker. Less than 0.05 gph is beyond the scope of measurement ability and the tank is considered tight. Present technology is imprecise in detecting leaks smaller than 0.05 gph. This standard is listed in NFPA 329, Final Test, now renamed Precision Test.

Tank Monitoring

There are four general methods for detecting leaks in USTs:

- Volumetric (quantitative) leak testing and leak rate measurement
- Non-volumetric (qualitative) leak testing

- Inventory monitoring
- Environmental monitoring

These methods can be used independently or in combination.

Figure 5.2.1 illustrates some of these leak detecting alternatives.

Regardless of the UST monitoring techniques used, the effects of major variables must be compensated for (Table 5.2.1). It is important for USTs to be tested under conditions close to normal operating conditions without uncovering the tanks. Uncovering tanks or piping is expensive and time-consuming, and can cause new leaks. Inadvertent pressurization during testing may rupture the tank and piping.

VOLUMETRIC LEAK TESTING

Table 5.2.2 summarizes common volumetric tank testing systems. These systems may be used to detect leaks and ascertain tightness of tanks and associated piping.

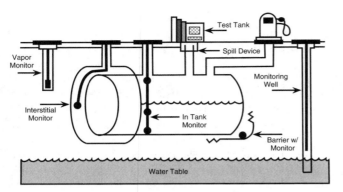

FIG. 5.2.1 Leak detection alternatives (EPA).

An experimental device that detects leaks based on laser interferometry is currently being developed by SRI International under contract to API (Figure 5.2.2). The device aims a laser beam at the underground tank and the beam is reflected to a detector that computes the liquid level in the tank. Test results to date indicate that this de- vice can detect liquid level changes in micro-inches. The API has specified that the device must instantly detect leaks as small as 0.05 gph.

NONVOLUMETRIC LEAK TESTING

Table 5.2.3 presents a number of nonvolumetric (qualita- tive) leak testing methods. If a leak is occurring, volumet- ric testing is used to determine the rate of the leak.

INVENTORY MONITORING

This involves thorough record keeping of product pur- chases and consumption, regular inspections, and recog- nition of conditions indicating leaks. This simple, low-cost leak-detecting method is applicable to any product stored or transported in pipelines. In addition, it does not require interruption of tank service or a set degree of tank full- ness. However, this method requires good bookkeeping and will not detect small leaks. Table 5.2.4 summarizes three common inventory monitoring techniques.

TABLE 5.2.1   MAJOR VARIABLES AFFECTING LEAK DETECTION

Variable	Impact
Temperature change	Expansion or contraction of a tank and its contents can mask leak and/or leak rate.
Water table	Hydrostatic head and surface tension forces caused by groundwater may mask tank leaks partially or completely.
Tank deformation	Changes or distortions of the tank due to changes in pressure or temperature can cause an apparent volume change when none exists.
Vapor pockets	Vapor pockets formed when the tank must be overfilled for testing can be released during a test or expand or contract from temperature and pressure changes and cause an apparent change in volume.
Product evaporation	Product evaporation can cause a decrease in volume that must be accounted for during a test.
Piping leaks	Leaks in piping can cause misleading results during a tank test because many test methods cannot differentiate between piping leaks and tank leaks.
Tank geometry	Differences between the actual tank specifications and nominal manufacturer's specification can affect the accuracy of change in liquid volume calculations.
Wind	When fill pipes or vents are left open, wind can cause an irregular fluctuation of pressure on the surface of the liquid and/or a wave on the liquid-free surface that may affect test results.
Vibration	Vibration can cause waves on the free surface of the liquid that can cause inaccurate test results.
Noise	Some nonvolumetric test methods are sound-sensitive, and sound vibrations can cause waves to affect volumetric test results.
Equipment accuracy	Equipment accuracy can change with the environment (e.g., temperature and pressure).
Operator error	The more complicated a test method, the greater the chance for operator error, such as not adequately sealing the tanks if required by the test method in use.
Type of liquid stored	The physical properties of the liquid (including effects of possible contaminants) can affect the applicability or repeatability of a detection method (e.g., viscosity can affect the sound characteristics of leaks in acoustical leak-detection methods).
Power vibration	Power vibration can affect instrument readings.
Instrumentation limitation	Instruments must be operated within their design range or accuracy will decrease.
Atmospheric pressure	A change in this parameter has the greatest effect when vapor pockets are in the tank, particularly for leak-rate determination.
Tank inclination	The volume change per unit of level change is different in an inclined tank than in a level one.

Source: Reprinted, from U.S. Environmental Protection Agency (EPA), 1986, *Underground tank leak, detection methods: a state-of-the-art review* (EPA, EPA 600–2–80–001, Washington, D.C.).

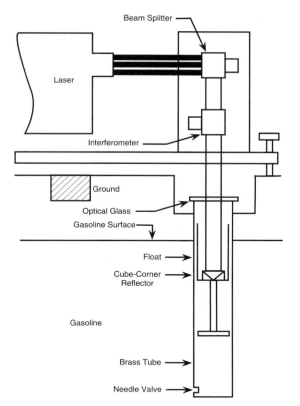

FIG. 5.2.2 Laser interferometer used to measure level changes. (Source: U.S. EPA.)

ENVIRONMENTAL MONITORING

Monitoring wells are the most prevalent form of environmental monitoring for USTs. With environmental effects monitoring, it is difficult to determine which tank is leaking when there is more than one tank. These methods do not provide information on leakage rates or the size of the leak; however, once installed, a leak effects monitoring system enables more frequent checking for leaking tanks than the other methods. Table 5.2.5 presents the principal environmental monitoring methods.

An early warning monitoring technique of double-walled tanks involves monitoring the space between the inner and outer walls of the tank, using either fluid sensors or pressure sensors. This is accurate and applicable with any double-walled tank.

Vapor wells may be used to detect hazardous gases or vapors released into the soil surrounding the UST. Gas detectors or portable gas sampling devices can be used to monitor for gaseous contaminants.

Groundwater monitoring wells may be used to detect or define the movement of leaked substances in a groundwater table. This typically entails drilling monitoring wells, installing monitoring casings, and performing chemical analyses. Table 5.2.6 presents a generalized groundwater sampling protocol.

Table 5.2.7 summarizes the advantages and disadvantages of the various leak monitoring techniques, including those presented in this section.

TABLE 5.2.2   VOLUMETRIC LEAK TESTING METHODS

Method	Principle	Claimed Accuracy, gal/h	Total Down-time for Testing	Requires Empty or Full Tank for Test
Ainlay tank integrity testing	Pressure measurement by a coil-type manometer to determine product level change in a propane bubbling system	0.02	10–12 h (filled a night before 1.5-h testing)	Full
ARCO HTC underground tank detector	Level change measurement by float and light-sensing system	0.05	4–6 h	No
Certi-Tec testing	Monitoring of pressure changes resulting from product level changes	0.05	4–6 h	Full
"Ethyl" tank sentry	Level change magnification by a J tube manometer	Sensitive to 0.02-in level change	Typically 10 h	No
EZY-CHEK leak detector	Pressure measurement to determine product level change in an air bubbling system	Less than 0.01	4–6 h (2 h waiting after fillup, 1-h test)	Full
Fluid-static (standpipe) testing	Pressurizing of system by a standpipe; keeping the level constant by product addition or removal; measuring rate of volume change	Gross	Several days	Full

(Continued on next page)

TABLE 5.2.2 *(Continued)*

Method	Principle	Claimed Accuracy, gal/h	Total Down-time for Testing	Requires Empty or Full Tank for Test
Heath Petro Tite tank and line testing (Kent-Moore)	Pressurizing of system by a standpipe; keeping the level constant by product addition or removal; measuring volume change; product circulation by pump	Less than 0.05	6–8 h	Full
Helium differential pressure testing	Leak detection by differential pressure change in an empty tank; leak rate estimation by Bernoulli's equation	Less than 0.05	Minimum 48 h	Empty
Mooney tank test detector	Measuring level change with a dip stick	0.02	14–16 h* (12 to 14 h waiting after fillup)	Full
PACE tank tester	Magnification of pressure change in a sealed tank by using a tube (based on manometer principle)	Less than 0.05	14 h	Full
PALD-2 leak detector	Pressurizing system with nitrogen at three different pressures; level measurement by an electrooptical device; estimate of leak rate based on the size of leak and pressure difference across the leak	Less than 0.05	14 h (preferably 1 day before, 1-h fill testing, includes sealing time)	Full
Pneumatic testing	Pressurizing system with air or other gas; leak rate measurement by change in pressure	Gross	Several hours	No
Tank auditor	Principle of buoyancy	0.00001 in the fill pipe; 0.03 at the center of a 10.5-ft-diameter tank	1.5–3 h	Typically full
Two-tube laser interferometer system	Measuring level change by laser beam and its reflection	Less than 0.05	4–5 h†	No (at existing level)

*Including the time for tank end stabilization when testing with standpipe.
†Including 1 to 2 h for reference tube temperature equilibrium.
Source: Reprinted from U.S. EPA, 1986.

Corrective Technologies

The most important considerations are the volume and type of substance released and constraining site features that can hinder or prevent effective implementation of a technology. To lesser extent, the financial ability of responsible parties to implement certain technologies and the impact on facility production or service operations also should be considered.

Table 5.2.8 summarizes potential applicable corrective action technologies commensurate with release volume and chemical characteristics. Applicable site data needs for potential technologies are presented in Table 5.2.9.

Initial Response Action. The first response action must minimize immediate risk to human health and the environment; all remaining product must be removed from leaky tanks. Table 5.2.10 lists potential situations and their associated initial corrective actions.

Permanent Response Action. Table 5.2.11 gives examples of permanent corrective actions for a variety of site-specific problems.

—*David H.F. Liu*

TABLE 5.2.3   NONVOLUMETRIC LEAK TESTING METHODS

Method	Principle	Claimed Accuracy, gal/h	Total Downtime for Testing	Requires Empty or Full Tank for Test
Acoustical Monitoring System (AMS)	Sound detection of vibration and elastic waves generated by a leak in a nitrogen-pressurized system; triangulation techniques to detect leak location	Does not provide leak rate; detects leaks as low as 0.01 gal/h	1–2 h	No
Leybold-Heraeus helium detector, Ultratest M2	Rapid diffusivity of helium; mixing of a tracer gas with products at the bottom of the tank; helium detected by a sniffer mass spectrometer	Does not provide leak rate; helium could leak through 0.005-in leak size	None	No
Smith & Denison helium test	Rapid diffusivity of helium; differential pressure measurement; helium detection outside a tank	Provides the maximum possible leak detection based on the size of the leak (does not provide leak rates); helium could leak through 0.05-in leak size	Few—24 h (excludes sealing time)	Empty
TRC rapid leak detector for underground tanks and pipes	Rapid diffusion of tracer gas; mixing of a tracer gas with product; tracer gas detected by a sniffer mass spectro-meter with a vacuum pump	Does not provide leak rate; tracer gas could leak through 0.005-in leak size	None	No
Ultrasonic leak detector (Ultrasound)	Vacuuming the system (5 lb/in^2); scanning entire tank wall by ultrasound device; noting the sound of the leak by headphones and registering it on a meter	Does not provide leak rate; a leak as small as 0.001 gal/h of air could be detected; a leak through 0.005-in could be detected	Few hours (includes tank preparation and 20-min test)	Empty
VacuTect (Tanknology)	Applying vacuum at higher than product static head; detecting bubbling noise by hydrophone; estimating approximate leak rate by experience	Provides approximate leak rate	1 h	No
Varian leak detector (SPY2000 or 938–41)	Similar to Smith & Denison	Similar to Smith & Denison	Few—24 h (excludes sealing time)	Empty

Source: Reprinted from U.S. EPA, 1986.

TABLE 5.2.4   INVENTORY MONITORING METHODS

Method	Principle	Claimed Accuracy
Gauge stick	Measuring project level with dip stick when station is closed	Gross
MFP-414 TLG leak detector	Monitoring product weight by measuring pressure and density at the top, middle, and bottom of tank	Sensitive to 0.1% of product height change
TLS-150	Using electronic level measurement device or programmed microprocessor inventory system	Sensitive to 0.1-in. level change

Source: Reprinted from U.S. EPA, 1986.

TABLE 5.2.5   ENVIRONMENTAL MONITORING METHODS

Method	Principle
Collection sumps	Using collection mechanism of product in collection sump through sloped floor under the storage tank
Dye method	Hydrocarbon detection by use of soluble dye through perforated pipe
Groundwater and soil sampling	Water and soil sampling
Interstitial monitoring in double-walled tanks	Monitoring in interstitial space between the walls of double-walled tanks with vacuum or fluid sensors
LASP	Diffusion of gas and vapor to a plastic material
Observation wells	Product sensing in liquid through monitoring wells at areas with high groundwater
Pollulert and Leak-X	Difference in thermal conductivity of water and hydrocarbons through monitoring wells
Remote infrared sensing	Determining soil temperature characteristic change due to the presence of hydrocarbons
Surface geophysical methods	Hydrocarbon detection by ground-penetrating radar, electromagnetic induction, or resistivity techniques
U-tubes	Product sensing in liquid; collection sump for product directed through a horizontal pipe installed under a tank
Vapor wells	Monitoring of vapor through monitoring well

Source: Reprinted from U.S. EPA, 1986.

TABLE 5.2.6   GENERALIZED GROUNDWATER SAMPLING PROTOCOL

Step	Goal	Recommendations
Hydrologic measurements	Establish nonpumping water level.	Measure the water level to ±0.01 ft (±0.3 cm)
Well purging	Remove or isolate stagnant H_2O which would otherwise bias representative sample.	Pump water until well purging parameters (e.g., pH, T, Ω^{-1}, Eh) stabilize to ±10% over at least two successive well volumes pumped.
Sample collection	Collect samples at land surface or in well-bore with minimal disturbance of sample chemistry.	Pumping rates should be limited to −100 mL/min for volatile organics and gas-sensitive parameters.
Filtration and preservation	Filtration permits determination of soluble constituents and is a form of preservation. It should be done in the field as soon as possible after collection.	Filter: Trace metals, inorganic anions and cations, alkalinity. Do not filter: TOC, TOX, volatile organic compound samples; other organic compound samples only when required.
Field determinations	Field analyses of samples will effectively avoid bias in determining parameters and constituents which do not store well; e.g., gases, alkalinity, pH.	Samples for determining gases, alkalinity, and pH should be analyzed in the field if at all possible.
Field blanks and standards	These blanks and standards will permit the correction of analytical results for changes which may occur after sample collection: preservation, storage, and transport.	At least one blank and one standard for each sensitive parameter should be made up in the field on each day of sampling. Spiked samples are also recommended for good QA/QC.
Sample storage and transport	Refrigerate and protect samples to minimize their chemical alteration prior to analysis.	Observe maximum sample holding or storage periods recommended by the Agency. Documentation of actual holding periods should be carefully performed.

TABLE 5.2.7 COMPARISON OF VARIOUS LEAK MONITORING TECHNIQUES

Approach	Description	Applications	Substances Detected	Relative Cost	Advantages/Disadvantages
Inventory Control	A system based on product record keeping, regular inspections, and recognition of the conditions which indicate leaks.	Any storage tanks and buried pipelines.	Any product stored or transported.	Low	The technique is widely applicable to any product stored or transported in pipelines. However, it requires good bookkeeping, and will not detect small leaks.
Thermal Conductivity Sensors	Uses a probe that detects the presence of stored product by measuring thermal conductivity.	Can monitor groundwater or normally dry areas.	Any liquid.	Medium	Primary advantage is early detection which makes it possible for leaks and spills to be corrected before large volumes of material are discharged. Typically requires $\frac{1}{4}$ inch of product on ground/water interface in wet (groundwater) applications.
Electric Resistivity Sensors	Consists of one or series of sensor cables that deteriorate in the presence of the stored product, thereby indicating a leak.	Can monitor groundwater or normally dry areas.	Any liquid.	Medium	Primary advantage is the early detection of spills. Once a leak or spill is detected the sensors must be replaced. Can detect small as well as large leaks.
Gas Detectors	Used to monitor the presence of hazardous gases in vapors in the soil.	Areas of highly permeable, dry soil, such as excavation backfill or other permeable soils, above ground-water table.	Highly volatile liquids, such as gasoline.	Medium	Once the contaminant is present and detected, gas detectors are no longer of use until contamination has been cleaned up.
Sampling	Grabbing soil or water samples from area for analysis.	Universal; primarily used to collect groundwater samples, as would be the case with tanks stored in high ground-water area.	Any substance	High	Highly accurate intermittent evaluation tool. However, does not provide continuous monitoring.
Interstitial Monitoring in Double-Walled Tanks	Monitors pressure level or vacuum in space between walls of a double-walled tank.	Double-walled tanks.	Pressure sensors monitor tank integrity and are applicable with any stored liquid. Fluid sensors monitor presence of any liquid in a normally dry area and are also applicable with any stored liquid.	High	Accurate technique which is applicable with any double-walled tanks.

(Continued on next page)

TABLE 5.2.7 (Continued)

Approach	Description	Applications	Substances Detected	Relative Cost	Advantages/Disadvantages
Groundwater Monitoring Wells (wet wells)	Wet wells are used to detect and determine the extent of contamination in groundwater tables.	Area-wide or local monitoring for groundwater contamination from underground storage tanks and pipelines. May be used for periodic sampling or may employ one of the sensors described above to detect leaks or spills.	Any hazardous liquids which can be detected by on-site instruments or laboratory analysis.	Medium to High	The type, number and location of wet wells depends upon the site's hydrogeology, the direction of groundwater flow, and the type of spill containment and spill collection systems used.
Vapor (sniff) Wells	Vapor wells are used to detect and monitor the presence of hazardous gases and vapors in the soil.	Area-wide or local monitoring of the soil surrounding underground storage tanks and pipe-lines.	Many different combustible and non-combustible gases and vapors.	Low	The type, number and location of vapor wells depend upon the extent of the spill, the volatility of the product, and the soil characteristics. Vapor wells are subject to contamination from surface spills and cannot be used at contaminated sites.
Dyes and Tracers	Substances with a characteristic color or other characteristics (e.g., radioactive tracers) that can be used to trace the origin of a spill.	Area-wide monitoring of underground tanks and buried pipelines.	Dye itself is detected visually or with the use of instruments.	Low Medium	Dye or tracer could be low in cost, but the time required to perform a study could be great. Also may require the drilling of observation wells to trace the dye of other material. Radioactive tracers require a license and approval from the Nuclear Regulatory Commission of the U.S. Department of Labor. Therefore they are generally discouraged.

TABLE 5.2.8   APPLICATIONS OF TYPICAL CORRECTIVE ACTION TECHNOLOGIES

Technology	Small- to moderate-volume recent gasoline or petroleum release (gas station or tank farms)	Large-volume or long-term chronic gasoline or petroleum release (gas station or tank farms)	Release from tanks containing hazardous substances (organic)	Release from tanks containing hazardous substances (inorganic)
Removal and excavation of tank, soil, and sediment				
Tank removal	●	●	●	●
Soil excavation	●	●	●	●
Sediment removal		●		●
On-site and off-site treatment and disposal of contaminants				
Solidification or stabilization		●	●	●
Landfilling				●
Landfarming	●	●	●	●
Soil washing				●
Thermal destruction		●	●	
Aqueous waste treatment	●	●	●	●
Deep well injection				
Free product recovery				
Dual pump systems	●	●	●	●
Floating filter pumps	●	●	●	●
Surface oil and water separators	●	●	●	
Groundwater recovery systems				
Groundwater pumping	●	●	●	●
Subsurface drains	●	●	●	●
Subsurface barriers				
Slurry walls		●	●	●
Grouting		●	●	●
Sheet piles				
Hydraulic barriers	●	●	●	●
In situ treatment				
Chemical treatment				
Physical treatment			●	●
Soil flushing	●	●	●	●
Biostimulation	●	●	●	●
Groundwater treatment				
Air stripping	●	●	●	
Carbon adsorption	●	●	●	
Biological treatment	●	●	●	
Precipitation, flocculation, sedimentation				●
Dissolved air flotation				●
Groundwater treatment				
Granular media filtration				●
Ion-exchange resin adsorption				●
Oxidation-reduction				●
Neutralization				●
Steam stripping			●	
Reverse osmosis				●
Sludge dewatering				●

(Continued on next page)

TABLE 5.2.8 *(Continued)*

Technology	Small- to moderate-volume recent gasoline or petroleum release (gas station or tank farms)	Large-volume or long-term chronic gasoline or petroleum release (gas station or tank farms)	Release from tanks containing hazardous substances (organic)	Release from tanks containing hazardous substances (inorganic)
Vapor migration control, collection, and treatment				
Passive collection systems		●	●	
Active collection systems		●	●	
Ventilation of structures	●	●	●	
Adsorption			●	
Flaring				
Surface water and drainage controls				
Diversion and collection systems	●	●	●	●
Grading	●	●	●	●
Capping		●	●	●
Revegetation	●	●	●	●
Restoration of contaminated water supplies and sewer lines				
Alternative central water supplies		●	●	●
Alternative point-of-use water supplies	●	●	●	●
Treatment of central water supplies		●	●	●
Treatment of point-of-use water supplies		●	●	●
Replacement of water and sewer lines		●	●	
Cleaning and restoration of water and sewer lines	●	●	●	●

TABLE 5.2.9 SITE INFORMATION FOR USE IN EVALUATING CORRECTIVE ACTION ALTERNATIVES

Technology	*Geographic and Topographic Characteristics*								*Land and Water Use Patterns*				*Hydrogeologic Characteristics*											
	Precipitation	Temperature	Evapotranspiration	Topography	Accessibility	Site Size	Proximity to Surface Water	Proximity to Human Interfaces	Current Water-use Patterns	Future Water-use Patterns	Current Land-use Patterns	Growth Projections	Soil Profiles	Soil Physical Properties	Soil Chemical Properties	Depth to Bedrock	Depth to Groundwater	Aquifer Physical Properties	Groundwater Flow Rate (Volume)	Groundwater Flow Direction	Recharge Areas	Recharge Rates	Aquifer Characteristics	Natural Groundwater Quality
Removal/excavation of tank, soil and sediment																								
Tank removal	•	•		•	•								•	•	•		•	•						
Soil excavation	•	•		•	•								•											
Sediment removal	•	•		•	•								•											
On-site and off-site treatment and disposal of contaminants																								
Solidification and stabilization	•	•		•	•	•	•	•	•	•	•		•	•	•	•	•	•	•	•	•	•	•	•
Landfilling	•			•	•	•	•	•	•	•	•		•	•	•	•	•	•	•	•	•	•	•	•
Landfarming	•	•	•	•	•	•	•	•	•	•	•		•	•	•	•	•	•	•	•	•	•	•	•
Soil washing	•		•	•	•	•	•	•	•	•	•		•	•	•	•	•	•	•	•	•	•	•	•
Thermal destruction	•	•		•	•	•	•	•	•	•	•		•		•	•	•	•	•	•	•	•	•	•
Aqueous waste treatment	•				•		•	•	•	•	•		•			•	•	•	•	•	•	•	•	•
Deep well injection	•			•	•	•		•					•			•	•	•	•	•	•	•	•	•
Free product recovery					•											•	•	•	•	•	•	•	•	•
Dual pump systems	•				•								•			•	•	•	•	•	•	•	•	•
Floating filter pumps					•																•			
Surface oil and water separators					•																•			
Groundwater recovery systems					•																•			
Groundwater pumping	•	•			•	•		•	•	•	•	•	•			•	•	•	•	•	•	•	•	•
Subsurface drains	•	•			•	•		•	•	•	•	•	•			•	•	•	•	•	•	•	•	•
Subsurface barriers					•																•			
Slurry walls	•	•		•	•	•		•					•	•	•	•	•	•	•	•	•	•	•	•
Grouting	•	•		•	•	•		•					•	•	•	•	•	•	•	•	•	•	•	•
Sheet piles		•		•	•	•		•					•	•	•	•	•	•	•	•	•	•	•	•
Hydraulic barriers	•			•	•	•		•						•	•	•	•	•	•	•	•	•	•	•
In situ treatment					•																•			
Soil flushing	•	•		•	•	•		•	•	•	•	•	•	•	•	•	•	•	•	•	•	•	•	•
Biostimulation	•	•		•	•	•		•	•	•	•	•	•	•	•	•	•	•	•	•	•	•	•	•
Chemical treatment	•	•		•	•	•		•	•	•	•	•	•	•	•	•	•	•	•	•	•	•	•	•
Physical treatment	•	•			•	•									•	•	•	•	•	•	•	•	•	•
Groundwater treatment					•																•			
Air stripping	•	•	•	•	•	•													•		•			•
Carbon adsorption	•		•	•	•	•													•		•			•

(Continued on next page)

TABLE 5.2.9 (Continued)

> Note: The following reconstructs a rotated multi-column applicability matrix. Dot (•) indicates the characteristic is relevant to the technology. Column groups: *Hydrogeologic Characteristics*, *Land and Water Use Patterns*, *Geographic and Topographic Characteristics*.

Technology	Natural Groundwater Quality	Aquifer Characteristics	Recharge Rates	Recharge Areas	Groundwater Flow Direction	Groundwater Flow Rate (Volume)	Aquifer Physical Properties	Depth to Groundwater	Depth to Bedrock	Soil Chemical Properties	Soil Physical Properties	Soil Profiles	Growth Projections	Current Land-use Patterns	Future Water-use Patterns	Current Water-use Patterns	Proximity to Human Interfaces	Proximity to Surface Water	Site Size	Accessibility	Topography	Evapotranspiration	Temperature	Precipitation
Groundwater treatment cont'd.																								
Biological treatment	•	•	•	•		•													•	•	•	•		•
Precipitation, flocculation, sedimentation	•	•	•	•		•													•	•	•	•		•
Dissolved air flotation	•	•	•	•		•													•	•	•	•		•
Granular media filtration	•	•	•	•		•													•	•	•	•		•
Ion-exchange resin adsorption	•	•	•	•		•							•	•	•	•	•	•	•	•	•	•		•
Oxidation-reduction	•	•	•	•		•							•	•	•	•	•	•	•	•	•	•		•
Neutralization	•	•	•	•		•							•	•	•	•	•	•	•	•	•	•		•
Steam stripping	•	•	•	•		•							•	•	•	•	•	•	•	•	•	•		•
Reverse osmosis	•	•	•	•		•							•	•	•	•	•	•	•	•	•	•		•
Sludge dewatering													•	•	•	•	•	•	•	•	•	•		•
Vapor migration control, collection, and treatment																								
Passive collection systems										•	•	•	•	•			•		•	•	•			•
Active collection systems										•	•	•	•	•			•		•	•	•			•
Ventilation of structures										•	•	•	•	•			•		•	•	•			•
Adsorption										•	•	•	•	•			•		•	•	•			•
Flaring										•	•	•	•	•			•		•	•	•			
Surface water and drainage controls																								
Diversion and collection systems											•	•	•	•	•	•	•	•	•	•	•	•		•
Grading											•	•	•	•	•	•	•	•	•	•	•	•		•
Capping											•	•	•	•	•	•	•	•	•	•	•	•		•
Revegetation											•	•	•	•	•	•	•	•	•	•	•	•		•
Restoration of contaminated water supplies and sewer lines																								
Alternative central water supplies	•	•		•									•	•	•	•	•	•	•	•	•	•	•	•
Alternative point-of-use water supplies	•	•		•									•	•	•	•	•	•	•	•	•	•	•	•
Treatment of central water supplies	•	•		•		•							•	•	•	•	•	•	•	•	•	•	•	•
Treatment of point-of-use water supplies	•	•		•		•							•	•	•	•	•	•	•	•	•	•	•	•
Replacement of water and sewer lines	•	•				•							•	•	•	•	•	•	•	•	•	•		•
Cleaning and restoration of water and sewer lines	•	•				•							•	•	•	•	•	•	•	•	•	•		•

Note: Technologies in italics are likely to be used in response to UST releases at gasoline stations.

120

TABLE 5.2.10 POTENTIAL INITIAL RESPONSE SITUATIONS AND ASSOCIATED CORRECTIVE ACTIONS

Situation	Tank repair or removal	Free product recovery	Groundwater recovery and treatment	Subsurface barriers	Soil excavation	Vapor migration control and collection	Sediment removal	Surface water diversion drainage	Alternative or treatment central water supply	Alternative or treatment point-of-use water supply	Restoration of utility, water, and sewer lines	Evacuation of nearby residents	Restricted egress or ingress
Groundwater contamination													
Existing public or private wells	●	●	●	●									
Potential future source of water supply	●	●	●	●	●								
Hydrologic connection to surface water	●	●	●	●									
Soil contamination													
Potential for direct human contact: nuisance or health hazard	●				●		●	●					
Agricultural use	●				●		●	●					
Potential source of future releases to ground water	●	●	●	●	●	●							
Surface water contamination													
Drinking water supply	●	●	●	●					●	●	●		
Source or irrigation water	●	●	●	●					●	●	●		
Water-contact recreation	●							●					●
Commercial or sport fishing	●							●					●
Ecological habitat	●							●					●
Other hazards													
Danger of fire or explosion	●					●						●	●
Property damage to nearby dwellings						●						●	●
Vapors in dwellings	●					●						●	●

Source: Reprinted from U.S. Environmental Protection Agency (EPA), 1987.

TABLE 5.2.11 POTENTIAL SITE-SPECIFIC PROBLEMS AND ASSOCIATED PERMANENT CORRECTIVE ACTIONS

	Removal or excavation of soil and sediments	On-site and off-site treatment and disposal of contaminants	Free product recovery	Groundwater recovery systems	Subsurface barriers	In situ treatment	Groundwater treatment	Vapor migration control, collection, and treatment	Surface water drainage	Restoration of contaminated water supplies and sewerlines
Volatilization of chemicals into air								●		
Hazardous particulates released to atmosphere	●							●		
Dust generation by heavy construction or other site activities										
Contaminated site runoff	●						●		●	
Erosion of surface by water									●	
Surface seepage of released substance	●		●						●	
Flood hazard or contact of surface water body with released substance	●								●	
Released substance migrating vertically or horizontally			●	●	●					
High water table which may result in groundwater contamination or interfere with other corrective action				●	●					
Precipitation infiltrating site and accelerating released substance migration					●					●
Explosive or toxic vapors migrating laterally underground					●			●		

(Continued on next page)

TABLE 5.2.11 *(Continued)*

	Removal or excavation of soil and sediments	On-site and off-site treatment and disposal of contaminants	Free product recovery	Groundwater recovery systems	Subsurface barriers	In situ treatment	Groundwater treatment	Vapor migration control, collection, and treatment	Surface water drainage	Restoration of contaminated water supplies and sewerlines
Contaminated surface water, groundwater, or other aqueous or liquid waste	●	●		●	●	●	●			
Contaminated soils	●	●				●				
Toxic and/or explosive vapors that have been collected		●								
Contaminated stream banks and sediments	●	●								
Contaminated drinking water distribution system		●		●	●		●			●
Contaminated utilities								●		●
Free product in groundwater and soils	●	●	●	●	●	●	●	●		

Source: Reprinted from U.S. EPA, 1987.

6
Radioactive Waste

6.1
PRINCIPLES OF RADIOACTIVITY

Radioactivity in the environment provokes public reaction faster than any other environmental occurrence. The mere word *radioactivity* evokes fear in most people, even trained and skilled workers in the field. This fear has been etched in the public mind by such names as Hiroshima, Three Mile Island, and Chernobyl. This legacy of fear has made it difficult for proponents of the usefulness of radioactivity to gain the public trust.

Handling radioactivity in the environment was formerly the territory of the nuclear engineer. This has changed dramatically during the past twenty-five years. Today, individuals working in the environmental arena may be required to deal with radiological issues as one part of a broad environmental program. The aim of this section, and the remainder of the chapter, is to help such individuals grasp the principles of radioactivity as they pertain to environmental engineering.

The following questions must always be answered when one encounters or suspects the presence of radiation:

- What type and how much radioactive material is present?
- How can it be handled safely?
- How can it be contained and/or disposed, including classification and transportation?

Radioactivity enters the environment from natural and man-made sources. Radioactivity can exist as gaseous, liquid or solid materials. Radon is a well-known example of a radioactive gas. Water often contains dissolved amounts of radium and uranium. Solid radioactive waste is produced from many sources, including the uranium and rare earth mining industries, laboratory and medical facilities, and the nuclear power industry.

The Environmental Protection Agency (EPA) has the authority to develop federal radiation protection guidelines for release of radioactivity into the general environment and for exposure of workers and the public. The Nuclear Regulatory Commission (NRC) and individual states authorized by the NRC, called agreement states, see Figure 6.1.1, implement the EPA's general environmental standards through regulations and licensing actions. These standards are usually based on recommendations developed by the International Atomic Energy Agency (IAEA).

Types of Radioactivity

Radioactivity is defined as the property possessed by some elements with spontaneously emitting alpha particles (α), beta particles (β), or sometimes gamma rays (γ) by the disintegration of the nuclei of atoms. It is a naturally occurring phenomenon, it can not be stopped, and it has been taking place since the beginning of time. The process of unstable nuclei giving off energy to reach a stable condition is called radioactive decay. This process produces nuclear radiation, and the emitting isotopes are called radionuclides (radio isotopes). All isotopes of elements with atomic numbers larger than 83 (Bismuth) are radioactive. A few elements with lower atomic numbers, such as potassium and rubidium, have naturally occurring isotopes that are also radioactive. The kind of ionizing radiation emitted, the amount of energy, and the period of time it takes to become stable differs for each radioactive isotope. The following three types of radiation can be emitted.

ALPHA PARTICLES

Emitted by many high-atomic-number natural and man-made radioactive elements such as thorium, uranium and plutonium, alpha decay does not lead directly to stable nuclei. Intermediate isotopes are first produced, then these undergo further decay. The relatively high mass of the alpha particle means that for a given energy, the velocity is relatively low. The heavy, slow-moving, highly charged particles are completely absorbed by a few centimeters of air. After absorption, they are released to the atmosphere as harmless helium gas.

BETA PARTICLES

Emitted by both high and low atomic weight radioactive elements, the beta particle is an electron possessing kinetic energy due to the speed with which it is emitted from the nucleus. The velocities of the more energetic betas approach the speed of light. Beta decay is the most common mode of radioactive decay among artificial and natural radioisotopes. The range of beta particles in air may be more than a meter for energetic betas, and such particles will penetrate several meters of aluminum.

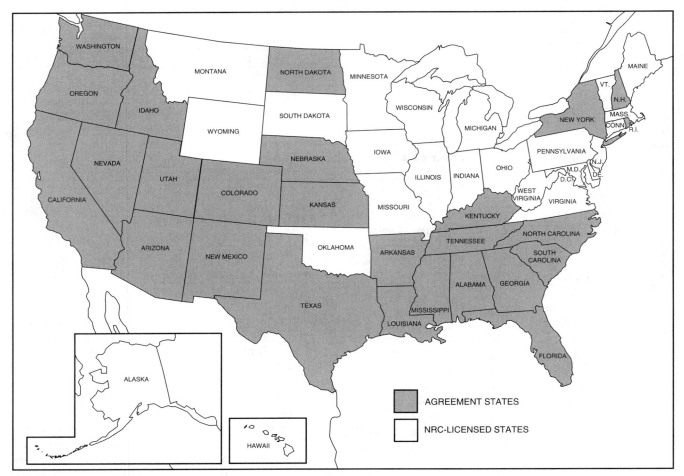

FIG. 6.1.1 States with NRC-licenses or agreements for possession of radioactive materials.

GAMMA RAYS

This type of emission consists not of particles but quanta of energy, similar to radio-waves, but containing much higher levels of energy. This emission is a secondary process following rapidly after certain alpha or beta decay events. The emission of these particles leaves the nucleus in an unstable state, and the excess energy is released as gamma radiation. Gamma rays are very penetrating: absorbing materials can not stop them completely, only reduce their intensity.

Half-Life and Decay of Radioisotopes

All decay processes result from energy changes that eventually result in the formation of a stable nucleus. For example, thorium decays to radium, which decays to actinium, which eventually produces non-radioactive lead 208. The unstable nucleus releases excess energy by one or more of these processes according to characteristic rates. All radioisotopes follow the same law of decay: a fixed fraction of the atoms present will decay in a unit of time. That is, for each isotope there is a period of time during which half of the atoms initially present will decay (Table 6.1.1). Each radioactive element has a constant speed of decay, so that each element can be characterized by the time it takes for half of the element to decay. This is called the half-life of the element. Some elements decay in seconds while others take thousands of years. Since the rate of radioactive decay is not dependent on physical variables such as temperature or pressure, the half-life of each radioisotope is constant. Specific activity relates the half-life of an element to its mass, and is conventionally used to characterize radionuclides.

The *specific activity* (SpA) of a radioisotope is the activity per gram of the pure radioisotope. The number of atoms of a pure radioisotope in one gram (N) is given by

$$N = \frac{N_A}{A} \qquad 6.1(1)$$

TABLE 6.1.1   HALF-LIFE AND DECAY MODE OF SELECTED RADIOISOTOPES

Element	Half-Life Duration	Type of Emission	Element	Half-Life Duration	Type of Emission
$^{14}_{6}C$	5770 y	(β^-)	$^{226}_{88}Ra$	1590 y	(α)
$^{13}_{7}N$	10.0 m	(β^+)	$^{228}_{88}Ra$	6.7 y	(β^-)
$^{24}_{11}Na$	15.0 h	(β^-)	$^{228}_{89}Ac$	6.13 h	(β^-)
$^{32}_{15}P$	14.3 d	(β^-)	$^{228}_{90}Th$	1.90 y	(α)
$^{40}_{19}K$	1.3×10^9 y	$(\beta^-$ or $E.C.)$	$^{232}_{90}Th$	1.39×10^{10} y	$(\alpha, \beta^-,$ or $S.F.)$
$^{60}_{27}Co$	5.2 y	(β^-)	$^{233}_{90}Th$	23 m	(β^-)
$^{87}_{37}Rb$	4.7×10^{10} y	(β^-)	$^{234}_{90}Th$	24.1 d	(β^-)
$^{90}_{38}Sr$	28 y	(β^-)	$^{223}_{91}Pa$	27 d	(β^-)
$^{115}_{49}In$	6×10^{14} y	(β^-)	$^{233}_{92}U$	1.62×10^5 y	(α)
$^{131}_{53}I$	8.05 d	(β^-)	$^{234}_{92}U$	2.4×10^5 y	$(\alpha$ or $S.F.)$
$^{142}_{58}Ce$	5×10^{15} y	(α)	$^{235}_{92}U$	7.3×10^8 y	$(\alpha$ or $S.F.)$
$^{198}_{79}Au$	64.8 h	(β^-)	$^{238}_{92}U$	4.5×10^9 y	$(\alpha$ or $S.F.)$
$^{208}_{81}Tl$	3.1 m	(β^-)	$^{239}_{92}U$	23 m	(β^-)
$^{210}_{82}Pb$	21 y	(β^-)	$^{239}_{93}Np$	2.3 d	(β^-)
$^{212}_{82}Pb$	10.6 h	(β^-)	$^{239}_{94}Pu$	24,360 y	$(\alpha$ or $S.F.)$
$^{214}_{82}Pb$	26.8 m	(β^-)	$^{240}_{94}Pu$	6.58×10^3 y	$(\alpha$ or $S.F.)$
$^{206}_{83}Bi$	6.3 d	$(\beta^+$ or $E.C.)$	$^{241}_{94}Pu$	13 y	$(\alpha$ or $\beta^-)$
$^{210}_{83}Bi$	5.0 d	(β^-)	$^{241}_{95}Am$	458 y	(α)
$^{212}_{83}Bi$	60.5 m	$(\alpha$ or $\beta^-)$	$^{242}_{96}Cm$	163 d	$(\alpha$ or $S.F.)$
$^{207}_{84}Po$	5.7 h	$(\alpha, \beta^+,$ or $E.C.)$	$^{243}_{97}Bk$	4.5 h	$(\alpha$ or $E.C.)$
$^{210}_{84}Po$	138.4 d	(α)	$^{245}_{98}Cf$	350 d	$(\alpha$ or $E.C.)$
$^{212}_{84}Po$	3×10^{-7} s	(α)	$^{253}_{99}Es$	20.0 d	$(\alpha$ or $S.F.)$
$^{216}_{84}Po$	0.16 s	(α)	$^{254}_{100}Fm$	3.24 h	$(S.F.)$
$^{218}_{84}Po$	3.0 m	$(\alpha$ or $\beta^-)$	$^{255}_{100}Fm$	22 h	(α)
$^{215}_{85}At$	10^{-4} s	(α)	$^{256}_{101}Md$	1.5 h	$(E.C.)$
$^{218}_{85}At$	1.3 s	(α)	$^{254}_{102}No$	3 s	(α)
$^{220}_{86}Rn$	54.5 s	(α)	$^{257}_{103}Lr$	8 s	(α)
$^{222}_{86}Rn$	3.82 d	(α)	$^{263}_{106}(106)$	0.9 s	(α)
$^{224}_{88}Ra$	3.64 d	(α)			

Symbol in parentheses indicates type of emission; $E.C.$ = K-electron capture, $S.F.$ = spontaneous fission; y = years, d = days, h = hours, m = minutes, s = seconds.

where:

N_A = Avogadro's number (6.0248×10^{23})/nuclidic mass
A = nuclidic mass

The specific activity (SpA) of a particular radioisotope is:

$$SpA \text{ (disintegration/sec)} = \frac{0.693\ N_A}{T_{1/2}\ A} \qquad 6.1(2)$$

or

$$SpA \text{ (Ci/gm)} = \frac{1.128 \times 10^{13}}{T_{1/2}\ A} \qquad 6.1(3)$$

where:

$T_{1/2}$ = half-life of the radioisotope in sec.

The curie (Ci) is thus the quantity of any radioactive material in which the number of disintegrations is 3.7×10^{10} per second. This is a rather large amount of radio-activity, and smaller quantities are expressed in such units as *millicuries* (1 mCi = 10^{-3} Ci), *microcuries* (1 μCi = 10^{-6} Ci), and *picocuries* (1 pCi = 10^{-12} Ci). Since the curie is a measure of the emission rate, it is not a satisfactory unit for setting safety standards for handling radioactive materials.

Radioactivity originates from natural and man-made sources. Man-made radioactive materials produce *artificial radioactivity*. The radioactivity produced from nuclear fission in a nuclear reactor is a classic example of artificial radioactivity.

Naturally occurring radioisotopes of higher atomic number elements belong to chains of successive disintegrations. The original element, which starts the whole decay series, is called the parent. The new elements formed are called daughters, and the whole chain is called a family. The parent of a natural radioactive series undergoes a series of disintegrations before reaching its stable form.

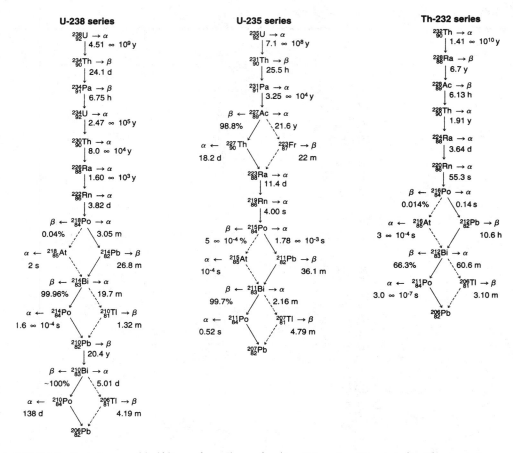

FIG. 6.1.2 Emissions and half-lives of members of radioactive series. (Reprinted, with permission, from R.C. Weast, ed. 1978. *Handbook of chemistry and physics*. CRC Press, Inc.)
[a]The abbreviations are y, year; d, day; m, minute; and s, second.

When a series is in secular equilibrium, one Ci of the parent will coexist with one Ci of each of the daughters. Three series, the uranium, actinium, and thorium, make up most of the naturally radioactive elements found in the periodic table (Figure 6.1.2).

—*Paul A. Bouis*

6.2
SOURCES OF RADIOACTIVITY IN THE ENVIRONMENT

Radioactivity in the environment comes from natural and man-made sources (Figure 6.2.1). Although natural radioactivity is the most likely to be encountered in the environment due to its widespread dispersal, man-made radioactivity poses the greatest environmental risk. Natural radioactivity harnessed by man and not properly disposed of is also a potential threat to the environment. There are five basic sources of radioactivity in the environment: the nuclear fuel cycle, mining activities, medical and laboratory facilities, nuclear weapons testing and seepage from natural deposits.

Nuclear Fuel Cycle

The nuclear fuel cycle is defined as the activities carried out to produce energy from nuclear fuel. These activities include, but are not limited to, mining of uranium-containing ores, enrichment of uranium to fuel grade specifications, fabrication and use of fuel rods, and isolation and storage of waste produced from power plants. The nuclear

fuel cycle is shown in Figure 6.2.2. Note that commercial fuel reprocessing, currently practiced in Europe, was discontinued in the United States in 1972 for safety and security reasons. The Department of Energy (DOE), however, does reprocess most of its spent fuels (U.S. DOE 1988).

Mining Activities

Mining, processing, and the use of coal, natural gas, phosphate rock, and rare earth deposits result in the concentration and release or disposal of large amounts of low–level radioactive material (UNSCEAR 1977). Coal–fired power plants release as much radioactivity (radon) to the environment as nuclear facilities, and the fly ash residue contains low levels of several natural radioisotopes. Natural gas is one of many radon sources in the environment. Phosphate rock always has associated natural radioisotopes; in many cases, tailings from phosphate operations have levels above those allowed by the

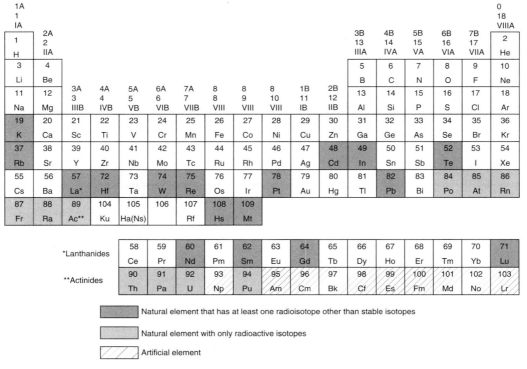

FIG. 6.2.1 Periodic table showing different types of radioisotopes.

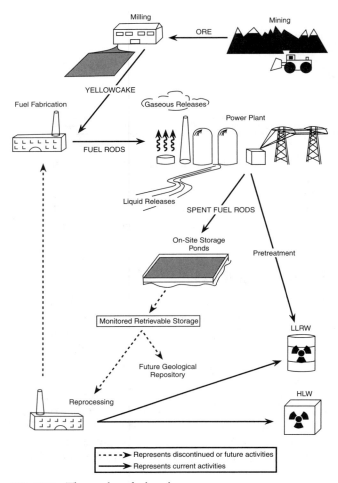

FIG. 6.2.2 The nuclear fuel cycle.

Medical and Laboratory Facilities

Radioisotopes are used extensively in medical facilities, biomedical research laboratories, and to a lesser extent in other types of laboratories (Table 6.2.2). Clinical use of radioisotopes is expanding rapidly in such areas as cancer treatment and diagnostic testing. The lack of waste management plans at many of these facilities results in frequent misclassification of materials as radioactive (Party & Gershey 1989). Relatively large amounts of radioisotopes are used in clinical procedures. Although most of these isotopes are strong emitters of gamma radiation, they have short half lives (Table 6.2.3).

Nuclear Weapons Testing

The use of nuclear devices in weapons is the primary cause of radioactive fallout, although the nuclear accident at Chernobyl and various volcanic eruptions have also contributed. Tritium (^{3}H) and several isotopes of iodine, cesium and strontium are found in the environment largely because of nuclear testing. In the United States, most radioactive waste is a by-product of nuclear weapons production. It is estimated that 70% of U.S. radioactive waste results from defense department activities (Eisenbud 1987). The DOE is currently trying to remedy many defense sites due to past poor waste management practices.

Natural Deposits

The majority of radioactivity in groundwater is due to seepage from natural deposits of uranium and thorium (Table 6.2.4). Strict guidelines for acceptable levels of radionuclides in drinking water exist (U.S. EPA 1986). The EPA has established maximum levels for radium (U.S. EPA 1976, 1980) to monitor for the presence of natural radionuclides (Table 6.2.5). Radon, a colorless, odorless, inert, radioactive gas that seeps out of the earth has been found at dangerously high levels in inadequately venti-

NRC for release to the environment. Processing rare-earth containing ores produces concentrated waste high enough in radioactivity to be disposed of as low-level radioactive waste. If monazite ore is the rare-earth source, nearly one ton in ten must be disposed of in this manner. Disposal costs have dramatically reduced imports of monazite (Table 6.2.1).

TABLE 6.2.1   U.S. IMPORTS FOR CONSUMPTION OF MONAZITE, BY COUNTRY

Country	1985 Quantity (Metric Tons)	1986 Quantity (Metric Tons)	1987 Quantity (Metric Tons)	1988 Quantity (Metric Tons)	1989 Quantity (Metric Tons)
Australia	5,694	2,660	—	382	180
India	—	300	—	—	—
Indonesia	—	—	—	1,144	794
Malaysia	—	—	527	197	—
Thailand	—	—	594	201	—
Total	5,694	2,960	1,121	1,924	974
RBO content[a]	3,132	1,628	617	1,058	536

[a]Estimated.
Source: Reprinted and adopted from U.S. Bureau of the Census.

TABLE 6.2.2   RADIOISOTOPES ENCOUNTERED IN LABORATORIES

Half-Life	Element and Symbol	Atomic Number	Mass Number	Gamma Radiation Energy (MeV)
88 days	Sulfur (S)	16	35	none
115 days	Tantalum (Ta)	73	182	0.068, .10, .15, .22, 1.12, 1.19, 1.22
120 days	Selenium (Se)	34	75	0.12, .14, .26, .28, .40
130 days	Thulium (Tm)	69	170	0.084
138 days	Polonium (Po)	84	210	0.80
165 days	Calcium (Ca)	20	45	none
245 days	Zinc (Zn)	30	65	1.12
270 days	Cobalt (Co)	27	57	0.12, .13
253 days	Silver (Ag)	47	110	0.66, .68, .71, .76, .81, .89, .94, 1.39
284 days	Cerium (Ce)	58	144	0.08, .134
303 days	Manganese (Mn)	25	54	0.84
367 days	Ruthenium (Ru)	44	106	none
1.81 yr	Europium (Eu)	63	155	0.09, .11
2.05 yr	Cesium (Cs)	55	134	0.57, .60, .79
2.6 yr	Promethium (Pm)	61	147	none
2.6 yr	Sodium (Na)	11	22	1.277
2.7 yr	Antimony (Sb)	51	125	0.18, .43, .46, .60, .64
2.6 yr	Iron (Fe)	26	55	none
3.8 yr	Thallium (Tl)	81	204	none
5.27 yr	Cobalt (Co)	27	60	1.3, 1.12
11.46 yr	Hydrogen (H)	1	3	none
12 yr	Europium (Eu)	63	152	0.12, .24, .34, .78, .96, 1.09, 1.11, 1.41
16 yr	Europium (Eu)	63	154	0.123, .23, .59, .72, .87, 1.00, 1.28
28.1 yr	Strontium (Sr)	38	90	none
21 yr	Lead (Pb)	82	210	0.047
30 yr	Cesium (Cs)	55	137	0.661
92 yr	Nickel (Ni)	28	63	none
1602 yr	Radium (Ra)	88	226	0.186
5730 yr	Carbon (C)	6	14	none
2.12×10^5 yr	Technetium (Tc)	43	99	none
3.1×10^5 yr	Chlorine (Cl)	17	36	none

Source: Reprinted, from U.S. Department of Health, Education and Welfare (HEW), 1970, *Radiological health handbook,* rev. ed. (HEW, Rockville, Md., [January]).

TABLE 6.2.3   PRINCIPAL CLINICALLY ADMINISTERED RADIOISOTOPES

Radionuclide	Principal Uses	Half-life	Typical Dose (mCi)	Number of Procedure[a]	Total Curies	% of Total
^{99m}Tc	Organ imaging	6.0 hr	4–25	8,040,000	116,580	96
^{309}Tl	Myocardial & parathyroid imaging	74.0 hr	2	960,000	1,920	2
Ga	Tumor/infection diagnosis	78.1 hr	5	600,000	3,000	2
I	Thyroid imaging	8.1 days	0.1	960,000	96	<0.1
Total				10,560,000	121,596	100

[a]Annual number of procedures in the United States.

TABLE 6.2.4 AVERAGED SOIL AND ROCK CONCENTRATION OF URANIUM AND THORIUM

	U	Th
	(mg kg^{-1})	
Rocks		
Igneous		
Silica (granites)	4.7	20.0
Intermediate (diorites)	1.8	8.0
Mafic (basalt)	0.9	2.7
Ultramafic (dunites)	0.03	6.0
Sedimentary		
Limestones	2.2	1.7
Carbonates	2.1	1.9
Sandstones	1.5	3.0
Shales	3.5	11.0
(Mean value in earth's crust)	3.0	11.4
Soils		
Typical range	1–4	2–12
World average	2	6.7
Average specific activity (pCi/kg^{-1})	670	650

Source: Reprinted, with permission, from M. Boyle, 1988, Radon testing of soils, *Environmental Science Technology* [22(12):1397–1399].

TABLE 6.2.5 CONCENTRATIONS OF RA^{226} AND DAUGHTERS IN CONTINENTAL WATERS (pCi per liter)

	Ra^{226}	Rn^{222}	Pb^{210}	Po^{210}
Deep wells	1–10	10^4–10^5	<0.1*	~0.02
Ground water	0.1*–1	10^2–10^3	<0.1*	~0.01
Surface water	<1	10	<0.5	—
Rainwater	—	10^3–10^5†	0.5–3	~0.5

*Below detection limits.
†As determined through presence of short-lived Rn^{222} daughters.

lated buildings. Radon originates from the radioactive decay of uranium, thorium and/or radium (Figure 6.2.3). The EPA states that levels above four pCi/L should be reviewed for possible corrective actions (Boyle 1988).

—Paul A. Bouis

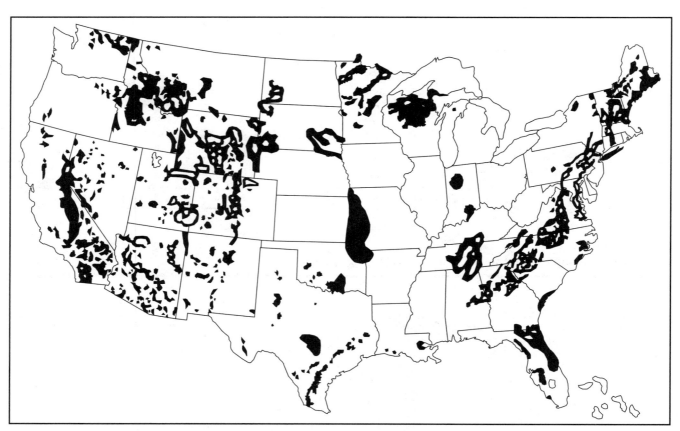

FIG. 6.2.3 Areas with potentially high radon levels. (Reprinted, with permission, from M. Boyle, 1988, Radon testing of soils, *Environmental Science Technology* [22(12):1397–1399].)

References

Boyle, M. 1988. Radon testing of soils, *Environmental Science Technology,* 22(12):1397–1399.

Eisenbud, M.E. 1987. *Environmental radioactivity from natural, industrial, and military sources.* 3rd ed. Orlando, Fla.: Academic Press.

Party, E.A., and E.L. Gershey. 1989. Recommendations for radioactive waste reduction in biomedical/academic institutions. *Health Physics* 56(4):571–572.

U.S. Department of Energy (DOE). 1988. *Database for 1988: spent fuel and radioactive waste inventories, projections, and characteristics.* DOE/RW–0006, Rev. 4. Washington, D.C.

United Nations Scientific Committee on the Effects of Atomic Radiation (UNSCEAR). 1977. *Sources and effects of ionizing radiation.* New York, N.Y.

U.S. Environmental Protection Agency (EPA). 1976. Drinking water regulations, radionuclides. *Federal Register* 41:28402.

———. 1980. *Prescribed procedures of measurement of radioactivity in drinking water.* EPA 600–4–80–032.

———. 1986. Water pollution control: radionuclides: advance notice of proposed rulemaking. 40 CFR Part 141, 34836: *Federal Register* 51:189.

6.3
SAFETY STANDARDS

Radioactivity presents special hazards because it cannot be detected by the normal human senses. Strict safety standards have been established by international organizations to ensure that exposure to workers is minimized and that the public is not exposed to radiation from other than the natural background (ICRP 1979; NCRPM 1959, 1987a,b). This background radiation, from naturally occurring radioisotopes and cosmic rays, is the base exposure level, and cannot be practically reduced.

All types of radiation share the property of losing energy by absorption in passing through matter. The degree of absorption depends upon the type of radiation, but all types are absorbed to some extent (Figure 6.3.1). The process of absorption always results in ionization. This process of stripping electrons from atoms causes damage to human tissues. It also allows for the design of instruments for detection and measurement of radioactivity. The properties of the various radiations determine the protective measures needed and the methods of measurement. Three types of radiation exist.

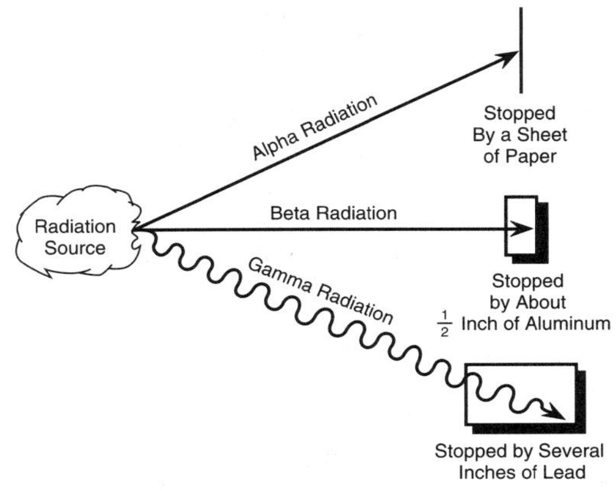

Alpha Radiation: Radiation from alpha particles loses energy very quickly when passing through matter. As a result, alpha radiation travels only a few inches in air and can easily be stopped by the outer layer of human skin. Alpha radiation sources are most harmful to humans if they are ingested. Alpha radiation can be very damaging to body organs, especially the lungs if the alpha source is inhaled as fine particles (BEIR 1988).

Beta Radiation: Radiation produced by beta particles travels much farther in air than alpha radiation, and can penetrate several layers of human skin. The human body can be damaged by being near a source of beta radiation for a long period of time or by ingesting a source of beta radiation. Beta radiation can be stopped by absorbing materials.

Gamma Radiation: Gamma radiation travels great distances and easily penetrates matter. It can pass completely through the human body, damaging cells en route, or be absorbed by tissue and bone. Three feet of concrete or two inches of lead are required to stop 90% of typical gamma radiation. Excessive external gamma radiation can cause serious internal damage to the human body, but cannot induce radioactivity in it.

The biological effect of radiation is measured in units called *rems.* A rem is the amount of beta/gamma radiation that transfers a specific quantity of energy to a kilogram of matter. A single exposure to 300 rems would result in death within thirty days for 50% of the persons exposed. The unit of dose is difficult to put into perspective, however, a comparison of the allowable doses helps. The permissi-

FIG. 6.3.1 Relative penetrating power of alpha, beta, and gamma radiation.

ble level for occupational radiation exposure is five rems per year to the whole body. It is believed that this level can be absorbed for a working lifetime without any sign of biological damage. Background radiation is measured in millirems (0.001 rem).

The average person is exposed to ionizing radiation from many sources. The environment, and even the human body, contains naturally occurring radioactive materials. Cosmic radiation contributes additional exposure. The use of x-rays and radioisotopes in medicine and dentistry adds to the public exposure. Table 6.3.1 shows the estimated average individual exposure in millirems from natural background and other sources.

Protection from Exposure

Maximum permissible levels of external and internal radiation (Table 6.3.2) have been set by the National Council on Radiation Protection (NCRP) and by the International Commission on Radiological Protection (ICRP). In addition, the practice of keeping exposures *As Low As Reasonably Achievable* (NRC 1976) is recommended by these and many other organizations. This means that every activity involving exposure to radiation should be planned to minimize unnecessary exposure to workers and the public. For further explanations of radiation protection the reader is referred to the many references available (Henry 1969, Olishifski 1981, Schapiro 1981).

Basic Radiation Safety

Safety practices for handling radioactive materials are aimed at protecting individuals from external and internal hazards.

TABLE 6.3.1   U.S. GENERAL POPULATION EXPOSURE ESTIMATES

Source	Average Individual Dose mrem/yr
Natural Background	100
Mining Releases	5
Medical	90
Nuclear Fallout	7
Nuclear Energy	0.3
Consumer Products	0.03
Total	≈200

TABLE 6.3.2   MAXIMUM PERMISSIBLE DOSE EQUIVALENT FOR OCCUPATIONAL EXPOSURE

Combined whole body occupational exposure	
Prospective annual limit	5 rems in any one year
Retrospective annual limit	10–15 rems in any one year
Long-term accumulation	(N–18) × 5 rems, where N is age in years
Skin	15 rems in any one year
Hands	75 rems in any one year (25/qtr)
Forearms	30 rems in any one year (10/qtr)
Other organs, tissues and organ systems	15 rems in any one year (5/qtr)
Fertile women (with respect to fetus)	0.5 rem in gestation period
Dose limits for the public, or	
occasionally exposed individuals:	
Individual or occasional	0.5 rem in any one year
Students	0.1 rem in any one year
Population dose limits	
Genetic	0.17 rem average per year
Somatic	0.17 rem average per year
Emergency dose limits—Life saving:	
Individual (older than 45, if possible)	100 rems
Hands and forearms	200 rems, additional (300 rems, total)
Emergency dose limits—Less urgent:	
Individual	25 rems
Hands and forearms	100 rems, total
Family of radioactive patients:	
Individual (under age 45)	0.5 rems in any one year
Individual (over age 45)	5 rems in any one year

Reprinted from National Committee on Radiation Protection (NCRP), 1975, *Review of the current state of radiation protection philosophy*. NCRP Publication No. 43.

TABLE 6.3.3   CLASSIFICATION OF ISOTOPES ACCORDING TO RELATIVE RADIOTOXICITY PER UNIT ACTIVITY (THE ISOTOPES IN EACH CLASS ARE LISTED IN ORDER OF INCREASING ATOMIC NUMBER)

CLASS 1 (very high toxicity)

Sr-90 + Y-90, *Pb-210 + Bi-210 (RaD + E), Po-210, At-211, Ra-226 + percent *daughter products, Ac-227, *U-233, Pu-239, *Am-241, Cm-242.

CLASS 2 (high toxicity)

Ca-45, *Fe-59, Sr-89, Y-91, Ru-106 + *Rh-106, *I-131, *Ba-140 + La-140, Ce-144 + *Pr-144, Sm-151, *Eu-154; *Tm-170, *Th-234 + *Pa-234, *natural uranium.

CLASS 3 (moderate toxicity)

*Na-22, *Na-24, P-32, S-35, Cl-36, *K-42, *Sc-46, Sc-47, *Sc-48, *V-48, *Mn-52, *Mn-54, *Mn-56, Fe-55, *Co-58, *Co-60, Ni-59, *Cu-64, *Zn-65, *Ga-72, *As-74, *As-76, *Br-82, *Rb-86, *Zr-95 − *Nb-95, *Nb-95, *Mo-99, Tc-98, *Rh-105, Pd-103 − Rh-103, *Ag-105, Ag-11, Cd-109 − *Ag-109, *Sn-113, *Te-127, *Te-129, *I-132, Cs-137 − *Ba-137, *La-140, Pr-143, Pm-147, *Ho-166, *Lu-177, *Ta-182, *W-181, *Re-183, *Ir-190, *Ir-192, Pt-191, *Pt-193, *Au-198, *Au-199, Tl-200, Tl-202, Tl-204, *Pb-203.

CLASS 4 (slight toxicity)

H-3, *Be-7, C-14, F-18, *Cr-51, Ge-71, *Tl-201.

*Gamma-emitters.

Source: International Atomic Energy Agency (IAEA), *Safe handling of radionuclides*, Safety Series No. 1 (Vienna, Austria: IAEA).

EXTERNAL RADIATION

Protection from external radiation is accomplished by adhering to the principles of maximizing distance, minimizing time, and shielding individuals from the radioactive source. Exposure levels are readily measured using conventional radiation-measuring devices (IAEA 1976). This allows the distance, time, and the necessary amount of shielding to be determined to minimize exposure.

INTERNAL RADIATION

The most frequent routes for radioactive materials intake are through inhalation or open wounds. Air monitoring in areas where radioactive materials are handled is always recommended. Simple methods and measuring devices exist. Periodic testing of urine, body fluids, and excrement is also recommended as a secondary means of determining if radioisotopes have been ingested. The toxicity of various isotopes is shown in Table 6.3.3.

—Paul A. Bouis

References

Committee on the Biological Effects of Ionizing Radiation (BEIR). 1988. *Health risks from radon and other internally deposited alpha emitters.* BEIR IV Report. Washington, D.C.: National Academy Press.

Henry, H.F. 1969. *Fundamentals of radiation protection.* New York, N.Y.: Wiley Interscience.

International Atomic Energy Agency (IAEA). 1976. Manual on radiological safety in uranium and thorium mines and mills. *Safety Series No. 43,* Vienna, Austria.

International Commission on Radiation Protection (ICRP). 1979. Limits for intake of radionuclides by workers. *ICRP Publication 30.* New York, N.Y.: Pergamon Press.

National Committee on Radiation Protection and Measurements (NCRP). 1959. Maximum permissible body burdens and maximum permissible concentrations of radionuclides in air and water for occupational exposure. *NBS Handbook No. 69.*

———. 1987a. Recommendations on limits for exposure to ionizing radiation. *NCRP Report No. 91.* Bethesda, Md.

———. 1987b. Ionizing radiation exposure of the population of the United States. *NCRP Report No. 93.* Bethesda, Md.

U.S. Nuclear Regulatory Commission (NRC). 1976. Operating philosophy for maintaining occupational radiation exposures as low as is reasonably achievable. *NRC Regulatory Guide 8.10.*

Olishifski, J.B. 1981. *Fundamentals of industrial hygiene.* Chicago, Il.: National Safety Council.

Schapiro, J. 1981. *Radiation protection: A guide for scientists and physicians,* 2d ed. Cambridge, Mass.: Harvard University Press.

gaseous radioisotopes such as radon. Radium 226 analysis by radon de-emanation into a Lucas cell is a classic use of the scintillation counter (Greenberg, Clesceri & Eaton 1992) (Figure 6.4.2). Calibration of each cell is required to obtain quantitative results.

Gamma Ray Spectroscopy

Simultaneous analysis of multiple specific radionuclides can be done using gamma ray spectroscopy (Heath 1974). This method is applicable to the analysis of gamma-emitting radionuclides with gamma energies ranging from 80 keV to approximately 2000 keV. The technique minimizes the sample preparation required to do radiochemical analysis. Using a NaI detector, it is possible to routinely analyze four to eight gamma-emitting radionuclides. Personal computer-based, high-resolution intrinsic germanium detector systems are now used almost exclusively in gamma ray spectroscopy. These systems can analyze an almost unlimited number of radionuclides, and are especially suited for low-level analysis of environmental samples. A comparison of the superior resolution of an intrinsic germanium detector is shown in Figure 6.4.3. The most frequently used photo-energy peaks for common radionuclides are shown in Table 6.4.1. Liquid and solid samples can be placed directly in a Marineli beaker for analysis. An efficiency calibration is required for quantitative analysis. The result of this calibration is an efficiency versus energy curve (Figure 6.4.4).

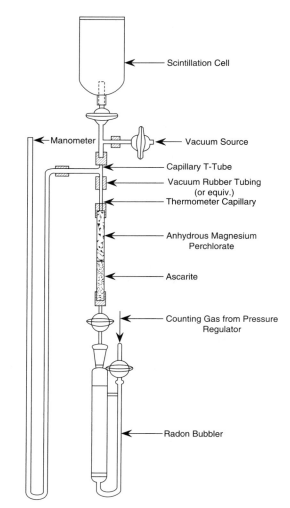

FIG. 6.4.2 De-emanation assembly.

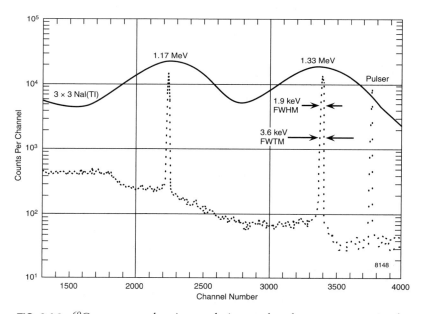

FIG. 6.4.3 ⁶⁰Co spectrum showing resolutions and peak-to-compton ratios for an intrinsic Ge detector and a NaI(Tl) detector.

TABLE 6.4.1   PRINCIPAL GAMMA-RAY PHOTO PEAKS USED FOR ANALYSIS OF SELECTED RADIONUCLIDES BY GAMMA-RAY SPECTROSCOPY

Radionuclide	Gamma Energy keV
Ra 226	186
Pb 212	239
Cr 51	321
Cs 134	605
Cs 137	662
Mn 54	835
Ac 228 (Ra 228)	911
Co 58	1100
Zn 65	1110
Co 60	1173
Co 58	1290
Co 60	1333
Ce 144	1387
Eu 152	1408

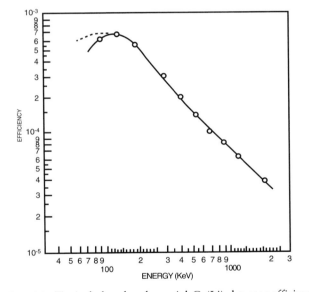

FIG. 6.4.4 Typical closed-end coaxial Ge(Li) detector efficiency calibration curve. The dashed curve indicates the increased low-energy efficiency of intrinsic Ge detectors.

ANALYTICAL METHODS

Gross Alpha-Beta

A proportional counter with heavy shielding is recommended for this method. The instrument is calibrated by adding radionuclide standards to a matrix similar to the sample. A standard solution of cesium 137 or strontium 90 certified by the National Institute of Standards and Technology (NIST) is suitable for gross beta analysis. A solution of natural uranium, thorium, plutonium 239 or americium 241 is recommended for gross alpha analysis. Gross alpha-beta results are always reported in comparison to a specific standard.

The sample, usually a liquid, is evaporated onto a *planchette*, to a thin film. The standard is prepared in the same manner. Counts from the sample are then compared to the standard. Samples suspected of containing fission or artificial radionuclides can be tested for gross beta using either the cesium or strontium standard. Environmental samples suspected of containing natural radionuclides can be tested for gross alpha using any of the alpha standards previously listed. Careful attention must be paid to self absorption of alpha and beta particles due to sample thickness on the planchette whenever test results are evaluated.

Radioactive Cesium

An extremely hazardous fission product, the interim EPA drinking water regulations limit cesium 134 to 80 pCi/L and cesium 137 to 200 pCi/L. Samples suspected of containing moderate or high levels of cesium can be tested by gamma ray spectroscopy. Low-level environmental samples can be purified and concentrated by co-precipitation with ammonium phosphomolyodate and analyzed either by gamma ray analysis or by beta counting (Kreiger 1976).

Radioactive Iodine

Radioiodine originates from nuclear weapons testing and from the nuclear fuel cycle. Fission products may contain iodines 129 through 135. The EPA drinking water maximum for iodine 135 is 3 pCi/L. Samples are preconcentrated either by precipitation as PdI_2, absorption on an anion exchange resin, or by distillation. The concentrated sample is then beta counted (U.S. EPA 1980).

Radium

The EPA has established strict limits on radium in public drinking waters. These regulations require that if the radium 226 activity exceeds 3 pCi/L, radium 228 activity must be measured. If the combined activities of these radioisotopes exceed 5 pCi/L, the water supply exceeds the EPA limit for radium in water. The standard methods of analyzing for radium involve either alpha counting a purified barium-radium sulfate co-precipitate, or measuring the radian de-emanated from radium 226-containing samples. An involved wet chemical procedure based on the ingrowth of actinium 228 has been published by the EPA for radium 228 analysis (U.S. EPA 1980). Simpler, more precise gamma ray spectroscopy methods have been developed, significantly lowering the detection limits (U.S. EPA 1980).

Strontium

Nuclear fission produces radioactive strontium isotopes. Strontium 90 is an extremely hazardous isotope. Upon ingestion it tends to concentrate in bone. Analysis of strontium involves tedious and complicated wet procedures of large samples. It is impossible to separate the isotopes of strontium, therefore strontium 90 is actually determined by measuring the amount of its daughter, yttrium 90. The final purified concentrate is beta counted using cesium 137 as the calibration standard.

Tritium

Tritium is found in the environment as a result of natural cosmic rays, nuclear weapons testing, and the nuclear fuel cycle. Tritium eventually decays by beta emission to helium. Analysis consists of an alkaline permanganate distillation, mixing with a liquid scintillator, and beta counting with a liquid scintillation spectrometer.

Uranium

Uranium is found in most drinking water supplies as a soluble carbonate. Uranium 238 is the primary isotope found in these waters. Standard uranium methods involve complicated wet procedures combined with ion exchange purification prior to alpha counting with a proportional counter (Barker 1965). A direct fluorescence analyzer is now commercially available, considerably simplifying this analysis.

—*Paul A. Bouis*

References

Analytical chemistry, lab guide edition, Vol. 65, No. 16. American Chemical Society.

American Laboratory. *Buyers' Guide Edition.* Vol. 26, No. 4. International Scientific Communications, yearly publication.

Barker, F.B., et al. 1965. *Determination of uranium in natural waters.* U.S. Geological Survey, Water Supply Paper 1696-C. Washington, D.C.: U.S. Government Printing Office.

Greenberg, A.E., L.S. Clesceri, and A.D. Eaton. 1992. *Standard methods for the examination of water and wastewater,* 18th ed. APHA. Washington, D.C.

Heath, R.L. 1974. *Gamma ray spectrum catalogue, Ge(Li) and Si(Li) spectrometry.* ANCR-1000-2. National Technical Information Service. Springfield, Va.

International Atomic Energy Agency (IAEA). 1976. Manual on radiological safety in uranium and thorium mines and mills. *Safety Series No. 43.* Vienna, Austria: IAEA.

Knoll, G.F. 1989. *Radiation detection and measurements.* New York, N.Y.: J. Wiley & Sons.

Kreiger, H.L. 1976. *Interim radiochemical methodology for drinking water.* EPA 600–4–75–008 (Revised). U.S. Environmental Protection Agency, Environmental Monitoring and Support Lab. Cincinnati, Oh.

National Council on Radiation Protection and Measurements (NCRP). 1976. Environmental radiation measurements. *NCRP Report No. 50.* Washington, D.C.

———. 1978. A handbook of radioactivity measurement procedures. *NCRP Report No. 58.* Washington, D.C.

U.S. Environmental Protection Agency (EPA). 1980. *Prescribed procedures for measurements of radioactivity in drinking water.* EPA 600–4–80–032. Environmental Monitoring and Support Lab. Cincinnati, Oh.

6.5
MINING AND RECOVERY OF RADIOACTIVE MATERIALS

The nuclear fuel cycle begins with the exploration and mining of uranium-containing ores. Although few active mining sites are currently in operation, there are numerous closed mines throughout the world. Many of the facilities in the United States, built and operated under contract to the DOE, have not been properly remediated. The proliferation of sites occurred during a period in the 1970s when uranium prices skyrocketed. Ores containing as little as 0.2% U_3O_8 were processed during this period.

A simplified schematic of a typical uranium mill is shown in Figure 6.5.1. Closure of mining and mill sites is strictly regulated by the NRC and its agreement states. Detailed criteria for closure can be found in the Code of Federal Regulations (10 CFR Part 40). The mining and milling of uranium ores produces large quantities of contaminated rock, sludge, gases, and liquids. These materials contain varying concentrations of the radioactive ore and its daughters such as radon, radium, lead, thorium,

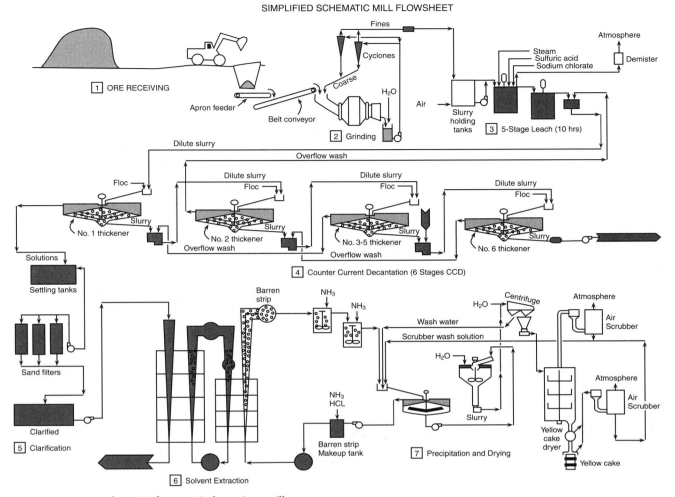

FIG. 6.5.1 Process schematic for a typical uranium mill.

and bismuth. Tailings and waste produced from the extraction or concentration of uranium or thorium from any ore processed primarily to recover *source material* is called by-product material. Source material is defined as any material containing more than 0.05% uranium and/or thorium by weight. Exemptions from most NRC regulations exist for many commercial uses of source material. By-product material is regulated as radioactive waste, including surface wastes from uranium solution extraction. Underground ore bodies depleted by such techniques are not considered by-product material.

Proper handling and disposal of waste classified as by-product material is a large part of any mining and milling operation (IAEA 1976). The typical process produces waste at almost every step. Most wastes are put into tailings ponds where uranium is periodically recovered. These ponds are normally highly acidic due to the large quantities of acid used in the ore leaching step. They may also be contaminated with organic solvents or ion exchange resins if these are used in the recovery step. Many of these

impoundments have contaminated local groundwaters (UNSCEAR 1977). Treatments for runoff from uranium mills have been developed; a typical treatment process is shown in Figure 6.5.2. This process is designed to produce an effluent that comes close to meeting the drinking water limits of 5 pCi/L total radium and 3 pCi/L of radium 226. The concentrated radioactive (radium) sludge (Tsivoglou & O'Connell 1965) is then handled as a low-level radioactive waste.

Tailings ponds, even after closure, are a constant source of radon from the decay of radium. Tailings are sometimes used as building materials, posing a potential health hazard from radon seepage.

Non-radioactive mining such as phosphate rock operations can produce tailings containing uranium, thorium, and radium at levels above those permissible for release to the environment. These tailings also often find their way into commerce as building materials.

—*Paul A. Bouis*

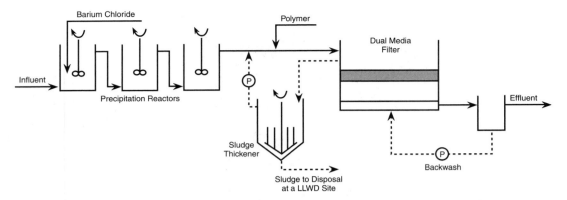

FIG. 6.5.2 Process schematic of a typical radium removal operation.

References

International Atomic Energy Agency (IAEA). 1976. Manual on radiological safety in uranium and thorium mines and mills. *Safety Series No. 43.* IAEA. Vienna, Austria.

Tsivoglou, E.C., and R.S. O'Connell. 1965. Nature, volume and activity of uranium mill wastes. *Radiological Health and Safety in Mining and Milling of Nuclear Materials.* IAEA. Vienna, Austria.

United Nations Scientific Committee on the Effects of Atomic Radiation (UNSCEAR). 1977. *Sources and effects of ionizing radiation.* New York, N.Y.

6.6
LOW-LEVEL RADIOACTIVE WASTE

Low-level radioactive waste is a general term for a wide range of materials contaminated with radioisotopes (Gershey, Klein, Party & Wilkerson 1990; Burns 1988). Industries and hospitals, medical, educational and research institutions, private and government laboratories, and nuclear fuel cycle facilities using radioactive materials generate low-level radioactive wastes as part of normal operations. These wastes are generated in many physical and chemical forms, and at many levels of contamination. Low-level radioactive waste (LLRW) accounts for only one percent of the activity (curies, bequerels) but eighty-five percent of the volume of radioactive waste generated in the United States. The NRC defines LLRW as "radioactive material subject to NRC regulations that is *not* high-level waste, spent nuclear fuel, or mill tailings and which NRC classifies in 10 CFR Part 61 as low-level radioactive waste."

Table 6.6.1 shows the origins of most radioactive wastes. Figure 6.6.1 shows general classifications for all radioactive wastes. Low-level wastes fall under four categories:

1. Below regulatory concern
2. Generator disposed
3. Class A, B, or C
4. Greater than class C

Approximately two million cubic feet of LLRW are disposed of annually at currently operating commercial disposal sites. The nuclear fuel cycle accounts for over fifty percent of this volume, and more than eighty percent of the activity.

Although contact with radioactive waste in the environment should be minimal, due to the highly regulated nature of the waste handling protocols, the ongoing design, operation, and maintenance of the numerous sites are ongoing activities requiring the expertise of environmental engineers and scientists.

Waste Classification

No worldwide agreement has been reached for classification of radioactive wastes. This is contrary to the rules established for release of radioactive materials to the environment and for protection of the general public and workers from radiological exposure. However, most countries agree that waste is best classified from the point of view of disposal. The NRC, in 10 CFR Part 61, classifies low-level radioactive waste based on its suitability for near surface disposal. According to the NRC, classifying radioactive waste involves two factors:

TABLE 6.6.1   ORIGINS, TYPES, QUANTITIES AND CHARACTERISTICS OF RADIOACTIVE WASTE GENERATED IN THE UNITED STATES

Waste	Principal Generators	Typical Nuclides	U.S. Inventory Curies	m	Surface Exposure	Hazard Duration (years)	Overall Hazard Potential
Spent fuel	Nuclear power plants, DOE activities	^{137}Cs, ^{60}Co, ^{235}U, ^{238}U, $^{239-242}$Pu	1.8×10^{10}	6.80×10^3	High	$>10^5$	Requires isolation in perpetuity
High-level	DOE reprocessing of spent fuels	^{90}Sr-^{90}Y, ^{137}Cs, ^{144}Ce, ^{106}Ru, $^{239-242}$Pu	1.3×10^9	3.82×10^5	High	$>10^5$	Requires long-term isolation
Transuranic	Plutonium production for nuclear weapons	$^{239-242}$Pu, ^{241}Am, ^{244}Cm	4.1×10^6	2.80×10^5	Moderate	$>10^5$	Soluble and respirable
Mill tailings	Mining and milling of uranium/thorium ores	^{235}U, ^{230}Th, ^{226}Ra	1.4×10^5	1.20×10^8	Low	$>10^4$	Hazard to worker
Greater than Class C	Nuclear power plants, users and manufacturers of sealed-source devices	^{60}Co, ^{137}Cs, ^{90}Sr, ^{241}Am	2.40×10^6	1.30×10^2	High	500	High
Low-level DOE	Various processes, including decontamination and remedial action cleanup projects	Fission products, ^{235}U, ^{230}Th, α-bearing waste, ^{3}H	1.4×10^7	2.40×10^6	Unknown	$>10^3$	High, poorly managed in the past
Low-level commercial Class A	Fuel cycle, power plants, industry, institutions		3.6×10^5	1.3×10^6	Low	200	Low
Class B	Principally power plants and industry		9.5×10^5	2.7×10^4	Moderate	$\sim 10^3$	Moderate
Class C	Power plants, some industry		2.5×10^6	6.5×10^3	High	$>10^5$	High

TABLE 6.6.2   OVERVIEW OF CLASSES A, B, AND C, WASTE CHARACTERISTICS

Characteristic	Class A Waste	Class B Waste	Class C Waste
Concentration	low concentrations of radionuclides	higher concentrations of radionuclides	highest concentration of radionuclides
Waste Form	must meet minimum waste form requirements does not require stabilization (but may be stabilized)	must meet minimum waste form requirements requires stabilization for 300 years	must meet minimum waste form requirements requires stabilization for 300 years
Examples	typically contaminated protective clothing, paper, laboratory trash	typically resins and filters from nuclear power plants	typically nuclear reactor components, sealed sources, high activity industrial waste
Intruder Protection	after 100 years, decays to acceptable levels to an intruder requires no additional measures to protect intruder	after 100 years, decays to acceptable levels to an intruder, provided waste is recognizable requires stabilization to protect intruder	after 500 years, decays to acceptable levels to an intruder requires stabilization and deeper disposal (or barriers) to protect intruder
Segregation	unstable Class A must be segregated from Classes B and C	need not be segregated from Class C	need not be segregated from Class B

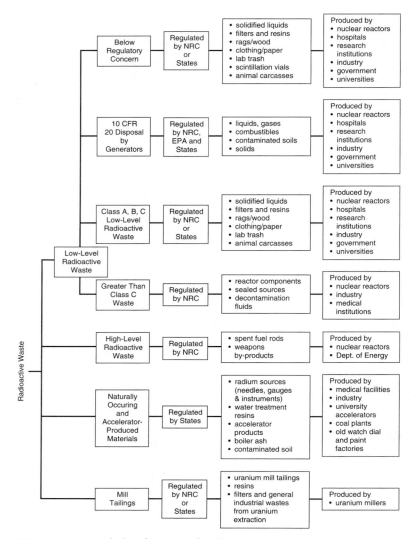

FIG. 6.6.1 General classifications of radioactive waste.

1. Long-lived radionuclide concentrations posing potential hazards that will persist long after such precautions as institutional controls, improved waste forms and deeper disposal have ceased to be effective
2. Shorter-lived radionuclide concentrations for which institutional controls, waste forms, and disposal methods are effective

Low-level radioactive waste is classified as Class A, B, and C waste. An overview of the characteristics of wastes in these classes is shown in Table 6.6.2. 10 CFR §61.54 defines these classes as follows:

1. Class A wastes are usually segregated from other waste classes at the disposal site. The physical form and characteristics must meet the minimum requirements set forth in these regulations (10 CFR §61.56[a]), e.g., contains less than 1% liquid by volume, etc. If Class A waste also meets the stability requirements set forth in 10 CFR §61.56(b), it is not necessary to segregate the waste for disposal.

2. Class B wastes must meet more rigorous waste form requirements to ensure stability after disposal.
3. Class C wastes must meet more rigorous waste form requirements, and also require additional measures at the disposal facility to protect against inadvertent intrusion.

Wastes with form and disposal methods more stringent than Class C are not acceptable for near surface disposal. These wastes must be disposed of in geological repositories.

Classification by specific long- and short-lived radionuclide concentrations is also given in 10 CFR §61.54. The reader is referred to this section for details.

The 10 CFR Part 61 radioactive waste classification is a systematic attempt to control the potential dose to man from disposed waste. System components include site characteristics, site design and operation, institutional controls, waste forms, and intruder barriers. The quantity and type of radionuclides permitted in each class are based on these various disposal components and on radioactive material concentrations expected in the waste and important for disposal. Since low-level radioactive waste typically contains short- and long-lived radionuclides, three time intervals, 100, 300, and 500 yr, are used to set waste classification limits (Table 6.6.3).

Sources of Low-Level Radioactive Waste

NUCLEAR FUEL CYCLE WASTE

Fuel cycle and utility wastes consist mostly of compacted trash and dry wastes, filters, tools, and ion-exchange resin. Many of these wastes are generated from systems designed to minimize escape of any radioactivity to the environment.

INDUSTRIAL WASTE

The industrial LLRW category encompasses wastes generated by private research and development companies, manufacturers, non-destructive testing, mining, fuel fabrication facilities, and radiopharmaceutical manufacturers. Most wastes are generated by manufacturing concerns producing radioactive materials for use in nuclear fuel and non-fuel cycles. Manufacturing companies also produce waste from consumer goods such as smoke detectors and luminous devices. An estimated ninety-five percent of waste is generated by one percent of the approximately 4000 industrial generators.

GOVERNMENT WASTE

Waste generated by state and federal agencies falls into this category. Waste from private facilities working under contract to the government is very often excluded from this category. Government waste is the most diverse since it is generated by so many different organizations. LLRW produced in defense-related areas is handled by the DOE and is not included in this category.

MEDICAL WASTE

Medical generators include hospitals and clinics, research facilities, and private medical offices. More than 120 million medical procedures using radioactive materials are conducted annually in the United States (SNM 1988). Relatively large doses of isotopes, frequently powerful gamma emitters with short half-lives, are used in clinical procedures. Medical waste volumes were historically too large due to improper classifications. Rising disposal costs have improved proper classification.

TABLE 6.6.3 MAXIMUM CONCENTRATION LIMITS FOR LOW-LEVEL RADIOACTIVE WASTE FROM 10 CFR PART 61

Radionuclide	Half-life (years)	Maximum Concentration Limits $(Ci/m^3)^a$		
		Class A	Class B	Class C
Nuclides with half-lives <5 years[a]	<5.0	700.000	NL[b]	...
^{60}Co	5.3	700.000	NL	...
^{3}H	11.3	40.000	NL	...
^{90}Sr	28.0	0.040	150.0	7000.00
^{137}Cs	30.0	1.000	44.0	4600.00
^{63}Ni	92.0	3.500	70.0	700.00
^{63}Ni in activated metal	92.0	35.000	700.0	7000.00
^{14}C	5,730.0	0.800	...	8.00
^{14}C in activated metal	5,730.0	8.000	...	80.00
^{94}Nb in activated metal	20,000.0	0.020	...	0.20
^{59}Ni in activated metal	80,000.0	22.000	...	220.00
^{99}Tc	212,000.0	0.300	...	3.00
^{129}I	17,000,000.0	0.008	...	0.08
α-emitting transuranic nuclides with half-lives <5 years	<5.0	10.000 nCi/g	...	100.00 nCi/g
^{242}Cm	0.45	2,000.000 nCi/g	...	20,000.00 nCi/g
^{241}Pu	13.2	350.000 nCi/g	...	3,500.00 nCi/g

[a]Including, but not limited to: ^{32}P, ^{35}S, ^{51}Cr, ^{54}Mn, ^{55}Fe, ^{58}Co, ^{59}Fe, ^{65}Zn, ^{67}Ga, ^{125}I, ^{131}I, ^{134}Cs, ^{144}Ce, and ^{192}Ir.
[b]No upper limit on concentration.

ACADEMIC WASTE

Academic waste includes university hospitals and university medical and nonmedical research facilities. It tends to be low in activity and relatively high in volume, often due to improper classification of some materials as radioactive waste.

GREATER THAN CLASS C WASTE

Greater than Class C (GTCC) wastes contain concentrations of radionuclides greater than Class C limits established in 10 CFR Part 61. These wastes, as mentioned earlier, cannot be disposed of as LLRW but must go to a geological repository. GTCC waste comes primarily from decontamination and decommissioning of nuclear power plants. Nonutility generators include manufacturers of sealed sources used as measuring devices. GTCC waste volume is projected to expand during the next twenty years as more nuclear plants are decommissioned.

BELOW REGULATORY CONCERN WASTE

Below regulatory concern (BRC) wastes have radioactive content so low that unregulated release does not pose an unacceptable risk to public health or safety (Table 6.6.4). This class was established to make practical, timely determinations of when wastes need to go to a licensed LLRW site. The low-level radioactive waste policy amendments act of 1985 established procedures for acting expeditiously on petitions to exempt specific radioactive waste streams from NRC regulations (NRC 1986). Petitions already filed could dramatically reduce the total LLRW needing disposal.

TABLE 6.6.4   EXAMPLES OF MATERIALS EXEMPT FROM LICENSING REQUIREMENTS UNDER 10 CFR PART 31

Product[a]	Permissible Activity ($\leq$)
Static-elimination devices	500 μCi ^{210}Po
Ion-generating tubes	500 μCi ^{210}Po or 50 mCi ^{3}H
Luminous devices in aircraft	10 Ci ^{3}H or 300 mCi ^{147}Pm
Calibration sources	5 μCi ^{241}Am
Ice-detection devices	50 μCi ^{90}Sr
Prepackaged in vitro/clinical	10 μCi ^{125}I/test
testing kits	10 μCi ^{131}I/test
	10 μCi ^{14}C
	50 μCi ^{3}H
	20 μCi ^{59}Fe
	10 μCi ^{55}Fe

[a]The use of thorium in gas mantles, vacuum tubes, welding rods, incandescent lamps, photographic films, and finished optical lenses is also not regulated. Naturally occurring radioactive materials (NORM) present in geologic specimens, petroleum drilling wastes, and rare earth minerals processing wastes (with the exception of uranium and thorium) are also not regulated.

MIXED WASTE

Mixed low-level radioactive waste contains both radioactive and hazardous components and meets, respectively, NRC's definition of low-level radioactive waste in 10 CFR Part 61, and the Environmental Protection Agency's definition of hazardous material in 40 CFR Part 261. Although any type of low-level waste may be "mixed," surveys of waste generators indicate that less than five percent of the wastes to be sent to commercial sites would be classified as mixed (Bowerman, Davis & Siskind 1986). An example of a mixed waste would be a contaminated flammable extraction solvent used in radioisotope recovery. NRC deregulation of scintillation fluids containing minimal quantities of ^{3}H and ^{14}C has eliminated the largest source of mixed waste from disposal as LLRW.

Quantities of LLRW Generated

Each year, the DOE national low-level waste management program publishes data on both national and state-specific LLRW commercially disposed of in the United States (Fuchs & McDonald 1993). Data are categorized by disposal site, generator category, waste class, volume, and radionuclide activity. A distinction is made between LLRW shipped directly for disposal by generators, and waste handled by an intermediary. Wastes are subdivided into five categories:

- Academic
- Government
- Industrial
- Medical
- Utility

The volume of LLRW disposed of at commercial sites exceeded 3,500,000 ft^3 in 1980 (LLWMP 1982). The volume of LLRW disposed of at these sites since that time has steadily declined. In 1992, commercial LLRW disposal facilities received a total volume of 1,743,279 ft^3 of waste containing an activity of 1,000,102 curies. Waste distribution by disposal site is presented in Table 6.6.5. Tables 6.6.6 and 6.6.7 provide typical radionuclide and waste forms associated with commercial LLRW. Table

TABLE 6.6.5   DISTRIBUTION OF LOW-LEVEL RADIOACTIVE WASTE RECEIVED AT DISPOSAL SITES IN 1992

Site	Volume (ft^3)	Percent of Total	Activity (curies)	Percent of Total
Barnwell	830,512	48	815,974	82
Beatty	514,726	29	90,205	9
Richland	398,041	23	93,923	9
Total	1,743,279	100	1,000,102	100

TABLE 6.6.6 REPORTED LOW-LEVEL RADIOACTIVE WASTE RADIONUCLIDES RECEIVED AT DISPOSAL SITES IN 1992 FOR DIRECT AND NONDIRECT SHIPMENTS IN ORDER OF HIGHEST TO LOWEST ACTIVITY LEVELS

Nondirect	Reactors	Academic	Medical	Industrial	Government
H-3	Fe-55	Pm-147	Cs-137	H-3	Sr-90
Cs-137	Co-60	H-3	Sr-90	Co-60	Co-60
Fe-55	Ni-63	Co-60	Ni-63	Cs-137	Fe-55
Co-60	Mn-54	I-129	Co-57	Fe-55	U-238
S-35	Ag-110m	S-35	Ba-133	S-35	Mn-54
Ni-63	Cs-137	Cr-51	Ra-226	Ir-192	Ni-63
Co-58	Co-58	P-32	Rn-222	U-238	Co-58
C-14	Cr-51	C-14		Sr-90	Ra-226
Mn-54	Cs-134	I-131		Th-232	C-14
Kr-85	H-3	Ca-45		Th-228	H-3
Cs-134	Cd-109	Ni-63		Ce-144	Eu-152
Sr-90	Sb-125	Ra-226		P-32	U-235
P-32	Sr-90	Co-57		Ni-63	Ni-59
Cr-51	Ni-59	I-123		Ag-110m	Co-57
Zn-65	Nb-95	Re-186		Sb-125	Tc-99
I-125	Zr-95	Cu-67		Ra-228	Am-241
Am-241	Fe-59	Cu-64			Eu-154
Ni-59	C-14	K-40			I-125
Sb-125		Sr-85			Cs-137
Fe-59		Zn-65			
Ra-226		Ag-108			

TABLE 6.6.7 TYPICAL WASTE FORMS BY GENERATOR CATEGORIES

Academic
Compacted trash or solids
Institutional laboratory or biological waste
Absorbed liquids
Animal carcasses

Government
Compacted trash or solids
Contaminated plant hardware
Absorbed liquids

Industrial
Depleted uranium
Compacted trash or solids
Contaminated plant hardware
Absorbed liquids
Sealed sources

Medical
Compacted trash or solids
Institutional laboratory or biological waste
Absorbed liquids
Sealed sources

Utilities
Spent resins
Evaporator bottoms and concentrated waste
Filter sludge
Dry compressible waste
Irradiated components
Contaminated plant hardware

6.6.8 shows volume and activity according to generator category.

LLRW Commercial Disposal Sites

There were only two low-level radioactive waste disposal sites operating in the United States in 1994. Located in Barnwell, South Carolina and Richland, Washington, these facilities handle all low-level waste generated in the United States. Beginning in 1993, federal law allowed these states to refuse to accept any low-level waste generated outside their borders. The low-level radioactive waste policy act of 1980 made each of the 50 states responsible for dis-

TABLE 6.6.8 LOW-LEVEL RADIOACTIVE WASTES RECEIVED AT COMMERCIAL DISPOSAL SITES IN 1992

Generator Category	Volume (ft^3)	Activity (curies)
Academic	44,322.34	1,724.39
Government	158,186.17	40,780.08
Industrial	908,451.86	100,089.80
Medical	26,251.32	397.80
Utility	606,066.85	857,110.38
Total	1,743,278.54	1,000,102.45

(Fortom and Goode 1986). This petition proposes on-site incineration of solid biomedical waste containing a maximum of one curie of ^{3}H and one hundred millicuries of ^{14}C per year. The resulting ash would be disposed of as sanitary waste. This petition estimates a 90% reduction in institutional waste presently sent to LLRW disposal sites. It has not been enacted, in part due to concern for clean air requirements not related to radioactivity.

DEWATERING

Radioactive waste dewatering is an effective and efficient method for volume reduction. In addition, radioactive waste must not contain more than 0.5% freestanding water to be accepted at LLRW disposal sites. Centrifugation, filtration, and evaporation are standard techniques used to dewater wastes.

COMPACTION

Compaction is the primary volume-reduction method. Uncompacted waste has a typical density of approximately 130 kg/m^3 and can be increased three- to fourfold using a standard (20,000 psi) compactor. Super compactors can increase the density by a factor of ten. Shredding waste prior to compaction can also reduce the final volume. Compaction methods cannot be applied to hard and dense waste items for which volume reduction would be minimal. During compacting, potentially contaminated gases, liquids, and particulates are expelled from the waste and must be trapped by an off-gas (scrubber) treatment system.

INCINERATION

A large portion of LLRW is combustible and suitable for incineration. Used in combination with compacting, one-hundred fold volume reductions can be achieved. Radioactive waste incineration is an expensive and potentially troublesome treatment technique. Most European countries incinerate combustible radioactive waste prior to disposal. In the United States, incineration is reserved for cases where maximum volume reduction is required, and/or sophisticated off-gas treatment is not necessary. Clean air requirements make it increasingly difficult to build commercial incinerators.

Several waste characteristics are important in relation to incinerator performance. With very compact materials, combustion may be incomplete. Certain materials such as plastics (PVC) produce corrosive (HCl) gases that can damage the incinerator and must be scrubbed prior to release to the environment. The correct temperature must be maintained to ensure complete combustion. Since furnace temperature is controlled by the calorific value of the waste, the moisture content, and the combustion rate, it is clear that the feed rate is critical to successful incineration. The use of supplemental fuel to control combustion is dis-

couraged unless it is already contaminated with radioactive materials.

Liquid and Gaseous Effluent Treatment

LIQUID EFFLUENTS

LLRW is produced from the clean-up of drainings and cooling water at nuclear power plants, manufacturing sites, and R&D laboratories where radioactive materials are handled. These low-activity wastes are usually treated to remove most radionuclides, then discharged to the environment. Low-activity wastes can be collected and mixed for a more uniform effluent or segregated to utilize specific treatments for the individual components. If the first approach is utilized, the usual wastewater treatments of flocculation, precipitation, absorption, filtration, and ion exchange can be adapted to radioactive wastes (Table 6.6.11). Provisions must be made for water discharging and for drying, compacting, and disposing of the solids produced. Presently, solids are sent to a LLRW disposal site. Radium removal, covered in the section on mining and milling, is a good example of a specific treatment process.

If the total solids content of the contaminated water is low, if the volume is small, or if a final polishing of effluents is necessary, ion exchange may be a suitable treatment method. At nuclear power plants, ion exchange, filtration, evaporation, and reverse osmosis are the major processes used for contaminated water treatment (Figure 6.6.3).

GASEOUS EFFLUENTS

The primary source of radioactive gaseous effluents to the environment is from nuclear power plants. Coal-fired power plants also emit particulate radionuclides and are treated by conventional stack gas technology. Effluents from nuclear reactors include noble gas isotopes, radioiodines, tritium and some fission products (heavy water reactors). Typical treatment processes are shown schematically in Figure 6.6.3.

Conditioning Techniques

Proper LLRW disposal is closely regulated by the NRC and its agreement states. The application for a license to handle radioactive materials requires a sound disposal plan for any radioactive waste produced. The NRC permits LLRW disposal via six methods outlined in 10 CFR Part 20.

1. Transfer of waste to an authorized recipient.
2. Disposal by release into a sanitary sewerage system (meets limits in 10 CFR §20.303).

TABLE 6.6.11   TREATMENT PROCESSES FOR REMOVAL OF RADIOACTIVE WASTES

Process	Decontamination Factor[a]	
	Individual Radionuclides	Mixed Fission Products[b]
Conventional		
Coagulation and settling	0–100+	2–9.1
Clay addition, coagulation and settling	0–100	1.1–6.2
Sand filtration	1–100	
Coagulation, settling and filtration	1–50	1.4–13.3
Lime-soda ash softening	2–100	
Ion exchange, cation	1.1–500	2.0–6.1
Ion exchange, anion	0–125	
Ion exchange, mixed bed	11–3300	50–100
Solids-contact clarifier	1.9–15	2.0–6.1
Evaporation	1.00–10,000	
Nonconventional		
Phosphate	1.2–1000	125–250
Metallic dusts	1.1–1000	1.1–8.6
Clay treatment	0–100+	
Diatomaceous earth	1.1–∞	
Sedimentation	<1.05	
Activated sludge	1.03–8.2	4.8–9.8
Trickling filter	1.05–37	3.5–6.1
Sand filter	8.3–100	1.9–50
Oxidation ponds	<1.1–20	

[a]Decontamination factor $= \dfrac{\text{initial concentration}}{\text{final concentration}}$

[b]Where no data are listed it implies lack of information and not the unsuitability of the process.

3. Release to the environment, if material is below the maximum permissible concentration (MPC) in 10 CFR Part 20 appendix B, Table II.
4. Disposal by incineration according to 10 CFR §20.305, especially for waste oils and scintillation fluids.
5. Disposal of certain specific waste without regard to its radioactivity 10 CFR §20.306 (e.g., 0.05 mCi ^{3}H or ^{14}C).
6. Specific procedures approved as part of licensing to handle radioactive materials.

Radioactive waste is normally disposed of as a solid, except for liquids released to sanitary sewers or other water systems when radioactivity levels are below the maximum permissible concentration (MPC). In contrast to other types of waste, where pollutants can be eliminated by treatment, radioactivity can only be reduced by decay time. Thus the disposal methods used at NRC-authorized disposal sites are for solids and are based on the decay time required to make them non-radioactive. The correct preparation of radioactive waste is the first step to ensure the waste is disposed of economically and according to all applicable regulations.

Conditioning of radioactive wastes can include segregation, pretreatment, processing, and packaging. These techniques are covered in other sections of this chapter. Here, conditioning refers only to the various immobilization techniques used to prevent radioisotopes leaching into the environment. Immobilization is often used to help meet the NRC stability requirements for Class B and C waste and even for some forms of Class A and mixed wastes. The principle immobilization techniques are cementation, bituminization, polymerization, and vitrification. All of these techniques will increase the volume of radioactive waste.

CEMENTATION

Cement is used to solidify liquid waste. Cementation is relatively inexpensive but prone to leaching. The radioactive waste reacts with the cement and is bound to it. Waste compatibility must be verified, and special cement formulations are sometimes required to insure the product sets. This technique is sometimes used to dry a solid waste so that it contains less than 0.5% freestanding liquid.

BITUMINIZATION

The use of bitumen or asphalt is a classic immobilization technique. The process, carried out at the relatively high

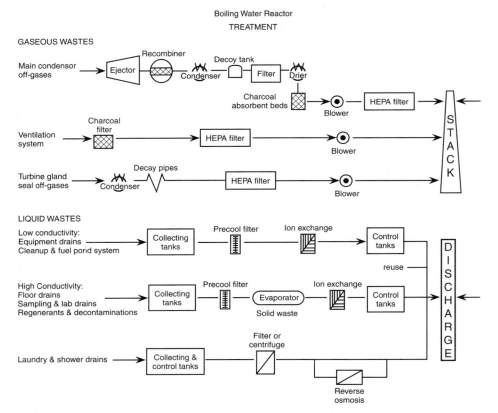

FIG. 6.6.3 Specific treatment and volume-reduction methods for nuclear plants. (Reprinted from International Atomic Energy Agency, 1986.)

temperature of $\geq 150°C$, is dangerous and requires specialized equipment. The product is less subject to normal leaching, but is susceptible to fire damage. The product also has a tendency to swell from the release of gases.

POLYMERIZATION

Polymerization of liquid and semi-liquid LLRW by in situ addition of monomers and initators is a relatively new technique. The process must be carefully adopted to the type of waste being immobilized. The product has shortcomings similar to bitumen waste.

VITRIFICATION

Vitrification in borosilicate waste is an expensive technique very rarely used in the immobilization of LLRW.

Disposal Techniques

Disposal of LLRW in the United States has been based on some form of land burial since ocean dumping was banned in the 1960s. The facilities must be on a site designed, operated, closed, and controlled after closure to meet all criteria in 10 CFR Part 61. Releases to the environment must

be as low as reasonably achievable (ALARA), and waste containment systems must be effective until the radioactivity has decayed to MPC levels.

SHALLOW LAND BURIAL

Shallow land burial (SLB) in trenches, often plastic lined, is the most economical disposal method. Prepackaged or preconditioned waste is carefully stacked into the trench, then covered with the excavated earth. Radioactivity can be successfully confined in the burial area if leaching of the waste by groundwater or rainwater can be reduced to negligible levels. Thus, careful geological, geochemical and hydrological studies must be made for burial site location.

DISPOSAL VAULTS

Below-ground vaults (BGV) and above ground vaults (AGV) are enclosed, engineered structures built to hold the most hazardous low-level radioactive wastes, such as Class C or greater than Class C (GTCC). The long-term effectiveness of this expensive solution has been questioned by proponents of SLB disposal (Gershey, Klein, Party & Wilkerson 1990).

EARTH-MOUNDED CONCRETE BUNKERS

Earth-mounded concrete bunkers (EMCB), a combination of trenches and vaults, are being strongly considered by many of the new state disposal sites mandated by Congress. EMCB disposal technology involves isolating low-level radioactive waste in an engineered vault located above or below the natural grade of the site. A multilayer, engineered earthen cover is positioned over the vault to provide an additional barrier. Depending on the design, Class A, B, or C wastes can be stored in these structures.

Other disposal methods for LLRW have been proposed, but at this time shallow land burial is the only successful and cost effective commercial method (Gershey, Klein, Party & Wilkerson 1990).

—Paul A. Bouis

References

Bowerman, B.S., R.E. Davis, and B. Siskind. 1986. *Document review regarding hazardous chemical characteristics of low-level waste.* NUREG, BNL. Upton, N.Y.

Burns, M.E. 1988. *Low-level radioactive waste regulations: science, politics and fear.* Chelsea, Mich.: Lewis Publishers.
Code of Federal Regulations. Title 10, Part 61.
Code of Federal Regulations. Title 40, Part 261.
Fortom, J.M., and D.J. Goode. 1986. *Deminimis waste impacts analysis methodology: impacts-BRC user's guide and methodology for radioactive wastes below regulatory concern.* NUREG/CR-3585, NRC, Washington, D.C.
Fuchs, R.L., and S.D. McDonald. 1993. *1992 state-by-state assessment of low-level radioactive waste received at commercial disposal sites.* DOE/LLN-181. Springfield, Va.: NTIS.
Gershey, E.L., R.C. Klein, E. Party, and A. Wilkerson. 1990. *Low-level radioactive waste: from cradle to grave.* New York, N.Y.: Van Nostrand Reinhold.
National Low-Level Radioactive Waste Management Program (LLWMP). 1982. *The 1980 state-by-state assessment of low-level radioactive waste received at commercial disposal sites.* DOE/LLWMP-11T. Springfield, Va.: NTIS.
Nuclear Regulatory Commission (NRC). 1986. Guideline for wastes below regulatory concern (BRC). *Federal Register 51,* 30839.
Society of Nuclear Medicine (SNM). 1988. *Nuclear medicine self-study program 1.* SNM, New York, N.Y.

6.7
HIGH-LEVEL RADIOACTIVE WASTE

High-level radioactive waste consists of spent fuel elements from nuclear reactors, waste produced from reprocessing, and waste generated from the manufacture of nuclear weapons. All these wastes are highly regulated and controlled due to the dangerously high levels of radiation and the security issues caused by their plutonium content. Strict licensing requirements for the storage of spent nuclear fuel and high-level radioactive waste are specified in 10 CFR Part 72.

Spent nuclear fuel has been withdrawn from a reactor, has undergone at least one year of decay since being used as an energy source in a power reactor, and has not undergone chemical reprocessing. Spent fuel is normally stored on-site at nuclear power plants in an independent spent fuel storage installation (ISFSI). An ISFSI is defined in 10 CFR Part 72 as a complex designed and constructed for the interim storage of spent nuclear fuel and other radioactive materials associated with spent fuel storage.

Spent fuel reprocessing was discontinued in the United States in 1972, except for the DOE, which continues to reprocess most of its spent fuel. France, Germany and several other major nuclear power producers also reprocess their spent fuel. Reprocessing improves the cost effectiveness of nuclear power by recycling recovered uranium and plutonium. The reprocessing of spent fuel, using the PUREX process developed in the United States, involves dissolution in large volumes of acid, liquid/liquid extraction, chemical reduction, and precipitation (Lanham & Runiou 1949, Flagg 1961, Koch 1979). The highly radioactive waste produced from reprocessing is classified by the NRC as a high-level radioactive waste or HLW in 10 CFR Part 72.

Spent fuel elements, HLW, and other highly radioactive wastes, such as transuranic wastes, require permanent containment. The disposal method must be designed to allow decay of the longest-lived radionuclides present in significant amounts in the waste. This means a time period of several hundreds of thousands of years.

Burial in engineered geological repositories is the only current option being seriously considered on a worldwide basis. Except for TRU waste, no site has been selected in the U.S., making it necessary for power plants and the DOE to continue storing waste on site. TRU waste generated by the DOE from various weapons programs is being disposed of at the waste isolation pilot plant (WIPP), a geological repository constructed in a bedded salt dome in New Mexico (Kohn 1987).

Many books and publications are available on the subject of HLW and the reader is referred to these for further details (Delange 1987, IAE 1981, Gertz 1989).

—*Paul A. Bouis*

References

Code of Federal Regulations. Title 10, Sec. 72.

Delange, M. 1987. LWR spent fuel reprocessing at La Hague: ten years on. Proc. Int. Conf. Nucl. Fuel Reprocc. *Waste Management.* Vol. 1, Societe Francaise d'Energie Nucleaire. Paris.

Flagg, J.F. 1961. *Chemical processing of reactor fuels.* London: Academic Press.

Gertz, C.P. 1989. Yucca Mountain, Nevada: is it a safe place for isolation of high-level radioactive waste? *Waste Management,* Vol. 1:9–11.

International Association of Energy. 1981. Underground disposal of radioactive wastes—basic guidance. *Safety Series No. 54.*

Koch, G. 1979. *Existing and projected reprocessing plants: a general review.* Atomkernenerg/Kerntech, Vol. 33:241.

Kohn, K. 1987. Kerntech, Vol. 51:157–160.

Lanham, W.B., and T.C. Runiou. 1949. *Purex process for plutonium and uranium recovery.* U.S. Atomic Energy Commission (USAEC) Report ORNZ-479.

Ullmann's encyclopedia of industrial chemistry. 1993. 5th ed. Vol. A22, pp. 499–591. Weinheim, Germany: VCH.

6.8
TRANSPORT OF RADIOACTIVE MATERIALS

Approximately 2,500,000 packages of radioactive materials are shipped per year in the United States. The vast majority of these shipments involves small or intermediate quantities of material in relatively small packages. The U.S. Department of Transportation (DOT) has regulatory responsibility for safety in the transportation of radioactive materials. The DOT updates transport regulations to keep pace with the changing transportation scene. The NRC has promulgated requirements, in 10 CFR Part 71, for licensees delivering radioactive materials for transport. The principle sources of federal regulations pertaining to transport of radioactive materials are listed in Table 6.8.1. An excellent review of DOT regulations is available from the U.S. Government Printing Office (DOT 1983).

Materials Subject to DOT Regulations

For transportation purposes, radioactive materials are defined as materials that emit ionizing radiation and have a specific activity greater than 0.002 mci/g are not regulated by the DOT or IAEA. The International Atomic Energy Agency (IAEA) has established international regulations

TABLE 6.8.1   SOURCES OF FEDERAL REGULATIONS

Title 49: U.S. Department of Transportation's Hazardous Materials Regulations, Parts 100–177 and 178–199
Main Headings
49 CFR 106—Rulemaking Procedures
49 CFR 107—Hazardous Materials Program Procedures
49 CFR 171—General Information, Regulations and Definitions
49 CFR 172—Hazardous Materials Tables and Hazardous Materials Communications Regulations
49 CFR 173—Shippers—General Requirements for Shipments and Packagings
49 CFR 174—Carriage by Rail
49 CFR 175—Carriage by Aircraft
49 CFR 176—Carriage by Vessel
49 CFR 177—Carriage by Public Highway
49 CFR 178—Shipping Container Specifications
49 CFR 179—Specifications for Tank Cars

Title 10: U.S. Nuclear Regulatory Commission
10 CFR 71—Packaging of Radioactive Materials for Transport and Transportation of Radioactive Materials Under Certain Conditions

Title 39: U.S. Postal Service
Domestic Mail Manual, U.S. Postal Service Regulations, Part 124. (Postal Regulations for Transport of Radioactive Matter are published in U.S. Postal Service Publication 6, and in the U.S. Postal Manual.)

and requirements (IAEA 1978). Materials not subject to DOT regulations may be subject to use or transfer regulations issued by the NRC or even the EPA.

REGULATIONS FOR SAFE TRANSPORT

A primary consideration in safe transportation of radioactive materials is the use of proper packaging for the specific radioactive material to be transported. In order to determine the packaging requirements, the following questions must be answered.

1. What radionuclides are being shipped? 49 CFR §173.435 contains a listing of over 250 specific radionuclides. Certain ground rules for dealing with unlisted or unknown radionuclides, or with mixtures of radionuclides, appear in 49 CFR §173.433.
2. What quantity of the radionuclides is being shipped? Packaging requirements are related to the activity of the material.
3. Is the radionuclide material *normal* or *special* form? Special form refers to materials that, if released from a package, would present a direct external radiation hazard, but not from contamination (Figure 6.8.1). Figure 6.8.2 details normal form materials that are, therefore, any radioactive materials that do not qualify as special form.

QUANTITY LIMITS AND PACKAGING

The quantity or specific activity of a radioactive material determines the packaging requirements. The regulations use A_1 and A_2 values as points of reference for quantity limitations for every radionuclide. Every radio-nuclide is assigned an A_1 and an A_2 value. These two values, in curies, are the maximum activity of that radio-nuclide that may be transported in a Type A package (Figure 6.8.3). Table 6.8.2 gives examples of A_1 and A_2 values for some typical radionuclides. Type B quantities (Figure 6.8.4) are

defined as exceeding the appropriate A_1 or A_2 value. Type B packages, highway route controlled quantities, and fissile radioactive materials are additionally controlled by the NRC regulations in 10 CFR Part 71.

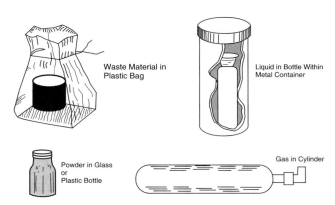

FIG. 6.8.2 Normal Forms of Radioactive Materials 49 CFR §173.403(s). Normal form materials may be solid, liquid or gaseous and include material that has not been qualified as special form. Type A Package Limits are A_2 Values.

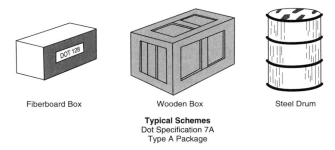

Typical Schemes
Dot Specification 7A
Type A Package

FIG. 6.8.3 Typical Type A Packaging. Package must withstand normal conditions (49 CFR §173.465) of transport, without loss or dispersal of radioactive contents.

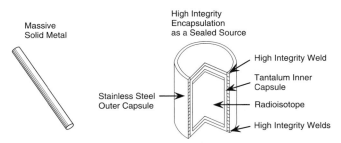

FIG. 6.8.1 *Special Form* R.A.M. (49 CFR §§173.403[z] and 173.469[a]). May present a direct radiation hazard if released from package, but presents little hazard due to contamination. Special form R.A.M. may be a *natural* characteristic, i.e., massive solid metal, or *acquired* through high integrity encapsulation.

TABLE 6.8.2   TYPE A PACKAGE QUANTITY LIMITS FOR SELECTED RADIONUCLIDES (ADDITIONAL RADIONUCLIDES ARE LISTED IN 49 CFR §173.435)

Symbol of Radionuclide	Element and Atomic Number	A_1 (Ci) (Special Form)	A_2 (Ci) (Normal Form)
^{14}C	Carbon (6)	1000	60
^{137}Cs	Cesium (55)	30	10
^{99}Mo	Molybdenum (42)	100	20
^{235}U	Uranium (92)	100	0.2
^{226}Ra	Radium (88)	10	0.05
^{201}Pb	Lead (82)	20	20

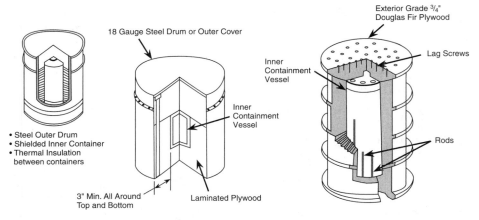

FIG. 6.8.4 Typical Type B Packagings. Package must stand both normal (49 CFR §173.465) and accident (10 CFR Part 71) test conditions without loss of contents.

TABLE 6.8.3   REMOVABLE EXTERNAL RADIOACTIVE CONTAMINATION: WIPE LIMITS

	Maximum Permissible Limits	
Contaminant	uCi/cm²	dpm/cm²
Beta/gamma-emitting radionuclides: all radionuclides with half-lives less than ten days; natural uranium; natural thorium; uranium-235; uranium-238; thorium-232; thorium-228 and thorium-230 when contained in ores or physical concentrates	10^{-5}	22
All other alpha-emitting radionuclides	10^{-6}	2.2

uCi/cm² = microcuries per square centimeter.
dpm/cm² = disintegrations per minute per square centimeter.

TABLE 6.8.4   RADIOACTIVE MATERIALS PACKAGES MAXIMUM RADIATION LEVEL LIMITATIONS (SEE SECTIONS 173.441(A) AND (B)

Radiation level (dose) rate at any point on external surface of any package of R.A.M. may not exceed:

A. 200 millirem per hr.
B. 10 millirem per hr at one meter (*transport index* may not exceed 10).

Unless the packages are transported in an *exclusive use* closed transport vehicle (aircraft prohibited), then the maximum radiation levels may be:

A. 1000 millirem per hr on the accessible external package surface.
B. 200 millirem per hr at external surface of the vehicle.
C. 10 millirem per hr at two meters from external surface of the vehicle.
D. 2 millirem per hr in any position of the vehicle which is occupied by a person.

EXTERNAL RADIATION AND CONTAMINATION LEVELS

Radiation levels may not exceed certain dose rates at any point from the package's external surface.

A. 200 millirems per hour at the surface
B. 10 millirems per hour at one meter from the surface.

If the package is transported in an "exclusive use" closed transport vehicle, the maximum radiation levels may be:

A. 1000 millirems per hr on the accessible surface of the package
B. 200 millirems per hr at the external surface of the transport vehicle
C. 10 millirems per hr at two meters from external surface of the vehicle
D. 2 millirems per hr in any position in the vehicle occupied by a person.

DOT regulations also prescribe limits for control of removable (non-fixed) radioactive contamination as shown in Table 6.8.3. Maximum levels for materials packages are covered in Table 6.8.4. A conversion chart (Table 6.8.5) and a list of NRC contacts are also provided for reference purposes.

—*Paul A. Bouis*

References

International Atomic Energy Agency (IAEA). 1978. Regulations for the safe transportation of radioactive materials. *Safety Series No. 6.* IAEA.
U.S. Department of Transportation (DOT). 1983. A review of the department of transportation regulations for transportation of radioactive materials. Washington, D.C.: U.S. Government Printing Office.

TABLE 6.8.5 CONVERSION FACTORS FOR IONIZING RADIATION

Quantity	Symbol for Quantity	Expression in SI Units	Expression in Symbols for SI Units	Special Name for SI Units	Symbols Using Special Names	Conventional Units	Symbol for Conventional Unit	Value of Conventional Unit in SI Units
Conversion Between SI and Other Units								
Activity	A	1 per second	s^{-1}	becquerel	Bq	curie	Ci	3.7×10^{10} Bq
Absorbed dose	D	joule per kilogram	$J\,kg^{-1}$	gray	Gy	rad	rad	0.01 Gy
Absorbed dose rate	$\dot{D}$	joule per kilogram second	$J\,kg^{-1}\,s^{-1}$		$Gy\,s^{-1}$	rad	$rad\,s^{-1}$	$0.01\,Gy\,s^{-1}$
Average energy per ion pair	W	joule	J			electronvolt	eV	1.602×10^{-19} J
Dose equivalent	H	joule per kilogram	$J\,kg^{-1}$	sievert	Sv	rem	rem	0.01 Sv
Dose equivalent rate	$\dot{H}$	joule per kilogram second	$J\,kg^{-1}\,s^{-1}$		$Sv\,s^{-1}$	rem per second	$rem\,s^{-1}$	$0.01\,Sv\,s^{-1}$
Electric current	I	ampere	A			ampere	A	1.0 A
Electric potential difference	U, V	watts per ampere	$W\,a^{-1}$	volt	V	volt	V	1.0 A
Exposure	X	coulomb per kilogram	$C\,kg^{-1}$			roentgen	R	2.58×10^{-4} C kg^{-1}
Exposure rate	$\dot{X}$	coulomb per kilogram second	$C\,kg^{-1}\,s^{-1}$			roentgen	$R\,s^{-1}$	2.58×10^{-4} C $kg^{-1}\,s^{-1}$
Fluence	ϕ	1 per meter squared	m^{-2}			1 per centimeter squared	cm^{-2}	$1.0 \times 10^{4}\,n^{-2}$
Fluence rate	Φ	1 per meter squared second	$m^{-2}\,s^{-1}$			1 per centimeter squared second	$cm^{-2}\,s^{-1}$	$1.0 \times 10^{4}\,m^{-2}\,s^{-1}$
Kerma	K	joule per kilogram	$J\,kg^{-1}$	gray	Gy	rad	rad	0.01 Gy
Kerma rate	$\dot{K}$	joule per kilogram second	$J\,kg^{-1}\,s^{-1}$		$Gy\,s^{-1}$	rad per second	$rad\,s^{-1}$	$0.01\,Gy\,s^{-1}$
Lineal energy	y	joule per meter	$j\,m^{-1}$			kiloelectron volt per micrometer	$keV\,\mu m^{-1}$	1.602×10^{-10} J m^{-1}
Linear energy transfer	L	joule per meter	$j\,m^{-1}$			kiloelectron volt per micrometer	$keV\,\mu m^{-1}$	1.602×10^{-10} J m^{-1}
Mass attenuation coefficient	μ/p	meter squared per kilogram	$m^2\,kg^{-1}$			centimeter squared per gram	$cm^2\,g^{-1}$	0.1 $m^2\,kg^{-1}$

Continued on next page)

TABLE 6.8.5 (*Continued*)

Quantity	Symbol for Quantity	Expression in SI Units	Special Name for SI Units	Symbols Using Special Names	Expression in Symbols for SI Units	Conventional Units	Symbol for Conventional Unit	Value of Conventional Unit in SI Units
Mass energy transfer coefficient	μ_e/ρ	meter squared per kilogram			$m^2\ kg^{-1}$	centimeter squared per gram	$cm^2\ g^{-1}$	$0.1\ m^2\ kg^{-1}$
Mass energy absorption coefficient	μ_{en}/ρ	meter squared per kilogram			$m^2\ kg^{-1}$	centimeter squared per gram	$cm^2\ g^{-1}$	$0.1\ m^2\ kg^{-1}$
Mass stopping power	S/ρ	joule meter squared per kilogram			$J\ m^2\ kg^{-1}$	MeV centimeter squared per gram	$MeV\ cm^2\ g^{-1}$	$1.602 \times 10^{-14}\ J\ m^2\ kg^{-1}$
Power	P	joule per second	watt	W	$J\ s^{-1}$	watt	W	$1.0\ W$
Pressure	P	newton per meter squared	pascal	Pa	$N\ m^{-2}$	torr	torr	$(101325/760)\ Pa$

Conversion Between SI and Other Units

Quantity	Symbol for Quantity	Expression in SI Units	Special Name for SI Units	Symbols Using Special Names	Expression in Symbols for SI Units	Conventional Units	Symbol for Conventional Unit	Value of Conventional Unit in SI Units
Radiation chemical yield	G	mole per joule			$mol\ J^{-1}$	molecules per 100 electron volts	molecules $(100\ eV)^{-1}$	$1.04 \times 10^{-7}\ mole\ J^{-1}$
Specific energy	z	joule per kilogram	gray	Gy	$J\ kg^{-1}$	rad	rad	$0.01\ Gy$

Converting SI Units/Non-SI Units

To Convert: From	To	Multiply By
becquerel (Bq)	curie	2.7×10^{-11}
curie (Ci)	becquerel	3.7×10^{10}
gray (Gy)	rad	100
rad (rad)	gray	0.01
sievert (Sv)	rem	100
rem (rem)	sievert	0.010

Taken from the National Council on Radiation Protection and Measurements Report No. 82. "SI Units in Radiation Protection and Measurements". Reproduced by permission of the copyright owner. Information regarding data in these tables is presented in the publication "NCRP Report No. 82" and is available from NCRP, 7910, Woodmont Avenue, Suite 1016, Bethesda, Maryland 20814.

NRC Contacts for Further Information

Division of Low-Level Waste Management and Decommissioning
NMSS
U.S. Nuclear Regulatory Commission
1555 Rockville Pike
Rockville, MD 20852
(301) 415-7000

Public Affairs
U.S. Nuclear Regulatory Commission
1555 Rockville Pike
Rockville, MD 20852
(301) 415-7715

State Liaison Officer
Region I
475 Allendale Road
King of Prussia, PA 19406
(610) 337-5246

State & Government Affairs Staff
Director
Region II
101 Marietta Street NW, Suite 2900
Atlanta, GA 30323
(404) 331-5597

State and Government Affairs
Director
Region III
801 Warnerville Road
LaSalle, IL 60532-4351
(630) 829-9500

State Liaison Officer
Region IV
Parkway Central Plaza Building
611 Ryan Plaza Drive, Suite 400
Arlington, TX 76011-8064
(817) 860-8100

State Liaison Officer
Region V
1450 Maria Lane, Suite 300
Walnut Creek, CA 94596-5368
(510) 975-0200

Bibliography

American Society for Testing and Materials (ASTM). 1983. *A standard guide for examining the incompatibility of selected hazardous waste based on binary chemical mixtures.* Philadelphia, Pa.

Berlin, R.E., and C.C. Stanton. 1989. *Radioactive waste management.* New York, N.Y.: Wiley.

Blackman, W.C., Jr. 1992. *Basic hazardous waste management.* Boca Raton, Fla.: Lewis Publishers.

Chapman, N.A., and I.G. McKinley. 1987. *The geological disposal of nuclear waste.* Chichester, Great Britain: John Wiley & Sons.

Chenoweth, D. 1995. DOT penalties for shipping container violations. *Chemical Processing* (April).

Cheremisinoff, P. 1990. Biological treatment of hazardous wastes, sludges, and wastewater. *Pollution Engineering* (May).

Cheremisinoff, P. 1992. *A guide to underground storage tanks: evolution, site assessment, and remediation.* Englewood Cliffs, N.J.: Prentice-Hall, Inc.

David, M.L. and D.A. Cornwell. 1991. *Introduction to environmental engineering.* New York, N.Y.: McGraw-Hill, Inc.

Gershey, E.L., R.C. Klein, E. Party, and A. Wilkerson. 1990. *Low-level radioactive waste: from cradle to grave.* New York, N.Y.: Van Nostrand Reinhold.

Gollnick, D.A. 1988. *Basic radiation protection technology.* 2nd ed. Pacific Radiation Corp. Altadena, Calif.

Greenberg, A.E., L.S. Clesceri, and A.D. Eaton. 1992. *Standard methods for the examination of water and wastewater.* 18th ed. APHA. Washington, D.C.

Harris, M. 1988. Inhouse solvent reclamation efforts in Air Force maintenance operations, of *Hazardous waste minimization within the Department of Defense,* edited by J.A. Kaminsky. Office of the Deputy Assistant Secretary of Defense (Environment) Washington, D.C.

International Atomic Energy Agency (IAEA). 1986. Assessment of the radiological impact of the transport of radioactive materials. *Technical Document 398.* Vienna, Austria.

International Atomic Energy Agency (IAEA). 1987. Safe management of wastes from the mining and milling of uranium and thorium ores. *Safety Series No. 85.* Vienna, Austria.

International Atomic Energy Agency (IAEA). 1988. Immobilization of low intermediate level radioactive wastes and polymers. *Technical Report 289.* Vienna, Austria.

International Atomic Energy Agency (IAEA). 1989. *Nuclear power and fuel cycle: status and trends.* Vienna, Austria.

Irvine, R.L. and L.H. Ketchum, Jr. 1988. Sequencing batch reactors for biological waste water treatment. *Critical Reviews in Environmental Control,* 18(4):255–294.

Leiter, J.L., ed. 1989. *Underground storage tank guide.* Salisbury, Md.: Thomas Publishing Group.

Maillet, J. and C. Sombret. 1988. High-level waste vitrification: the state-of-the-art in France. *Waste Management* 88(2):165–172.

Mattus, A.J., R.D. Doyle, and D.P. Swindlehurst. 1988. Asphalt solidification of mixed wastes. *Waste Management* 89(1):229–234.

Murray, R.L. 1989. *Understanding radioactive waste.* 3rd ed. Columbus, Oh.: Battelle Press.

National Council on Radiation Protection and Measurements (NCRP). 1987. Ionizing radiation exposure of the population of the United States. *NCRP Report No. 93.* Bethesda, Md.

National Low-Level Waste Management Program. 1993. *The 1992 state-by-state assessment of low-level radioactive wastes received at commercial disposal sites.* DOE/LLW-181. Washington, D.C.

Party, E.P., and E.L. Gershey. 1989. Recommendations for radioactive waste reduction in biomedical/academic institutions. *Health Physics* 56(4):571–572.

Snelgrove, W.L. and B.O. Paul. 1995. The chemical industry and tank car development. Chemical Processing (April).

Theodore, L. and Y.C. McGuinn. 1992. *Pollution Prevention.* New York, N.Y.: Van Nostrand Reinhold.

Ullmann's encyclopedia of industrial chemistry, 5th ed. 1993. Vol. A22:499–591. Weinheim, Germany: VCH.

United Nations Scientific Committee on the Effects of Atomic Radiation (UNSCEAR). 1977. *Sources and effects of ionizing radiation.* New York, N.Y.

U.S. Department of Transportation (DOT). 1983. *A review of the*

Department of Transportation regulations for transportation of radioactive materials. Washington, D.C.

U.S. Environmental Protection Agency (EPA). 1985. Minimum technology guidance on double liner systems for landfills and surface impoundments, design, construction, and operation. *Report No. PB87-151072.* National Technical Information Service. Springfield, Va.

———. 1986. *RCRA orientation manual.* Office of Solid Waste. Washington, D.C.

———. 1987. Underground storage tank corrective action technologies. EPA 625–6–87–015. Washington, D.C.

———. 1987. Handbook—Groundwater. EPA 625–6–87–016. Office of Research and Development. Cincinnati, Oh.

———. 1987. *A compendium of technologies used in the treatment of hazardous wastes.* EPA 625–8–87–014 (September).

———. 1988. Musts for USTs. *Report No. 530–UST–88–008.* Office of Underground Storage. Washington, D.C. (September).

———. 1989. *Hazardous waste incineration measurement guidance manual.* EPA 625–6–89–021 (June).

———. 1990. *Guidance on PIC controls for hazardous waste incinerators.* EPA 625–530–SW–90–040 (April).

———. 1990. *Engineering Bulletin: Soil washing treatment.* Office of Research and Development. Cincinnati, Oh. (September).

———. 1990. *RCRA orientation manual.* 1990 edition. Superintendent of Documents. Washington, D.C.: U.S. Government Printing Office.

———. 1992. *A citizen's guide to soil washing.* EPA 524–F–92–003. Office of Solid Waste and Emergency Response. Washington, D.C. (March).

Wagner, H.N., and L.E. Ketchem. 1989. *Living with radiation.* Baltimore, Md.: The Johns Hopkins University Press.

Wentz, C. 1989. *Hazardous waste management.* McGraw-Hill, Inc.

Solid Waste

R.C. Bailie | J.W. Everett | Béla G. Lipták | David H.F. Liu |
F. Mack Rugg | Michael S. Switzenbaum

7
Source and Effect

7.1
DEFINITION

For practical purposes, the term *waste* includes any material that enters the waste management system. In this chapter, the term *waste management system* includes organized programs and central facilities established not only for final disposal of waste but also for recycling, reuse, composting, and incineration. Materials enter a waste management system when no one who has the opportunity to retain them wishes to do so.

Generally, the term *solid waste* refers to all waste materials except hazardous waste, liquid waste, and atmospheric emissions. *CII waste* refers to wastes generated by commercial, industrial, and institutional sources. Although most solid waste regulations include hazardous waste within their definition of solid waste, *solid waste* has come to mean *nonhazardous solid waste* and generally excludes hazardous waste.

This section describes the types of waste that are detailed in this chapter.

Waste Types Included

This chapter focuses on two major types of solid waste: municipal solid waste (MSW) and bulky waste. MSW comprises small and moderately sized solid waste items from homes, businesses, and institutions. For the most part, this waste is picked up by general collection trucks, typically compactor trucks, on regular routes.

Bulky waste consists of larger items of solid waste, such as mattresses and appliances, as well as smaller items generated in large quantity in a short time, such as roofing shingles. In general, regular trash collection crews do not pick up bulky waste because of its size or weight.

Bulky waste is frequently referred to as C&D (construction and demolition) waste. The majority of bulky waste generated in a given area is likely to be C&D waste. In areas where regular trash collection crews take anything put out, the majority of bulky waste arriving separately at disposal facilities is C&D waste. In areas where the regular collection crews are less accommodating, however, substantial quantities of other types of bulky waste, such as furniture and appliances, arrive at disposal facilities in separate loads.

Waste Types Not Included

In a broad sense, the majority of nonhazardous solid waste consists of industrial processing wastes such as mine and mill tailings, agricultural and food processing waste, coal ash, cement kiln dust, and sludges. The waste management technologies described in this chapter can be used to manage these wastes; however, this chapter focuses on the management of MSW and the more common types of bulky waste in most local solid waste streams.

—*F. Mack Rugg*

7.2
SOURCES, QUANTITIES, AND EFFECTS

This section identifies the sources of solid waste, provides general information on the quantities of solid waste generated and disposed of in the United States, and identifies the potential effects of solid waste on daily life and the environment.

Sources

The primary source of solid waste is the production of commodities and byproducts from solid materials. Everything that is produced is eventually discarded. A secondary source of solid waste is the natural cycle of plant growth and decay, which is responsible for the portion of the waste stream referred to as yard waste or vegetative waste.

The amount a product contributes to the waste stream is proportional to two principal factors: the number of items produced and the size of each item. The number of items produced, in turn, is proportional to the useful life of the product and the number of items in use at any one time. Newspapers are the largest contributor to MSW because they are larger than most other items in MSW, they are used in large numbers, and they have a useful life of only one day. In contrast, pocket knives make up a negligible portion of MSW because relatively few people use them, they are small, and they are typically used for years before being discarded.

MSW is characterized by products that are relatively small, are produced in large numbers, and have short useful lives. Bulky waste is dominated by products that are large but are produced in relatively small numbers and have relatively long useful lives. Therefore, a given mass of MSW represents more discreet acts of discard than the same mass of bulky waste. For this reason, more data are required to characterize bulky waste to within a given level of statistical confidence than are required to characterize MSW.

Most MSW is generated by the routine activities of everyday life rather than by special or unusual activities or events. On the other hand, activities that deviate from routine, such as trying different food or a new recreational activity, generate waste at a higher rate than routine activities. Routinely purchased items tend to be used fully, while unusual items tend to be discarded without use or after only partial use.

In contrast to MSW, most bulky waste is generated by relatively infrequent events, such as the discard of a sofa or refrigerator, the replacement of a roof, the demolition of a building, or the resurfacing of a road. Therefore, the composition of bulky waste is more variable than the composition of MSW.

In terms of generation sites, the principal sources of MSW are homes, businesses, and institutions. Bulky waste is also generated at functioning homes, businesses, and institutions; but the majority of bulky waste is generated at construction and demolition sites. At each type of generation site, MSW and bulky waste are generated under four basic circumstances:

Packaging is removed or emptied and then discarded. This waste typically accounts for approximately 35 to 40% of MSW prior to recycling. Packaging is generally less abundant in bulky waste.

The unused portion of a product is discarded. In MSW, this waste accounts for all food waste, a substantial portion of wood waste, and smaller portions of other waste categories. In bulky waste, this waste accounts for the majority of construction waste (scraps of lumber, gypsum board, roofing materials, masonry, and other construction materials).

A product is discarded, or a structure demolished, after use. This waste typically accounts for 30 to 35% of MSW and the majority of bulky waste.

Unwanted plant material is discarded. This waste is the most variable source of MSW and is also a highly variable source of bulky waste. Yard wastes such as leaves, grass clippings, and shrub and garden trimmings commonly account for as little as 5% or as much as 20% of the MSW generated in a county-sized area on an annual basis. Plant material can be a large component of bulky waste where trees or woody shrubs are abundant, particularly when lots are cleared for new construction.

Packaging tends to be concentrated in MSW because many packages destined for discard as MSW contain products of which the majority is discarded in wastewater or enters the atmosphere as gas instead of being discarded as MSW. Such products include food and beverages, cleaning products, hair- and skin-care products, and paints and other finishes.

Quantities

The most important parameter in solid waste management is the quantity to be managed. The quantity determines the size and number of the facilities and equipment required to manage the waste. Also important, the fee collected for each unit quantity of waste delivered to the facility (the tipping fee) is based on the projected cost of operating a facility divided by the quantity of waste the facility receives.

The quantity of solid waste can be expressed in units of volume (typically cubic yards or cubic meters) or in units of weight (typically short, long, or metric tons). In this chapter, the word ton refers to a short ton (2000 lb). Although information about both volume and weight are important, using weight as the master parameter is generally preferable in record keeping and calculations.

The advantage of measuring quantity in terms of weight rather than volume is that weight is fairly constant for a given set of discarded objects, whereas volume is highly variable. Waste set out on the curb on a given day in a given neighborhood occupies different volumes on the curb, in the collection truck, on the tipping floor of a transfer station or composting facility, in the storage pit of a combustion facility, or in a landfill. In addition, the same waste can occupy different volumes in different trucks or landfills. Similarly, two identical demolished houses occupy different volumes if one is repeatedly run over with a bulldozer and the other is not. As these examples illustrate, the phrases "a cubic yard of MSW" and "a cubic yard of bulky waste" have little meaning by themselves; the phrases "a ton of MSW" and "a ton of bulky waste" are more meaningful.

Franklin Associates, Ltd., regularly estimates the quantity of MSW generated and disposed of in the United States under contract to the U.S. Environmental Protection Agency (EPA). Franklin Associates derives its estimates from industrial production data using the *material flows methodology*, based on the general assumption that what is produced is eventually discarded (see "Estimation of Waste Quantity" in Section 8.2). Franklin Associates estimates that 195.7 million tons of MSW were generated in the United States in 1990. Of this total, an estimated 33.4 million tons (17.1%) were recovered through recycling and composting, leaving 162.3 million tons for disposal (Franklin Associates, Ltd. 1992).

The quantity of solid waste is often expressed in pounds per capita per day (pcd) so that waste streams in different areas can be compared. This quantity is typically calculated with the following equation:

$$pcd = 2000T/365P \qquad 7.2(1)$$

where:

pcd = pounds per capita per day
T   = number of tons of waste generated in a year
P   = population of the area in which the waste is generated

Unless otherwise specified, the tonnage T includes both residential and commercial waste. With modification the equation can also calculate pounds per employee per day, residential waste per person per day, and so on.

Franklin Associates's (1992) estimate of MSW generated in the United States in 1990, previously noted, equates to 4.29 lb per person per day. This estimate is probably low for the following reasons:

Waste material is not included if Franklin Associates cannot document the original production of the material.
Franklin's material flows methodology generally does not account for moisture absorbed by materials after they are manufactured (see "Combustion Characteristics" in Section 8.1).

Table 7.2.1 shows waste quantities reported for various counties and cities in the United States. All quantities are given in pcd. Reports from the locations listed in the table indicate an average generation rate for MSW of 5.4 pcd, approximately 25% higher than the Franklin Associates estimate. Roughly 60% of this waste is generated in residences (residential waste) while the remaining 40% is generated in commercial, industrial, and institutional establishments (CII waste). The percentage of CII waste is usually lower in suburban areas without a major urban center and higher in urban regional centers.

Table 7.2.1 also shows generation rates for solid waste other than MSW. The quantity of other waste, most of which is bulky waste, is roughly half the quantity of MSW. The proportion of bulky and other waste varies, however, and is heavily influenced by the degree to which recycled bulky materials are counted as waste. The quantities of bulky waste shown for Atlantic and Cape May counties, New Jersey, include large amounts of recycled concrete, asphalt, and scrap metal. See also "Component Composition of Bulky Waste" in Section 8.1.

Franklin Associates (1992) projects that the total quantity of MSW generated in the United States will increase by 13.5% between 1990 and 2000 while the population will increase by only 7.3%. On a per capita basis, therefore, MSW generation is projected to grow 0.56% per year. No comparable projections have been developed for bulky waste. Table 7.2.2 shows the potential effect of this growth rate on MSW generation rates and quantities.

Effects

MSW has the following potential negative effects:

- Promotion of microorganisms that cause diseases
- Attraction and support of disease vectors (rodents and insects that carry and transmit disease-causing microorganisms)
- Generation of noxious odors
- Degradation of the esthetic quality of the environment
- Occupation of space that could be used for other purposes
- General pollution of the environment

TABLE 7.2.1   SOLID WASTE GENERATION RATES IN THE UNITED STATES

Location	Year	Residential Fraction of MSW (%)	Commercial/Industrial Fraction of MSW (%)	Total MSW (pcd)	Bulky Waste (pcd)	Other Solid Waste (pcd)[a]	Total Solid Waste (pcd)
Atlantic County, NJ	1991	—	—	6.0	5.9	0.3	12.2
Bexar County, TX	1990	—	—	—	—	—	6.5
Cape May County, NJ	1990	—	—	6.6	6.0	0.6	13.2
Delaware (state)	1990	—	—	—	—	—	7.1
Fairfax County, VA	1991	55	45	4.8	1.3	0.0	6.1
Marion County, FL	1989	—	—	5.4	—	—	—
Middlesex County, NJ	1988	—	—	4.4	2.1	1.6	8.2
Minnesota Metro Area	1991	—	—	6.5	2.6	0.0	9.1
Monmouth County, NJ	1987	75	25	4.8	2.7	0.0	7.5
Monroe County, NY	1990	—	—	5.7	—	—	—
Rhode Island (state)	1985	52	48	4.9	—	—	—
San Diego, CA	1985	—	—	—	—	—	8.0
Sarasota County, FL	1989	—	—	—	—	—	9.2
Seattle, WA	1987	37	63	7.6	—	—	—
Somerset County, NJ	1989	—	—	4.2	1.5	0.6	6.3
Warren County, NJ	1989	—	—	3.2	0.4	0.9	4.5
Wichita, KA	1990	61	39	6.6	1.1	0.0	7.7
	Average[b]	56	44	5.4	2.6	0.5	8.1
	Minimum	37	25	3.2	0.4	0.0	4.5
	Maximum	75	63	7.6	6.0	1.6	13.2
USA (Franklin Associates)	1990	62	38	4.3	—	—	—

Sources: Data from references listed at the end of this section.

Note: pcd = pounds per capita per day

[a]Most waste in this category falls within the definition of either MSW or bulky waste. Specific characteristics vary from place to place.

[b]Because different information is available from different locations, the overall average is not the sum of the averages for the individual waste types.

Bulky waste also has the potential to degrade esthetic values, occupy valuable space, and pollute the environment. In addition, bulky waste may pose a fire hazard.

MSW is a potential source of the following useful materials:

- Raw materials to produce manufactured goods
- Feed stock for composting and mulching processes
- Fuel

Bulky waste has the same potential uses except for composting feed stock.

The fundamental challenge of solid waste management is to minimize the potential negative effects while maximizing the recovery of useful materials from the waste at a reasonable cost.

Conformance with simple, standard procedures for the storage and handling of MSW largely prevents the promotion of disease-causing microorganisms and the attrac-

TABLE 7.2.2   PROJECTED GENERATION OF MSW IN THE UNITED STATES IN THE YEAR 2000

Year	Population (in millions)	MSW Quantity Projected by Franklin Associates (millions of tons)	Per Capita Generation Based on Franklin Associates (lb/day)	Average Annual Growth of Per Capita Generation Represented (%)	Per Capita Generation Based on Average in Table 7.2.1 (lb/day)	MSW Quantity Based on Average in Table 7.2.1 (millions of tons)
1990	249.9	195.7	4.3	—	5.4	247.6
2000	268.3	222.1	4.5	0.56	5.7	281.0

Source: Data from Franklin Associates, Ltd., 1992, *Characterization of municipal solid waste in the United States: 1992 Update* (EPA/530-R-92-019, NTIS PB92-207-166, U.S. EPA).

Note: Derived from Table 7.2.1.

tion and support of disease vectors. Preventing the remaining potential negative effects of solid waste remains a substantial challenge.

Solid waste can degrade the esthetic quality of the environment in two fundamental ways. First, waste materials that are not properly isolated from the environment (e.g., street litter and debris on a vacant lot) are generally unsightly. Second, solid waste management facilities are often considered unattractive, especially when they stand out from surrounding physical features. This characteristic is particularly true of landfills on flat terrain and combustion facilities in nonindustrial areas.

Solid waste landfills occupy substantial quantities of space. Waste reduction, recycling, composting, and combustion all reduce the volume of landfill space required (see Sections 9.1 to 10.6).

Land on which solid waste has been deposited is difficult to use for other purposes. Landfills that receive unprocessed MSW typically remain spongy and continue to settle for decades. Such landfills generate methane, a combustible gas, and other gases for twenty years or more after they cease receiving waste. Whether the waste in a landfill is processed or unprocessed, the landfill generally cannot be reforested. Tree roots damage the impermeable cap applied to a closed landfill to reduce the production of leachate.

Solid waste generates odors as microorganisms metabolize organic matter in the waste, causing the organic matter to decompose. The most acute odor problems generally occur when waste decomposes rapidly, consuming available oxygen and inducing anaerobic (oxygen deficient) conditions. Bulky waste generally does not cause odor problems because it typically contains little material that decomposes rapidly. MSW, on the other hand, typically causes objectionable odors even when covered with dirt in a landfill (see Section 10.5).

Combustion facilities prevent odor problems by incinerating the odorous compounds and the microorganisms and organic matter from which the odorous compounds are derived (see Section 10.1). Composting preserves organic matter while reducing its potential to generate odors. However, the composting process requires careful engineering to minimize odor generation during composting (see Section 10.6).

In addition to odors, solid waste can cause other forms of pollution. Landfill leachate contains toxic substances that must be prevented from contaminating groundwater and surface water (see Section 10.5). Toxic and corrosive products of solid waste combustion must be prevented from entering the atmosphere (see Section 10.1). The use

of solid waste compost must be regulated so that the soil is not contaminated (see Section 10.6).

While avoiding the potential negative effects of solid waste, a solid waste management program should also seek to derive benefits from the waste. Methods for deriving benefits from solid waste include recycling (Section 9.2), composting (Section 10.6), direct combustion with energy recovery (Section 10.1), processing waste to produce fuel (Sections 9.3 and 10.4), and recovery of landfill gas for use as a fuel (Section 10.5).

— *F. Mack Rugg*

References

Cal Recovery Systems, Inc. 1990. *Waste characterization for San Antonio, Texas.* Richmond, Calif. (June).
Camp Dresser & McKee Inc. 1990a. *Marion County (FL) solid waste composition and recycling program evaluation.* Tampa, Fla. (April).
———. 1990b. *Sarasota County waste stream composition study.* Draft report (March).
———. 1991a. *Cape May County multi-seasonal solid waste composition study.* Edison, N.J. (August).
———. 1991b. *City of Wichita waste stream analysis.* Wichita, Kans. (August).
———. 1992. *Atlantic County (NJ) solid waste characterization program.* Edison, N.J. (May).
Cosulich, William F., Associates, P.C. 1988. *Solid waste management plan, County of Monroe, New York: Solid waste quantification and characterization.* Woodbury, N.Y. (July).
Delaware Solid Waste Authority. 1992. *Solid waste management plan.* (17 December).
Franklin Associates, Ltd. 1992. *Characterization of municipal solid waste in the United States: 1992 update.* U.S. EPA, EPA/530-R-92-019, NTIS no. PB92-207 166 (July).
HDR Engineering, Inc. 1989. *Report on solid waste quantities, composition and characteristics for Monmouth County (NJ) waste recovery system.* White Plains, N.Y. (March).
Killam Associates. 1989; 1991 update. *Middlesex County (NJ) solid waste weighing, source, and composition study.* Millburn, N.J. (February).
———. 1990. *Somerset County (NJ) solid waste generation and composition study.* Millburn, N.J. (May). Includes data for Warren County, N.J.
Minnesota Pollution Control Agency and Metropolitan Council. 1993. *Minnesota solid waste composition study, 1991–1992 part II.* Saint Paul, Minn. (April).
Rhode Island Solid Waste Management Corporation. 1987. *Statewide resource recovery system development plan.* Providence, R.I. (June).
San Diego, City of, Waste Management Department. 1988. *Request for proposal: Comprehensive solid waste management system.* (4 November).
SCS Engineers. 1991. *Waste characterization study—solid waste management plan, Fairfax County, Virginia.* Reston, Va. (October).
Seattle Engineering Department, Solid Waste Utility. 1988. *Waste reduction, recycling and disposal alternatives: Volume II—Recycling potential assessment and waste stream forecast.* Seattle (May).

8
Characterization

8.1
PHYSICAL AND CHEMICAL CHARACTERISTICS

This section addresses the characteristics of solid waste including fluctuations in quantity; composition, density, and other physical characteristics; combustion characteristics; bioavailability; and the presence of toxic substances.

Fluctuations in Solid Waste Quantities

Weakness in the economy generally reduces the quantity of solid waste generated. This reduction is particularly true for commercial and industrial MSW and construction and demolition debris. Data quantifying the effect of economic downturns on solid waste quantity are not readily available.

The generation of solid waste is usually greater in warm weather than in cold weather. Figure 8.1.1 shows two month-to-month patterns of MSW generation. The less variable pattern is a composite of data from eight locations with cold or moderately cold winters (Camp Dresser & McKee Inc. 1992, 1991; Child, Pollette, and Flosdorf 1986; Cosulich Associates 1988; HDR Engineering, Inc. 1989; Killam Associates 1990; North Hempstead 1986; Oyster Bay 1987). Waste generation is relatively low in the winter but rises with temperature in the spring. The surge of waste generation in the spring is caused both by increased human activity, including spring cleaning, and renewed plant growth and associated yard waste. Waste generation typically declines somewhat after June but remains above average until mid to late fall. In contrast, Figure 8.1.1 also shows the pattern of waste generation in Cape May County, New Jersey, a summer resort area (Camp Dresser & McKee Inc. 1991). The annual influx of tourists overwhelms all other influences of waste generation.

Areas with mild winters may display month-to-month patterns of waste generation similar to the cold-winter pattern shown in Figure 8.1.1 but with a smaller difference between the winter and spring/summer rates. On the other hand, local factors can create a distinctive pattern not generally seen in other areas, as in Sarasota, Florida (Camp Dresser & McKee Inc. 1990). The surge of activity and plant growth in the spring is less marked in mild climates,

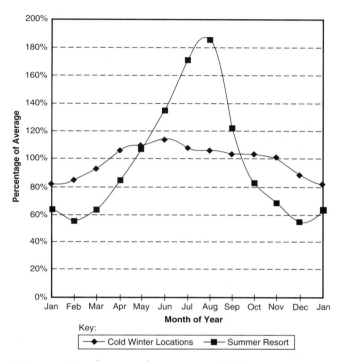

FIG. 2.1.1 Month-to-month variation in MSW generation rate.

and local factors can cause the peak of waste generation to occur in any season of the year.

Component Composition of MSW

Table 8.1.1 lists the representative component composition for MSW disposed in the United States and adjacent portions of Canada and shows ranges for individual components. Materials diverted from the waste stream for recycling or composting are not included. The table is based on the results of twenty-two field studies in eleven states plus the Canadian province of British Columbia. The ranges shown in the table are annual values for county-sized areas. Seasonal values may be outside these ranges, especially in individual municipalities.

TABLE 8.1.1 REPRESENTATIVE COMPONENT COMPOSITION OF MSW

Waste Category	Representative Composition (%)[b]	Range of Reasonable Reported Values (%)[b]
Organics/Combustibles	86.6	—
Paper	39.8	—
Newspaper	6.8	4.0–13.1
Corrugated	8.6	3.5–14.8
Kraft	1.5	0.5–2.3
Corrugated & kraft	10.1	5.4–15.6
Other paper[a]	22.9	17.6–30.6
High-grade paper	1.7	0.6–3.2
Other paper[a]	21.2	16.9–25.4
Magazines	2.1	1.0–2.9
Other paper[a]	19.1	12.5–23.7
Office paper	3.4	2.5–4.5
Magazines & mail	4.0	3.6–5.7
Other paper[a]	17.2	—
Yard waste	9.7	2.8–19.6
Grass clippings	4.0	0.3–6.5
Other yard waste	5.7	—
Food waste	12.0	6.8–17.3
Plastic	9.4	6.3–12.6
Polyethylene terephthalate (PET) bottles	0.4	0.1–0.5
High-density polyethylene (HDPE) bottles	0.7	0.4–1.1
Other plastic	8.3	5.8–10.2
Polystyrene	1.0	0.5–1.5
Polyvinyl chloride (PVC) bottles	0.06	0.02–0.1
Other plastic[a]	7.2	5.3–9.5
Polyethylene bags & film	3.7	3.5–4.0
Other plastic[a]	3.5	2.8–4.4
Other organics	15.7	—
Wood	4.0	1.0–6.6
Textiles	3.5	1.5–6.3
Textiles/rubber/leather	4.5	2.6–9.2
Fines	3.3	2.8–4.0
Fines <½ inch	2.2	1.7–2.8
Disposable diapers	2.5	1.8–4.1
Other organics	1.4	—
Inorganics/Noncombustibles	13.4	—
Metal	5.8	—
Aluminum	1.0	0.6–1.2
Aluminum cans	0.6	0.3–1.2
Other aluminum	0.4	0.2–0.9
Tin & bimetal cans	1.5	0.9–2.7
Other metal[a]	3.3	1.1–6.9
Ferrous metal	4.5	2.8–5.5
Glass	4.8	2.3–9.7
Food & beverage containers	4.3	2.0–7.7
Other glass	0.5	—
Batteries	0.1	0.04–0.1
Other Inorganics		
With noncontainer glass	3.2	1.9–4.9
Without noncontainer glass	2.7	1.8–3.8

[a]Each "other" category contains all material of its type except material in the categories above it.
[b]Weight percentage

Residential MSW contains more newspaper; yard waste; disposable diapers; and textiles, rubber, and leather. Nonresidential MSW contains more corrugated cardboard, high-grade paper, wood, other plastics, and other metals.

The composition of MSW varies from one CII establishment to another. However, virtually all businesses and institutions generate a variety of waste materials. For example, offices do not generate only paper waste, and restaurants do not generate only food waste.

Component Composition of Bulky Waste

Fewer composition data are available for bulky waste than for MSW. Table 8.1.2 shows the potential range of compositions. The first column in the table shows the composition of all bulky waste generated in two adjacent counties in southern New Jersey, including bulky waste reported as recycled. The third column shows the composition of bulky waste disposed in the two counties, and the middle column shows the estimated recycling rate for each bulky waste component based on reported recycling and disposal. Note that the estimated overall recycling rate is almost 80%.

The composition prior to recycling is dramatically different from the composition after recycling. For example, inorganic materials account for roughly three quarters of the bulky waste before recycling but little more than one quarter after recycling. Depending on local recycling practices, the composition of bulky waste received at a disposal facility in the United States could be similar to the first column of Table 8.1.2, similar to the third column, or anywhere in between.

The composition of MSW does not change dramatically from season to season. Even the most variable component, yard waste, may be consistent in areas with mild climates. In areas with cold winters, generation of yard waste generally peaks in the late spring, declines gradually through the summer and fall, and is lowest in January and February. A surge in yard waste can occur in mid to late fall in areas where a large proportion of tree leaves enter the solid waste stream and are not diverted for composting or mulching.

Density

As discussed in Section 7.2, the density of MSW varies according to circumstance. Table 8.1.3 shows representative density ranges for MSW under different conditions. The density of mixed MSW is influenced by the degree of compaction, moisture content, and component composition. As shown in the table, individual components of MSW have different bulk densities, and a range of densities exists within most components.

TABLE 8.1.2 COMPONENT COMPOSITION OF BULKY WASTE AND THE POTENTIAL IMPACT OF RECYCLING

Waste Category	Composition of all Bulky Waste Generated (%)[a]	Composition of Bulky Waste Recycled (%)[a]	Composition of Bulky Waste Landfilled (%)[a]
Organics/Combustibles	24.7	37.9	73.4
Lumber	13.1	47.2	33.0
Corrugated cardboard	0.7	2.5	3.1
Plastic	1.0	18.8	3.7
Furniture	1.3	0.0	6.3
Vegetative materials	3.8	73.0	4.9
Carpet & padding	0.7	0.0	3.2
Bagged & miscellaneous	2.1	0.0	10.2
Roofing materials	1.2	0.4	5.9
Tires	0.3	100.0	0.0
Other	0.6	0.0	3.1
Inorganics/Noncombustibles	75.3	92.6	26.6
Gypsum board & plaster	1.8	3.9	8.3
Metal & appliances	15.4	92.5	5.5
Dirt & dust	1.2	0.0	5.8
Concrete	26.5	96.7	4.2
Asphalt	28.7	99.9	0.1
Bricks & blocks	1.3	81.8	1.1
Other	0.3	0.0	1.6
Overall	100.0	79.1	100.0

Sources: Data from Camp Dresser & McKee, 1992, *Atlantic County (NJ) Solid Waste Characterization Program* (Edison, N.J. [May]) and *Idem*, 1991, *Cape May County Multi-Seasonal Solid Waste Composition Study* (Edison, N.J. [August]).

[a]Weight percentage

TABLE 8.1.3 DENSITY OF MSW AND COMPONENTS

Material and Circumstance	Density (lb/cu yd)
Mixed MSW	
Loose	150–300
In compactor truck	400–800
Dumped from compactor truck	300–500
Baled	800–1600
In landfill	800–1400
Loose Bulk Densities	
Aluminum cans (uncrushed)	54–81
Corrugated cardboard	50–135
Dirt, sand, gravel, concrete	2000–3000
Food waste	800–1500
Glass bottles (whole)	400–600
Light ferrous, including cans	100–250
Miscellaneous paper	80–250
Stacked high-grade paper	400–600
Plastic	60–150
Rubber	200–400
Textiles	60–180
Wood	200–600
Yard waste	100–600

Within individual categories of MSW, bulk density increases as physical irregularity decreases. Compaction increases density primarily by reducing irregularity. Some compaction occurs in piles, so density tends to increase as the height of a pile increases. In most cases, shredding and other size reduction measures also increase density by reducing irregularity. The size reduction of regularly shaped materials such as office paper, however, can increase irregularity and decrease density.

Particle Size, Abrasiveness, and Other Physical Characteristics

Figure 8.1.2 shows a representative particle size distribution for MSW based on research by Hilton, Rigo, and Chandler (1992). Environmental engineers generally estimate size distribution by passing samples of MSW over a series of screens, beginning with a fine screen and working up to a coarse screen. As shown in the figure, MSW has no characteristic particle size, and most components of MSW have no characteristic particle size.

MSW does not flow, and piles of MSW have a tendency to hold their shape. Loads of MSW discharged from compactor trucks often retain the same shape they had in-

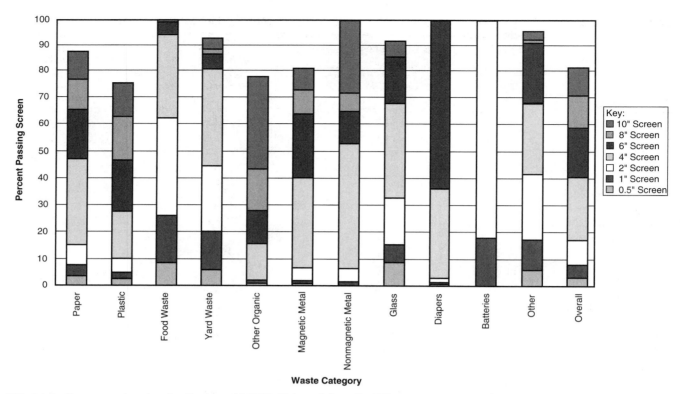

FIG. 8.1.2 Representative size distribution of MSW. (Adapted from D. Hilton, H.G. Rigo, and A.J. Chandler, 1992, Composition and size distribution of a blue-box separated waste stream, presented at *SWANA's Waste-to-Energy Symposium, Minneapolis, MN, January 1992.*)

side the truck. When MSW is removed from one side of a storage bunker at an MSW combustion facility, the waste on the other side generally does not fall into the vacated space. This characteristic allows the side on which trucks dump waste be kept relatively empty during the hours when the facility receives waste.

MSW tends to stratify vertically when mixed, with smaller and denser objects migrating toward the bottom and lighter and bulkier objects moving toward the top. However, MSW does not stratify much when merely vibrated.

Although MSW is considered soft and mushy, it contains substantial quantities of glass, metal, and other potentially abrasive materials.

Combustion Characteristics

Most laboratory work performed on samples of solid waste over the years has focused on parameters related to combustion and combustion products. The standard laboratory tests in this category are proximate composition, ultimate composition, and heat value.

PROXIMATE COMPOSITION

The elements of proximate composition are moisture, ash, volatile matter, and fixed carbon. The moisture content of

solid waste is defined as the material lost during one hour at 105°C. Ash is the residue remaining after combustion. Together, moisture and ash represent the noncombustible fraction of the waste.

Volatile matter is the material driven off as gas or vapor when waste is subjected to a temperature of approximately 950°C for 7 min but is prevented from burning because oxygen is excluded. Volatile matter should not be confused with *volatile organic compounds* (VOCs). VOCs are a small component of typical solid waste. In proximate analysis, any VOCs present tend to be included in the result for moisture.

Conceptually, fixed carbon is the combustible material remaining after the volatile matter is driven off. Fixed carbon represents the portion of combustible waste that must be burned in the solid state rather than as gas or vapor. The value for fixed carbon reported by the laboratory is calculated as follows:

$$\text{\% fixed carbon} = 100\% - \text{\% moisture} - \text{\% ash} - \text{\% volatile matter} \quad \textbf{8.1(1)}$$

Table 8.1.4 shows a representative proximate composition for MSW. The values in the table are percentages based on dry (moisture-free) MSW. Representative moisture values are also provided. These moisture values are for MSW and components of MSW as they are received at a disposal facility. Because of a shortage of data for the proximate composition of noncombustible materials, these materials are presented as 100% ash.

The dry-basis values in Table 8.1.4 can be converted to as-received values by using the following equation:

$$A = D(100\% - M) \qquad 8.1(2)$$

where:

A = value for waste as received at the solid waste facility
D = dry-basis value
M = percent moisture for waste received at the solid waste facility

Between initial discard at the point of generation and delivery to a central facility, moisture moves from wet materials to dry and absorbent materials. The largest movement of moisture is from food waste to uncoated paper discarded with food waste. This paper includes newspaper, kraft paper, and a substantial portion of other paper from residential sources as well as corrugated cardboard from commercial sources.

Other sources of moisture in paper waste include water absorbed by paper towels, napkins, and tissues during use, and precipitation. Absorbent materials frequently exposed to precipitation include newspaper and corrugated cardboard. Many trash containers are left uncovered, and precipitation is absorbed by the waste. Standing water in dumpsters is often transferred to the collection vehicle.

The value of proximate analysis is limited because (1) it does not indicate the degree of oxidation of the combustible waste and (2) it gives little indication of the quantities of pollutants emitted during combustion of the waste. Ultimate analysis supplements the information provided by proximate analysis.

TABLE 8.1.4 REPRESENTATIVE PROXIMATE AND ULTIMATE COMPOSITION OF MSW

Waste Category	Proximate Composition—Dry Basis			Ultimate Composition—Dry Basis[a]						Moisture (%)
	Ash (%)	Volatile Matter (%)	Fixed Carbon (%)	Carbon (%)	Hydrogen (%)	Nitrogen (%)	Chlorine (%)	Sulfur (%)	Oxygen (%)	
Organics/Combustibles	7.7	82.6	9.6	48.6	6.8	0.94	0.69	0.22	35.0	32.5
Paper	6.3	83.5	10.1	43.0	6.0	0.36	0.17	0.17	43.8	24.0
Newspaper	5.2	83.8	11.1	43.8	5.9	0.29	0.14	0.24	44.4	23.2
Corrugated & kraft paper	2.2	85.8	12.1	46.0	6.4	0.28	0.14	0.22	44.8	21.2
High-grade paper	9.1	83.4	7.5	38.1	5.6	0.15	0.12	0.07	46.9	9.3
Magazines	20.4	71.8	7.9	35.0	5.0	0.05	0.07	0.08	39.4	8.6
Other paper	6.9	83.8	9.3	42.7	6.1	0.50	0.22	0.14	43.3	28.7
Yard waste	9.6	73.0	17.4	45.0	5.6	1.5	0.31	0.17	37.7	53.9
Grass clippings	9.7	75.6	14.7	43.3	5.9	2.6	0.60	0.30	37.6	63.9
Leaves	7.3	72.7	20.1	50.0	5.7	0.82	0.10	0.10	36.0	44.0
Other yard waste	12.5	70.5	17.0	40.7	5.0	1.3	0.26	0.10	40.0	50.1
Food waste	11.0	79.0	10.0	45.4	6.9	3.3	0.74	0.32	32.3	65.4
Plastic	5.3	93.0	1.3	76.3	11.5	0.26	2.4	0.20	4.4	13.3
PET bottles	1.3	95.0	3.6	68.5	8.0	0.16	0.08	0.08	21.9	3.6
HDPE bottles	2.4	97.4	0.2	81.6	13.6	0.10	0.18	0.20	1.9	7.0
Polystyrene	1.8	97.8	0.4	86.3	7.9	0.28	0.12	0.30	3.4	10.8
PVC bottles	0.6	46.2	3.2	44.2	5.9	0.26	40.1	0.89	7.6	3.2
Polyethylene bags & film	8.8	90.1	1.1	77.4	12.9	0.10	0.09	0.12	1.8	19.1
Other plastic	4.2	94.1	1.7	72.9	11.4	0.45	5.3	0.24	5.5	10.5
Other Organics	11.3	77.8	10.9	46.2	6.1	1.9	1.0	0.36	33.3	27.3
Wood	2.8	83.0	14.1	46.7	6.0	0.71	0.12	0.16	43.4	14.8
Textiles/rubber/leather	6.6	84.0	9.4	50.3	6.4	3.3	1.8	0.33	31.3	12.4
Fines	25.3	64.7	10.0	37.3	5.3	1.6	0.54	0.45	29.5	41.1
Disposable diapers	4.1	87.1	8.7	48.4	7.6	0.51	0.23	0.35	38.8	66.9
Other organics	31.3	58.8	9.9	44.2	5.3	1.8	2.2	0.81	14.4	8.0
Inorganics/Noncombustibles[b]	100	0	0	0	0	0	0	0	0	0
Overall	24.9	67.2	7.8	39.5	5.6	0.76	0.56	0.18	28.5	28.2

[a]Also includes ash values from first column of proximate analysis.
[b]Values assumed for the purpose of estimating overall values.

ULTIMATE COMPOSITION

Moisture and ash, as previously defined for proximate composition, are also elements of ultimate composition. In standard ultimate analysis, the combustible fraction is divided among carbon, hydrogen, nitrogen, sulfur, and oxygen. Ultimate analysis of solid waste should also include chlorine. The results are more useful if sulfur is broken down into organic sulfur, sulfide, and sulfate; and chlorine is broken down into organic (insoluble) and inorganic (soluble) chlorine (Niessen 1995).

Carbon, hydrogen, nitrogen, sulfur, and chlorine are measured directly; calculating oxygen requires subtracting the sum of the other components (including moisture and ash) from 100%. Table 8.1.4 shows a representative ultimate composition for MSW. The dry-basis values shown in the table can be converted to as-received values with use of Equation 8.1(2).

The ultimate composition of MSW on a dry basis reflects the dominance of six types of materials in MSW: cellulose, lignins, fats, proteins, hydrocarbon polymers, and inorganic materials. Cellulose is approximately 42.5% carbon, 5.6% hydrogen, and 51.9% oxygen and accounts for the majority of the dry weight of MSW. The cellulose content of paper ranges from approximately 75% for low grades to approximately 90% for high-grade paper. Wood is roughly 50% cellulose, and cellulose is a major ingredient of yard waste, food waste, and disposable diapers. Cotton, the largest ingredient of the textile component of MSW, is approximately 98% cellulose (Masterton, Slowinski, and Stanitski 1981).

Despite the abundance of cellulose, MSW contains more carbon than oxygen due to the following factors:

- Most of the plastic fraction of MSW is composed of polyethylene, polystyrene, and polypropylene, which contain little oxygen.
- Synthetic fibers (textiles category) contain more carbon than oxygen, and rubber contains little oxygen.
- The lower grades of paper contain significant quantities of lignins, which contain more carbon than oxygen.
- Fats contain more carbon than oxygen.

The nitrogen in solid waste is primarily in organic form. The largest contributors of nitrogen to MSW are food waste (proteins), grass clippings (proteins), and textiles (wool, nylon, and acrylic). Chlorine occurs in both organic and inorganic forms. The largest contributor of organic chlorine is PVC or vinyl. Most of the PVC is in the other plastic and textiles components. The largest source of inorganic chlorine is sodium chloride (table salt). Sulfur is not abundant in any category of combustible MSW but is a major component of gypsum board. The sulfur in gypsum is largely noncombustible but not entirely so. In Table 8.1.4, gypsum board is included in the Inorganics/ Noncombustibles category, which is shown as 100% ash because of a lack of data on the ultimate composition.

The inorganic (noncombustible) waste categories contribute most of the ash in MSW. Additional ash is contributed by the inorganic components of combustible materials, including clay in glossy and high-grade paper, dirt in yard waste, bones and shells in food waste, asbestos in vinyl–asbestos floor coverings, fiberglass in reinforced plastic, and grit on roofing shingles.

HEAT VALUE

Table 8.1.5 shows the heat value of typical MSW based on the results of laboratory testing of MSW components. Calculations of the heat value based on energy output measurements at operating combustion facilities generally yield lower values (see Section 8.3).

The heat value shown for solid waste and conventional fuels in the United States, Canada, and the United Kingdom is typically the higher heating value (HHV). The HHV includes the latent heat of vaporization of the water created during combustion. When this heat is deducted, the result is called the lower heating value (LHV). For additional information see Niessen (1995).

The as-received heat value is roughly proportional to the percentage of waste that is combustible (i.e., neither moisture nor ash) and to the carbon content of the combustible fraction. The heat values of the plastics categories are highest because of their high carbon content, low ash content, and low-to-moderate moisture content. Paper categories have intermediate heat values because of their intermediate carbon content, moderate moisture content, and low-to-moderate ash content. Yard waste, food waste, and disposable diapers have low heat values because of their high moisture levels.

Bioavailability

Because microorganisms can metabolize paper, yard waste, food waste, and wood, this waste is classified as *biodegradable*. Disposable diapers and their contents are also largely biodegradable, as are cotton and wool textiles.

Some biodegradable waste materials are more readily metabolized than others. The most readily metabolized materials are those with high nitrogen and moisture content: food waste, grass clippings, and other green, pulpy yard wastes. These wastes are *putrescible* and have high *bioavailability*. Leaf waste generally has intermediate bioavailability. Wood, cotton and wool, although biodegradable, have relatively low bioavailability and are considered noncompostable within the context of solid waste management.

Toxic Substances in Solid Waste

Solid waste inevitably contains many of the toxic substances manufactured or extracted from the earth. Most

TABLE 8.1.5   REPRESENTATIVE HEAT VALUES OF MSW[a]

Waste Category	Dry-Basis Heat Value (HHV in Btu/lb)	Moisture Content (%)	As-Received Heat Value (HHV in Btu/lb)
Organics/Combustibles	9154	32.5	6175
Paper	7587	24.0	5767
Newspaper	7733	23.2	5936
Corrugated & kraft	8168	21.2	6435
High-grade paper	6550	9.3	5944
Magazines	5826	8.6	5326
Other paper	7558	28.7	5386
Yard waste	7731	53.9	3565
Grass clippings	7703	63.9	2782
Leaves	8030	44.0	4499
Other yard waste	7387	50.1	3689
Food waste	8993	65.4	3108
Plastic	16,499	13.3	14,301
PET bottles	13,761	3.6	13,261
HDPE bottles	18,828	7.0	17,504
Polystyrene	16,973	10.8	15,144
PVC bottles	10,160	3.2	9838
Polyethylene bags & film	17,102	19.1	13,835
Other plastic	15,762	10.5	14,108
Other organics	8698	27.3	6322
Wood	8430	14.8	7186
Textiles/rubber/ leather	9975	12.4	8733
Fines	6978	41.1	4114
Disposable diapers	9721	66.9	3222
Other organics	7438	8.0	6844
Inorganics/ Noncombustibles[b]	0	0.0	0
Overall	7446	28.2	5348

[a]Values shown are HHV. In HHV measurements, the energy required to drive off the moisture formed during combustion is not deducted.
[b]Values assumed for the purpose of estimating overall values.

toxic material in solid waste is in one of three categories:

- Toxic metals
- Toxic organic compounds, many of which are also flammable
- Asbestos

The results of studies of toxic metals in solid waste vary. Table 8.1.6 summarizes selected results of two comprehensive studies performed in Cape May County, New Jersey (Camp Dresser & McKee Inc. 1991a) and Burnaby, British Columbia (Chandler & Associates, Ltd. 1993; Rigo, Chandler, and Sawell 1993). Reports of both studies contain data for additional metals and materials, and the Burnaby reports contain results for numerous subcategories of the categories in the table. The Burnaby reports also analyze the behavior of specific metals from waste components during processing in an MSW incinerator.

Franklin Associates, Ltd. (1989) provided extensive information on sources of lead and cadmium in MSW, and Rugg and Hanna (1992) compiled detailed information on sources of lead in MSW in the United States.

Most MSW referred to as *household hazardous waste* is so classified because it contains toxic organic compounds. Large quantities of toxic organic materials from commercial and industrial sources were once disposed in MSW landfills in the United States, and many of these landfills are now officially designated as hazardous waste sites. The large-scale disposal of toxic organics in MSW landfills has been largely eliminated, but disposal of household hazardous waste remains a concern for many. Generally, household hazardous waste refers to whatever toxic materials remain in MSW, regardless of the source.

Estimates of the abundance of household hazardous waste vary. Reasons for the lack of consistency from one

TABLE 8.1.6 REPORTED METAL CONCENTRATIONS IN COMPONENTS OF MSW[a]

Waste Category	Arsenic CM	Arsenic BC	Cadmium CM	Cadmium BC	Chromium CM	Chromium BC	Copper CM	Copper BC	Lead CM	Lead BC	Mercury CM	Mercury BC	Nickel CM	Nickel BC	Zinc CM	Zinc BC
Organics/Combustibles																
Paper																
Newspaper	0.1	0.7	ND[b]	0.1	ND	49	17	18	ND	7	0.3	2	ND	28	58	21
Corrugated cardboard	0.2	0.6	ND	0.1	ND	2	13	3	19	4	0.2	0.1	6	4	56	10
Kraft paper	0.3	0.8	ND	0.1	5	5	11	11	15	9	0.1	0.5	ND	8	30	22
High-grade paper	0.7	1	ND	0.1	ND	3	7	8	ND	5	0.1	0.3	ND	8	28	208
Magazines	0.4	1	ND	0.2	4	11	46	32	ND	3	0.09	0.3	ND	13	88	27
Other	0.4	1	ND	1	4	27	52	25	9	182	0.07	0.3	ND	7	58	71
Yard waste	0.9	6	ND	5	4	87	10	571	14	137	0.1	1	3	21	89	321
Food waste	0.1	1	ND	2	ND	23	9	43	ND	72	0.02	0.3	2	5	20	186
Plastic																
PET	ND	0.8	ND	5	15	17	30	31	59	62	0.07	0.2	ND	8	21	97
HDPE	0.2	0.5	ND	3	52	15	14	24	211	61	0.1	0.2	ND	7	58	142
Film	0.5	0.6	ND	5	100	102	25	23	450	325	0.1	0.2	ND	7	120	658
Other	0.4	0.7	8	82	7	279	8	58	19	342	0.04	0.3	ND	40	69	231
Other organics																
Wood	34	24	ND	0.4	52	77	32	68	108	408	2	0.3	ND	3	205	174
Textiles & footwear	0.8	0.4	19	4	387	619	25	62	48	129	0.3	1	5	1	666	222
Fines	3	7	1	4	14	115	179	243	273	259	0.2	1	18	54	352	654
Disposable diapers	0.1	—	ND	—	1	—	2	—	ND	—	0.02	—	ND	—	28	—
Inorganics/Noncombustibles																
Metal																
Ferrous food & beverage cans	4	7	16	43	527	191	375	104	350	342	0.8	6	133	161	145	1552
Aluminum beverage cans	ND	0.4	ND	5	72	91	107	1105	30	41	0.7	0.4	54	21	80	229
Other metal	9	280	22	25	4702	768	6816	2082	1279	95	0.7	0.4	411	24	1675	199,000
Glass food & beverage containers	ND	2	ND	4	ND	91	ND	26	84	103	0.2	0.2	ND	15	ND	71
Household batteries																
Carbon-zinc & alkaline batteries[c]	7	2	53	1027	45	57	8400	6328	236	94	2900	136	—	512	180,000	103,000
Nickel-cadmium batteries	—	4	175,000	120,000	—	64	—	53	—	113	—	0.3	240,000	315	—	685
Other inorganics	1	12	ND	8	21	91	13	113	50	607	0.9	0.2	5	73	21	1997

Source: Data adapted from Camp Dresser & McKee Inc., 1991a, *Cape May County multi-seasonal solid waste composition study* (Edison, N.J. [August]); A.J. Chandler & Associates, Ltd. et al., 1993, *Waste analysis, sampling, testing and evaluation (WASTE) program: Effect of waste stream characteristics on MSW incineration: The fate and behaviour of metals. Final report of the mass burn MSW incineration study (Burnaby, B.C.), Vol. 1, Summary report* (Toronto [April]); and H.G. Rigo, A.J. Chandler, and S.E. Sawell, 1993, Debunking some myths about metals, in *Proceedings of the 1993 International Conference on Municipal Waste Combustion* (Williamsburg, Va. [30 March–2 April]).

[a]All values in mg/kg on an as-received basis. Values presented are based on reported results from studies in Cape May County, New Jersey and Burnaby, British Columbia. CM indicates Cape May, and BC indicates Burnaby.

[b]ND = Not detected.

[c]Current values for mercury are close to or below the Burnaby value.

study to another include the following:

Some quantity estimates include less toxic materials such as latex paint.

Most quantity estimates include the weight of containers, and many estimates include the containers even if they are empty.

Some quantity estimates include materials that were originally in liquid or paste form but have dried, such as dried paint and adhesives. Toxic substances can still leach from these dried materials, but drying reduces the potential leaching rate.

Strongly toxic organic materials, excluding their containers, appear to constitute well under 0.5% of MSW, and the toxic material is usually dispersed. Bulky waste typically contains no more toxic organic material than MSW, but bulky waste is more likely to contain concentrated pockets of toxic substances.

A statewide waste characterization study in Minnesota (Minnesota Pollution Control Agency 1992; Minnesota Pollution Control Agency and Metropolitan Council 1993) provides a detailed accounting of the household hazardous waste materials encountered.

Most of the asbestos in normal solid waste is in old vinyl–asbestos floor coverings and asbestos shingles. Asbestos in these forms is generally not a significant hazard.

—*F. Mack Rugg*

References

Camp Dresser & McKee Inc. 1990. *Sarasota County waste stream composition study*. Draft report (March).
———. 1991a. *Cape May County multi-seasonal solid waste composition study*. Edison, N.J. (August).
———. 1991b. *Cumberland County (NJ) waste weighing and composition analysis*. Edison, N.J. (January).
———. 1992. *Atlantic County (NJ) solid waste characterization program*. Edison, N.J. (May).
Chandler, A.J., & Associates, Ltd. et al. 1993. *Waste analysis, sampling, testing and evaluation (WASTE) program: Effect of waste stream characteristics on MSW incineration: The fate and behaviour of metals. Final report of the mass burn MSW incineration study (Burnaby, B.C.). Volume I, Summary report*. Toronto (April).
Child, D., G.A. Pollette, and H.W. Flosdorf. 1986. Waste stream analysis. *Waste Age* (November).
Cosulich, William F., Associates, P.C. 1988. *Solid waste management plan, County of Monroe, New York: Solid waste quantification and characterization*. Woodbury, N.Y. (July).
Franklin Associates, Ltd. 1989. *Characterization of products containing lead and cadmium in municipal solid waste in the United States, 1970 to 2000*. U.S. EPA (January).
HDR Engineering, Inc. 1989. *Report on solid waste quantities, composition and characteristics for Monmouth County (NJ) waste recovery system*. White Plains, N.Y. (March).
Killam Associates. 1990. *Somerset County (NJ) solid waste generation and composition study*. Millburn, N.J. (May).
Masterton, W.L., E.J. Slowinski, and C.L. Stanitski. 1981. *Chemical principles*. 5th ed. Philadelphia: Saunders College Publishing.
Minnesota Pollution Control Agency. 1992. *Minnesota solid waste composition study, 1990–1991 part I*. Saint Paul, Minn. (November).
Minnesota Pollution Control Agency and Metropolitan Council. 1993. *Minnesota solid waste composition study, 1991–1992 part II*. Saint Paul, Minn. (April).
Niessen, W.R. 1995. *Combustion and incineration processes: Applications in environmental engineering*. 2d ed. New York: Marcel Dekker, Inc.
North Hempstead, Town of (NY), transfer station scalehouse records, August 1985 through July 1986. 1986.
Oyster Bay, Town of (NY), transfer station scalehouse records, September 1986 through August 1987. 1987.
Rigo, H.G., A.J. Chandler, and S.E. Sawell. Debunking some myths about metals. In *Proceedings of the 1993 International Conference on Municipal Waste Combustion, Williamsburg, VA, March 30–April 2, 1993*.
Rugg, M. and N.K. Hanna. 1992. Metals concentrations in compostable and noncompostable components of municipal solid waste in Cape May County, New Jersey. *Proceedings of the Second United States Conference on Municipal Solid Waste Management, Arlington, VA, June 2–5, 1992*.

8.2
CHARACTERIZATION METHODS

This section describes and evaluates methods for estimating the characteristics of solid waste. The purposes of waste characterization are identified; and methods for estimating quantity, composition, combustion characteristics, and metals concentrations are addressed.

Purposes of Solid Waste Characterization

The general purpose of solid waste characterization is to promote sound management of solid waste. Specifically, characterization can determine the following:

The size, capacity, and design of facilities to manage the waste.
The potential for recycling or composting portions of the waste stream.
The effectiveness of waste reduction programs, recycling programs, or bans on the disposal of certain materials.
Potential sources of environmental pollution in the waste.

In practice, the immediate purpose of most waste characterization studies, including many extensive studies, is to comply with specific regulatory mandates and to provide information for use by vendors in preparing bids to design, construct, and operate solid waste management facilities.

The purposes of a waste characterization program determine the design of it. If all waste is to be landfilled, the characterization program should focus on the quantity of waste, its density, and its potential for compaction. The composition of the waste and its chemical characteristics are relatively unimportant. If all waste is to be incinerated, the critical parameters are quantity, heat value, and the percentage of combustible material in the waste. If recycling and composting are planned or underway, a composition study can identify the materials targeted for recovery, estimate their abundance in the waste, and monitor compliance with source separation requirements.

Basic Characterization Methods

Environmental engineers use one of two fundamental methods to characterize solid waste. One method is to collect and analyze data on the manufacture and sale of products that become solid waste after use. The method is called material flows methodology. The second method is a direct field study of the waste itself. Combining these two

fundamental methods creates hybrid methodologies (for example, see Gay, Beam, and Mar [1993]).

The direct field study of waste is superior in concept, but statistically meaningful field studies are expensive. For example, a budget of $100,000 is typically required for a detailed estimate of the composition of MSW arriving at a single disposal facility, accurate to within 10% at 90% confidence. A skilled and experienced team can often provide additional information at little additional cost, including an estimated composition for bulky waste based on visual observation.

The principal advantage of the material flows methodology is that it draws on existing data that are updated regularly by business organizations and governments. This method has several positive effects. First, the entire waste stream is measured instead of samples of the waste, as in field studies. Therefore, the results of properly conducted material flows studies tend to be more consistent than the results of field studies. Second, updates of material flows studies are relatively inexpensive once the analytical structure is established. Third, material flows studies are suited to tracking economic trends that influence the solid waste stream.

The principal disadvantages of material flows methodology follow.

Obtaining complete production data for every item discarded as solid waste is difficult.
Although data on food sales are available, food sales bear little relation to the generation of food waste. Not only is most food not discarded, but significant quantities of water are added to or removed from many food items between purchase and discard. These factors vary from one area to another based on local food preferences and eating patterns.
Material flows methodology cannot measure the generation of yard waste.
Material flows methodology does not account for the addition of nonmanufactured materials to solid waste prior to discard, including water, soil, dust, pet droppings, and the contents of used disposable diapers.
Some of the material categories used in material flows studies do not match the categories of materials targeted for recycling. For example, advertising inserts in newspapers are typically recycled with the newsprint, but in material flows studies the inserts are part of a separate commercial printing category.

In performing material flows studies for the U.S. EPA, Franklin Associates bases its estimates of food waste, yard waste, and miscellaneous inorganic wastes on field studies in which samples of waste were sorted. Franklin Associates (1992) also adjusts its data for the production of disposable diapers to account for the materials added during use.

In general, the more local and the more detailed a waste characterization study is to be, the greater are the advantages of a direct field study of the waste.

Estimation of Waste Quantity

The best method for estimating waste quantity is to install permanent scales at disposal facilities and weigh every truck on the way in and again on the way out. An increasing number of solid waste disposal facilities are equipped with scales, but many landfills still are not.

In the United States, facilities without scales record incoming waste in cubic yards and charge tipping fees by the cubic yard. Since estimating the volume of waste in a closed or covered vehicle or container is difficult, the volume recorded is usually the capacity of the vehicle or container. Because this estimation creates an incentive to deliver waste in full vehicles, the recorded volumes tend to be close to the actual waste volumes.

For the reasons previously stated, expressing waste quantity in tons is preferable to cubic yards. This conversion is conceptually simple, as shown in the following equation:

$$M = VD/2000 \qquad 8.2(1)$$

where:

M = mass of waste in tons
V = volume of waste in cubic yards
D = density of waste in pounds per cubic yard

If the density is expressed in tons per cubic yard, dividing by 2000 is unnecessary. In the United States, however, the density of solid waste is usually expressed in pounds per cubic yard.

Although simple conceptually, converting cubic yards to tons can be difficult in practice. The density of solid waste varies from one type of waste to another, from one type of vehicle to another, and even among collection crews. In small waste streams, local conditions can cause the overall density of MSW, as received at disposal facilities, to vary from 250 to 800 lb/cu yd. A conversion factor of 3.0 to 3.3 cu yd/tn (600 to 667 lb/cu yd) is reasonable for both MSW and bulky waste in many large waste streams; however, this conversion factor may not be reasonable for a particular waste stream.

At disposal facilities without permanent scales, environmental engineers can use portable scales to develop a better estimate of the tons of waste being delivered. Selected trucks are weighed, and environmental engineers use the results to estimate the overall weight of the waste stream.

Portable truck scales are available in three basic configurations: (1) platform scales designed to accommodate entire vehicles (or trailers), (2) axle scales designed to accommodate one axle or a pair of tandem axles at a time, and (3) wheel scales designed to be used in pairs to accommodate one axle or a pair of tandem axles at a time. Axle scales can be used singly or in pairs. Similarly, either one or two pairs of wheel scales can be used. When a single axle scale or a single pair of wheel scales is used, adding the results for individual axles yields the weight of the vehicle.

Platform scales are the easiest to use, but the cost can be prohibitive. The use of wheel scales tends to be difficult and time consuming. The cost of axle scales is similar to that of wheel scales, and axle scales are easier to use than wheel scales. The use of a pair of portable axle scales is recommended in the *Municipal solid waste survey protocol* prepared for the U.S. EPA by SCS Engineers (1979). Regardless of what type of scale is used, a solid base that does not become soft in wet weather is required.

Truck weighing surveys, like other waste characterization field studies, are typically conducted during all hours that a disposal facility is open during a full operating week. A full week is used because the variation in waste characteristics is greater among the hours of a day and among the days of a week than among the weeks of a month. Also, spreading the days of field work out over several weeks is substantially more expensive.

A truck weighing survey should be conducted during at least two weeks—one week during the period of minimum waste generation and one week during the period of maximum waste generation (see Section 8.1). One week during each season of the year is preferable. Holiday weeks should be avoided.

Weighing all trucks entering the disposal facility is rarely possible, so a method of truck selection must be chosen. A conceptually simple approach is to weigh every nth truck (for example, every 5th truck) that delivers waste to the facility. This approach assumes that the trucks weighed represent all trucks arriving at the facility. The total waste tonnage can be estimated with the following equation:

$$W = T(w/t) \qquad 8.2(2)$$

where:

W = the total weight of the waste delivered to the facility
T = the total number of trucks that delivered waste to the facility
w = the total weight of the trucks that were weighed
t = the number of trucks that were weighed

This approach is suited to a facility that receives a fairly constant flow of trucks. Unfortunately, the rate at which trucks arrive at most facilities fluctuates during the operating day. A weighing crew targeting every nth truck will miss trucks during the busy parts of the day and be idle at other times. Missing trucks during the busy parts of the day can bias the results; the trucks that arrive at these times tend to be curbside collection trucks, which have a distinctive range of weights. Also, having a crew and its equipment stand idle at slow times while waiting for the nth truck to arrive reduces the amount of data collected, which reduces the statistical value of the overall results.

A better approach is to weigh as many trucks as possible during the operating day, keeping track of the total number of trucks that deliver waste during each hour. A separate average truck weight and total weight is calculated for each hour, and the hourly totals are added to yield a total for the day. For this purpose, Equation 8.2(2) is modified as follows:

$$W = T_1(w_1/t_1) + T_2(w_2/t_2) \cdots + T_n(w_n/t_n) \qquad 8.2(3)$$

where:

  W = the total weight of the waste delivered to the facility
  T_1 = the number of trucks that delivered waste to the facility in the first hour
  T_2 = the number of trucks that delivered waste to the facility in the second hour
  T_n = the number of trucks that delivered waste to the facility in the last hour of the operating day
  w_1 = the total weight of the trucks that were weighed in the first hour
  w_2 = the total weight of the trucks that were weighed in the second hour
  w_n = the total weight of the trucks that were weighed in the last hour of the operating day
  t_1 = the number of trucks that were weighed in the first hour
  t_2 = the number of trucks that were weighed in the second hour
  t_n = the number of trucks that were weighed in the last hour of the operating day

Estimating the statistical precision of the results is complex when the ratio of the weighed trucks to the unweighed trucks varies from hour to hour. (Klee [1991, 1993] provides a discussion of this statistical problem.)

Sampling MSW to Estimate Composition

As in all statistical exercises based on sampling, the acquisition of samples is a critical step in estimating the composition of MSW. The principal considerations in collecting samples are the following:

Each pound of waste in the waste stream to be characterized must have an equal opportunity to be represented in the final results.
The greater the number of samples, the more precise the results.

The greater the variation between samples, the more samples must be sorted to achieve a given level of precision.
The greater the time spent collecting the samples, the less time is available to sort the samples.
The more the waste is handled prior to sorting, the more difficult and less precise the sorting.

A fundamental question is the time period(s) over which to collect the samples. One-week periods are generally used because most human activity and most refuse collection schedules repeat on a weekly basis. Sampling during a week in each season of the year is preferable. Spring sampling is particularly important because generation of yard waste, the most variable waste category, is generally least in the winter and greatest in the spring.

Another fundamental question is whether to collect the samples at the places where the waste is generated or at the solid waste facilities where the waste is taken. Sampling at solid waste facilities is generally preferred. Collecting samples at the points of generation may be necessary under the following circumstances, however:

The primary objective is to characterize the waste generated by certain sources, such as specific types of businesses.
The identity of the facilities to which the waste is taken is not known or cannot be predicted with confidence for any given week.
The facilities are widely spaced, increasing the difficulty and cost of the sampling and sorting operation.
Access to the facilities cannot be obtained.
Sufficient space to set up a sorting operation is not available at the facilities.
Appropriate loads of waste (e.g., loads from the geographic area to be characterized) do not arrive at the facilities frequently enough to support an efficient sampling and sorting operation.

Sampling at the points of generation tends to be more expensive and less valid than sampling at solid waste facilities. The added expense results from the increased effort required to design the sampling protocol and the travel time involved in collecting the samples.

The decreased validity of sampling at the points of generation has two principal causes. First, a significant portion of the waste is typically inaccessible. Waste can be inaccessible because it is on private property to which access is denied or because it is in trash compactors. Some waste is inaccessible during the day because it is not placed in outdoor trash containers until after business hours and it is picked up early in the morning. The second major cause of inaccuracy is that the relative portion of the waste stream represented by each trash receptacle is unknown because the frequency of pickup and the average quantity in the receptacle at each pickup are unknown. Random selection of receptacles to be sampled results in under-sampling of the more active receptacles, which represent more waste.

These problems are generally less acute for residential MSW than for commercial or institutional MSW. Residential MSW is usually accessible for sampling from the curb on collection day or from dumpsters serving multifamily residences. Because households generate similar quantities of waste, random selection of households for sampling gives each pound of waste a similar probability of being included in a sample. In addition, because waste characteristics are more consistent from household to household than from business to business, flaws in a residential sampling program are generally less significant than flaws in a commercial sampling program.

A universal protocol for sampling solid waste from the points of generation is impossible to state because circumstances vary greatly from place to place and from study to study. The following are general principles to follow:

Collect samples from as many different sectors of the target area as possible without oversampling relatively insignificant sectors.

If possible, collect samples from commercial locations in proportion to the size of the waste receptacles used and the frequency of pickup.

Collect samples from single-family and multifamily residences in proportion to the number of people living in each type of residence (unless a more sophisticated basis is readily available). The required population information can be obtained from U.S. census publications.

Give field personnel no discretion in selecting locations at which to collect samples. For example, field personnel should not be told to collect a sample from Elm Street but rather to collect a sample from the east side of Elm Street, starting with the second house (or business) north from Park Street.

To the extent feasible, add all waste from each selected location to the sample before going on to the next location. This practice reduces the potential for sampling bias.

Collecting samples at solid waste facilities is less expensive than collecting them at the points of generation and is more likely to produce valid results. Sample collection at facilities is less expensive because no travel is required. Samples collected at facilities are more likely to represent the waste being characterized because they are typically selected from a single line of trucks of known size that contain the entire waste stream.

Collecting samples at solid waste facilities has two stages: selecting the truck from which to take the sample and collecting the sample from the load discharged from the selected truck.

SELECTING SAMPLES

Environmental engineers usually select individual trucks in the field to sample, but they can select trucks in advance to ensure that specific collection routes are represented in the samples. Possible methods for selecting trucks in the field include the following:

- Constant interval
- Progress of sorters
- Random number generator
- Allocation among waste sources

The American Society for Testing and Materials (1992) *Standard test method for determination of the composition of unprocessed municipal solid waste* (ASTM D 5231) states that any random method of vehicle selection that does not introduce a bias into the selection process is acceptable.

Possible constant sampling intervals include the following in which n is any set number:

- Every nth truck
- Every nth ton of waste
- Every nth cubic yard of waste
- A truck every n minutes

Collecting a sample from every nth truck is relatively simple but causes the waste in small trucks and partially full trucks to be overrepresented in the samples. Collecting a sample from the truck containing every nth ton of waste is ideal but is difficult in practice because the weight of each truck is not apparent from observation. Collecting a sample from the truck containing every nth cubic yard of waste is more feasible because the volumetric capacity of most trucks can be determined by observation. However, basing the sampling interval on volumetric capacity tends to cause uncompacted waste and waste in partially full trucks to be overrepresented in the samples.

Basing the sampling interval on either a set number of trucks or a set quantity of waste causes the pace of the sampling operation to fluctuate during each day of field work. This fluctuation can result in inefficient use of personnel and deviations from the protocol when targeted trucks are missed at times of peak activity.

Collecting a sample from a truck every n minutes is convenient for sampling personnel but causes the waste in small trucks and partially full trucks to be overrepresented and the waste in trucks that arrive at busy times to be underrepresented in the samples. This approach also causes overrepresentation of waste arriving late in the day because the time interval between trucks tends to lengthen toward the end of the day and because trucks arriving late tend to be partially full, especially if the facility charges by the ton rather than by the cubic yard.

Obtaining samples as they are needed for sorting is similar to collecting a sample every n minutes and has the same disadvantages. Regardless of the sampling protocol used, however, the sorters should be kept supplied with waste to sort even if the available loads do not fit the protocol. Having more data is better.

ASTM D 5231 specifically identifies the use of a random number generator as an acceptable method for random selection of vehicles to sample. A random number generator can provide random intervals corresponding to each of the predetermined intervals just discussed. For example, if a facility receives 120 trucks per day and 12 are to be sampled, one can either sample every 10th truck or use the random number generator to generate 12 random numbers from 1 to 120. Similarly, random intervals of waste tonnage, waste volume, or elapsed time can be generated.

Random sampling intervals have the same disadvantages as the corresponding constant sampling intervals plus the following additional disadvantages:

Random sampling intervals increase the probability that the field crew is idle from time to time.
Random sampling intervals increase the probability that the field crew has to work overtime.
Random sampling intervals increase the probability that targeted trucks are missed when too many randomly selected trucks arrive within too short a time period.

In many cases, sampling by waste source minimizes the problems associated with these types of interval sampling. Sources of waste from which samples can be selected include individual municipalities, individual waste haulers, specific collection routes, waste generation sectors such as the residential sector and the commercial sector, and specific sources such as restaurants or apartment buildings. In general, sampling by source makes sense if adequate information is available on the quantity of waste from each source to be sampled. Samples can be collected from each source in proportion to the quantity of waste from each source, or the composition results for the various sources can be weighted based on the quantity from each source.

In the best case, the solid waste facility has a scale and maintains a computer database containing the following information for each load of waste: net weight, type of waste, type of vehicle, municipality of origin, hauler, and a number identifying the individual truck that delivered the waste. This information, combined with information on the hauling contracts in effect in each municipality, is usually sufficient to estimate the quantity of household and commercial MSW from each municipality.

The municipality is often the hauler for household waste, and, in those municipalities, private haulers usually handle commercial waste. In other cases, the municipality has a contract with a private hauler to collect household waste and discourages the hauler from using the same vehicles to service private accounts. Household and commercial waste can also be distinguished by the types of vehicles in which they are delivered. Dominant vehicle types vary from one region to another.

If the solid waste facility has no scale, environmental engineers can use records of waste volumes in designing a sampling plan but must differentiate between compacted and uncompacted waste. Many facilities receive little uncompacted MSW, while others receive substantial quantities.

Because per capita generation of household waste is relatively consistent, environmental engineers can use population data to allocate samples of household waste among municipalities if the necessary quantity records are not available.

Field personnel must interview private haulers arriving at the solid waste facility to learn the origins of the load of waste. Information provided by the haulers is often incomplete. In some cases this information can be supplemented or corrected during sorting of the sample.

McCamic (1985) provides additional information.

COLLECTING SAMPLES

Most protocols, including ASTM D 5231, state that each selected truck should be directed to discharge its load in an area designated for sample collection. This provision is convenient for samplers but is not necessary if a quick and simple sampling method is used. ASTM D 5231 states that the surface on which the selected load is discharged should be clean, but in most studies preventing a sample from containing a few ounces of material from a different load of waste is unnecessary.

Understanding the issues involved in selecting a sampling method requires an appreciation of the nature of a load of MSW discharged from a standard compactor truck onto the surface of a landfill or a paved tipping floor. Rather than collapsing into a loose pile, the waste tends to retain the shape it had in the truck. The discharged load can be 7 or 8 ft high. In many loads, the trash bags are pressed together so tightly that pulling material for the sample out of the load is difficult. Some waste usually falls off the top or sides of the load, but this loose waste should not be used as the sample because it can have unrepresentative characteristics.

In general, one sample should be randomly selected from each selected truck, as specified in ASTM D 5231. If more than one sample must be taken from one load, the samples should be collected from different parts of the load.

A threshold question is the size of the sample collected from each truck. Various sample sizes have been used, ranging from 50 lb to the entire load. Large samples have the following advantages:

The variation (standard deviation) between samples is smaller, so fewer samples are required to achieve a given level of precision.
The distribution of the results of sorting the samples is closer to a normal distribution (bell-shaped curve).
The boundary area between the sample and the remainder of the load is smaller in proportion to the volume of the sample, making the sampler's decisions on

whether to include bulky items from the boundary area less significant.

Small samples have a single advantage: shorter collection and sorting time.

A consensus has developed (SCS Engineers 1979; Klee and Carruth 1970; Britton 1971) that the optimum sample size is 200 to 300 lb (91 to 136 kg). This size range is recommended in ASTM D 5231. The advantages of increasing the sample size beyond this range do not outweigh the reduced number of samples that can be sorted. If the sample size is less than 200 lb, the boundary area around the sample is too large compared to the volume of the sample, and the sampler must make too many decisions about whether to include boundary items in the sample.

Environmental engineers use several general procedures to obtain samples of 200 to 300 lb from loads of MSW, including the following:

Assembling a composite sample from material taken from predetermined points in the load (such as each corner and the middle of each side)
Coning and quartering
Collecting a grab sample from a randomly selected point using a front-end loader
Manually collecting a column of waste from a randomly selected location

Numerous variations and combinations of these general procedures can also be used.

The primary disadvantage of composite samples is the same as that for small samples: the large boundary area forces the sampler to make too many decisions about whether to include items of waste in the sample. A composite sample tends to be a judgement sample rather than a random sample. A secondary disadvantage of composite samples is that they take longer to collect than grab samples or column samples.

A variation of composite sampling is to assemble each sample from material from different loads of waste. This approach has the same disadvantages as composite sampling from a single load of waste and is even more time-consuming.

In coning and quartering, samplers mix a large quantity of waste to make its characteristics more uniform, arrange the mixed waste in a round pile (coning), and randomly select a portion—typically one quarter—of the mixed waste (quartering). The purpose is to combine the statistical advantages of large samples with the reduced sorting time of smaller samples. The coning and quartering process can begin with the entire load of waste or with a portion of the load and can be performed once or multiple times to obtain a single sample. ASTM D 5231 specifies one round of coning and quartering, beginning with approximately 1000 lb of waste, to obtain a sample of 200 to 300 lb.

Coning and quartering has the following disadvantages and potential difficulties compared to grab sampling or column sampling:

Substantially increases sampling time
Requires more space
Requires the use of a front-end loader for relatively long periods. Many solid waste facilities can make a front-end loader and an operator available for brief periods, but some cannot provide a front-end loader for the longer periods required for coning and quartering.
Tends to break trash bags, making the waste more difficult to handle
Increases sorting time by breaking up clusters of a category of waste
Reduces accuracy of sorting by increasing the percentage of food waste adhering to or absorbed into other waste items
Promotes loss of moisture from the sample
Promotes stratification of the waste by density and particle size. The biasing potential of stratification is minimized if the quarter used as the sample is a true pie slice, with its sides vertical and its point at the center of the cone. This shape is difficult to achieve in practice.

The advantage of coning and quartering is that it reduces the variation (the standard deviation) among the samples, thereby reducing the number of samples that must be sorted. Coning and quartering is justified if it reduces the standard deviation enough to make up for the disadvantages and potential difficulties. If coning and quartering is done perfectly and completely, sorting the final sample is equivalent to sorting the entire cone of waste, and the standard deviation is significantly reduced. Since the number of samples that must be sorted to achieve a given level of precision is proportional to the square of the standard deviation, coning and quartering can substantially reduce the required number of samples. Note, however, that the more thoroughly coning and quartering is performed, the more pronounced are each of the disadvantages and potential difficulties associated with this method.

A more common method of solid waste sampling is collecting a grab sample using a front-end loader. This method is relatively quick and can often be done by facility personnel without unduly disrupting normal facility operations. Sampling by front-end loader reduces the potential impact of the personal biases associated with manual sampling methods but introduces the potential for other types of bias, including the following:

Like shovel sampling, front-end loader sampling tends to favor small and dense objects over large and light objects. Large and light objects tend to be pushed away or to fall away as the front-end loader bucket is inserted, lifted, or withdrawn.

On the other hand, the breaking of trash bags as the front-end loader bucket penetrates the load of waste tends to release dense, fine material from the bags, reducing the representation of this material in the sample.

Front-end loader samples taken at ground level favor waste that falls off the top and sides of the load, which may not have the same characteristics as waste that stays in place. On dirt surfaces, front-end loader samples taken at ground level can be contaminated with dirt.

The impact of these biasing factors can be reduced if the sampling is done carefully and the sampling personnel correct clear sources of bias, such as bulky objects falling off the bucket as it is lifted.

In front-end loader sampling, sampling personnel can use different sampling points for different loads to ensure that the various horizontal and vertical strata of the loads are represented in the samples. They can vary the sampling point either randomly or in a repeating pattern. The extent of the bias that could result from using the same sampling point for each load is not known.

An inherent disadvantage of front-end loader sampling is the difficulty in estimating the weight of the samples. Weight can only be estimated based on volume, and samples of equal volume have different weights.

A less common method of solid waste sampling is manually collecting a narrow column of waste from a randomly selected location on the surface of the load, extending from the bottom to the top of the load. This method has the following advantages:

- No heavy equipment is required.
- Sampling time is relatively short.
- Because different horizontal strata of the load are sampled, the samples more broadly represent the load than grab samples collected using a front-end loader. Note, however, that loads are also stratified from front to back, and column samples do not represent different vertical strata.
- The narrowness of the target area within the load minimizes the discretion of the sampler in choosing waste to include in the sample.

The major disadvantage of column sampling is that manual extraction of waste from the side of a well-compacted load is difficult, and the risk of cuts and puncture wounds from pulling on the waste is substantial.

Of the many hybrid sampling procedures that combine features of these four general procedures, two are worthy of particular note. First, in the sampling procedure specified in ASTM D 5231, a front-end loader removes at least 1000 lb (454 kg) of material along one entire side of the load; and this waste is mixed, coned, and quartered to yield a sample of 200 to 300 lb (91 to 136 kg). Compared to grab sampling using a front-end loader, the ASTM method has the advantage of generating samples more broadly representative of the load but has the disadvantage of increasing sampling time.

In a second hybrid sampling procedure, a front-end loader loosens a small quantity of waste from a randomly selected point or column on the load, and the sample is collected manually from the loosened waste. This method is safer than manual column sampling and provides more control over the weight of the sample than sampling by front-end loader. This method largely avoids the potential biases of front-end loader sampling but tends to introduce the personal biases of the sampler.

Number of Samples Required to Estimate Composition

The number of samples required to achieve a given level of statistical confidence in the overall results is a function of the variation among the results for individual samples (standard deviation) and the pattern of the distribution of the results. Neither of these factors can be known in advance, but both can be estimated based on the results of other studies.

ASTM D 5231 prescribes the following equation from classical statistics to estimate the number of samples required:

$$n = (t^*s/ex)^2 \qquad 8.2(4)$$

where:

n = required number of samples
t^* = student t statistic corresponding to the level of confidence and a preliminary estimate of the required number of samples
s = estimated standard deviation
e = level of precision
x = estimated mean

Table 8.2.1 shows representative values of the coefficient of variation and mean for various solid waste components. The coefficient of variation is the ratio of the standard deviation to the mean, so multiplying the mean by the coefficient of variation calculates the standard deviation. Table 8.2.2 shows values of the student t statistic.

Table 8.2.1 shows the coefficients of variation rather than standard deviations because the standard deviation tends to increase as the mean increases, while the coefficient of variation tends to remain relatively constant. Therefore, the standard deviations for sets of means different from those in the table can be estimated from the coefficients of variation in the table.

The confidence level is the statistical probability that the true mean falls within a given interval above and below the mean, with the mean as the midpoint (the confidence interval or confidence range). A confidence level of 90% is generally used in solid waste studies. The confidence interval is calculated based on the results of the study (see Table 8.2.3 later in this section).

TABLE 8.2.1   REPRESENTATIVE MEANS AND COEFFICIENTS OF VARIATION FOR MSW COMPONENTS

Waste Category	Mean (%)	Coefficient of Variation[a] (%)
Organics/Combustibles	86.6	10
Paper	39.8	30
Newspaper	6.8	80
Corrugated	8.6	95
Kraft	1.5	120
Corrugated & kraft	10.1	85
Other paper[b]	22.9	40
High-grade paper	1.7	230
Other paper[b]	21.2	40
Magazines	2.1	160
Other paper[b]	19.1	40
Office paper	3.4	—
Magazines & mail	4.0	90
Other paper[b]	17.2	40
Yard waste	9.7	160
Grass clippings	4.0	300
Other yard waste	5.7	180
Food waste	12.0	70
Plastic	9.4	40
PET bottles	0.40	100
HDPE bottles	0.70	95
Other plastic	8.3	50
Polystyrene	1.0	95
PVC bottles	0.06	200
Other plastic[b]	7.2	50
Polyethylene bags & film	3.7	45
Other plastic[b]	3.5	80
Other organics	15.7	55
Wood	4.0	170
Textiles	3.5	—
Textiles/rubber/leather	4.5	110
Fines	3.3	70
Fines <$\frac{1}{2}$ inch	2.2	80
Disposable diapers	2.5	110
Other organics	1.4	160
Inorganics/Noncombustibles	13.4	60
Metal	5.8	70
Aluminum	1.0	70
Aluminum cans	0.6	95
Other aluminum	0.4	120
Tin & bimetal cans	1.5	70
Other metal[b]	3.3	130
Ferrous metal	4.5	85
Glass	4.8	70
Food & beverage containers	4.3	85
Batteries	0.1	160
Other inorganics		
With noncontainer glass	3.2	160
Without noncontainer glass	2.7	200

[a]Standard deviation divided by the mean, based on samples of 200 to 300 pounds.

[b]Each "other" category contains all material of the previous type except material in those categories.

TABLE 8.2.2   VALUES OF STUDENT *t* STATISTIC

Number of Samples (n)	Student t Statistic	
	90% Confidence	95% Confidence
2	6.314	12.706
3	2.920	4.303
4	2.353	3.182
5	2.132	2.776
6	2.015	2.571
7	1.943	2.447
8	1.895	2.365
9	1.860	2.306
10	1.833	2.262
12	1.796	2.201
14	1.771	2.160
17	1.746	2.120
20	1.729	2.093
25	1.711	2.064
30	1.699	2.045
41	1.684	2.021
51	1.676	2.009
61	1.671	2.000
81	1.664	1.990
101	1.660	1.984
141	1.656	1.977
201	1.653	1.972
Infinity	1.645	1.960

The desired level of precision is the maximum acceptable error, expressed as a percentage or decimal fraction of the estimated mean. Note that a lower precision level indicates greater precision. A precision level of 10% (0.1) is frequently set as a goal but is seldom achieved.

After a preliminary value for n based on a preliminary value for t* is calculated, the calculation is repeated with the value of t* corresponding to the preliminary value for n.

Equation 8.2(4) assumes that the values for each variable to be measured (in this case the percentages of each solid waste component in the different samples) are normally distributed (conform to the familiar bell-shaped distribution curve, with the most frequent value equaling the mean). In reality, solid waste composition data are not normally distributed but are moderately to severely skewed right, with numerous values several times higher than the mean. The most frequent value is invariably lower than the mean, and in some cases is close to zero. The greater the number of waste categories, the more skewed the distributions of individual categories are.

Klee (1991; 1993) and Klee and Carruth (1970) have suggested equations to account for the effect of this skewness phenomenon on the required number of samples. Use of these equations is problematic. Like Equation 8.2(4), they are designed for use with one waste category at a time. For waste categories for which the mean is large compared to the standard deviation, the equations yield higher numbers of samples than Equation 8.2(4). This result is intuitively satisfying because more data should be needed to quantify a parameter whose values do not follow a predefined, normal pattern of distribution. For waste categories for which the mean is less than twice as large as the standard deviation, however, these equations tend to yield numbers of samples smaller than Equation 8.2(4). This result is counterintuitive since no reason is apparent for why an assumption of nonnormal distribution should decrease the quantity of data required to characterize a highly variable parameter.

An alternative method of accounting for skewness is to select or develop an appropriate equation for each waste category based on analysis of existing data for that category. Hilton, Rigo, and Chandler (1992) provide the results of a statistical analysis of the skewness of individual waste categories.

Equation 8.2(4) gives divergent results for different solid waste components. Based on the component means and coefficients of variation shown in Table 8.2.1 and assuming a precision of 10% at 90% confidence, the number of samples given by Equation 8.2(4) is 45 for paper other than corrugated, kraft, and high-grade; almost 700 for all yard waste; and more than 2400 for just grass clippings. The value of Equation 8.2(4) alone as a guide in designing a sampling program is therefore limited.

An alternative method is to estimate the number of samples required to achieve a weighted-average precision level equal to the required level of precision. The weighted-average precision level is the average of the precision levels for individual waste categories weighted by the means for the individual waste categories. The precision level for individual waste categories can be estimated with the following equation, which is Equation 8.2(4) solved for e:

$$e = t*s/xn^{1/2} \qquad 8.2(5)$$

The precision level for each category is multiplied by the mean for that category, and the results are totaled to yield the weighted-average precision level. The number of samples (n) is adjusted by trial and error until the weighted-average precision level matches the required value.

Calculation of the weighted-average precision level is shown in Table 8.2.3 later in this section. Figure 8.2.1 shows the relationship of the weighted-average precision level to the number of samples and the number of waste categories based on the values in Table 8.2.1. Overall precision improves as the number of samples increases and as the number of waste categories decreases. This statement does not mean that studies involving greater number of categories are inferior; it simply means that determining a few things precisely is easier than determining many things precisely.

Sorting and Weighing Samples of MSW

In most cases, sorting solid waste should be viewed as an industrial operation, not as laboratory research. While accuracy is essential, the appropriate measure of accuracy is ounces rather than grams or milligrams. Insistence on an excessive level of accuracy slows down the sorting process, reducing the number of samples that can be sorted. This

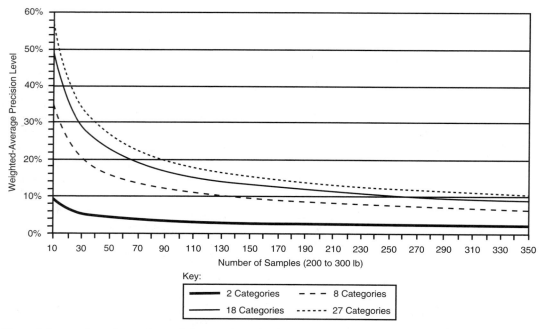

FIG. 8.2.1 Effect of the number of samples and the number of waste categories on weighted-average precision level (derived from Table 8.2.1).

reduction, in turn, reduces the statistical precision of the results. In the context of an operation in which a 10% precision level is a typical goal, inaccuracy of 1% is relatively unimportant.

The principles of industrial operations apply to solid waste sorting, including minimization of motion and maintenance of worker comfort and morale.

SORTING AREAS

A sorting area is established at the beginning of the field work and should have the following characteristics:

- A paved surface approximately 1000 sq ft in area and at least 16 ft wide
- Accessibility to vehicles
- Protection from precipitation and strong winds
- Heating in cold weather
- Separation from traffic lanes and areas where heavy equipment is used but within sight of arriving trucks

A typical sorting operation might use two sorting boxes and a crew of ten to twelve. The crew includes two sorting teams of four or five persons each, a supervisor, and a utility worker. The basic sorting sequence, starting when collection of the sample is complete, is as follows:

1. The sample is transported from the sampling point to the sorting area. A pickup truck or front-end loader can be used for this purpose.
2. The sampler gives the sorting supervisor a copy of a data form.
3. The sample is unloaded onto the surface of the sorting area.
4. Large items (e.g., corrugated cardboard and wood) and bags containing a single waste category (most often yard waste) are removed from the sample and set aside for weighing, bypassing the sorting box.
5. The remainder of the sample is transferred by increments into the sorting box, using broad-bladed shovels to transfer loose material.
6. The waste is sorted into the containers surrounding the sorting box.
7. The containers are brought to the scale, checked for accuracy of sorting, and weighed.
8. The gross weight of the waste and container and a letter symbol indicating the type of container are recorded on the data form.
9. If required, the waste in the containers is subsampled for laboratory analysis.
10. The containers are dumped in a designated receptacle or location.

The supervisor must ensure that each sample remains matched with the correct data form and that waste does not cross between samples.

SORTING CONTAINERS

Use of a counter-height sorting box speeds sorting, decreases worker fatigue, and encourages interaction among the sorters. All of these factors help build and sustain the morale of the sorters.

The following sorting box design has proven highly effective. The box is 4 ft wide, 6 ft long, 1 ft deep, and open at the top. It is constructed of $\frac{3}{8}$-in or $\frac{1}{2}$-in plywood with an internal frame of 2-by-3s or 2-by-4s. The long framing pieces extend 1 foot beyond the ends of the box at each bottom corner, like the poles of a stretcher. These framing pieces facilitate handling and extend the overall dimensions of the box to 4 ft by 8 ft by 1 ft. The box can lie flat within the bed of a full-sized pickup truck or standard cargo van.

A screen of $\frac{1}{2}$-in hardware cloth (wire mesh with $\frac{1}{2}$-in square openings) can be mounted in the bottom of the sorting box, $1\frac{1}{2}$ in from the bottom (the thickness of the internal framing pieces). If the screen is included, one end of the box must be open below the level of the screen to allow dumping of the fine material that falls through the screen. By allowing fine material to separate from the rest of the sample, the screen facilitates sorting of small items and makes dangerous items such as hypodermic needles easier to spot.

To facilitate dumping of the fines and to save space during transportation and storage, the sorting box is built without legs. During sorting, the sorting box is placed on a pair of heavy-duty sawhorses, 55-gal drums, or other supports. A support height of 32 in works well for a mixed group of male and female sorters. Fifty-five-gal drums are approximately 35 in high, approximately 3 in higher than optimum, and because of their size are inconvenient to store and transport.

The containers into which the waste is sorted should be a combination of 30-gal plastic trash containers and 5-gal plastic buckets. The 5-gal buckets are used for low-volume waste categories. Containers larger than 30 gal occupy too much space around the sorting box for efficient sorting and can be heavy when full. In a typical study with twenty-four to twenty-eight waste categories, each sorting crew should be equipped with approximately two dozen 30-gal containers and one dozen 5-gal buckets. In addition, each sorting crew should have several shallow plastic containers approximately 18 in wide, 24 in long, and 6 in deep.

For optimum use of space, the 30-gal containers should have rectangular rims. They should also have large handles to facilitate dumping. Recessed handholds in the bottom of the container are also helpful. In general, containers of heavy-duty HDPE are best. Because of their molded rims, these containers can be inverted and banged against pavement, the rim of a rolloff container, or the rim of a matching container to dislodge the material adhering to the inside of the container. The containers need not have wheels. Plastic containers slide easily across almost any flat surface.

Substantial field time can be saved when the containers of each type have fairly uniform weights so that each type of container can be assigned a tare weight rather than each container. When container weights are recorded on the data form after sorting, recording a letter code that refers to the type of container is faster than reading an individual tare weight on the container and recording it on the data form.

Assigning individual tare weights to containers weighing 2% more or less than the average weight for the container type is unnecessary. Batches of 5-gal buckets generally meet this standard, but many 30-gal containers do not. Ensuring that tare weights are consistent requires using portable scale when shopping for containers.

CONTAINER LABELING

Most sorting protocols, including ASTM D 5231, call for labeling each container to indicate which waste category is to be placed in it. When a sorting box is used, however, unlabeled containers have the following advantages:

The sorters are encouraged to establish a customary location for each waste category and sort by location, which is faster than sorting by labels.
When sorting is done by location rather than by labels, the containers can be placed closer to the sorters, which further speeds the sorting process.
Less time is required to arrange unlabeled containers around the sorting box after the sorted material from the previous sample has been weighed and dumped.
Keeping the containers unlabeled increases the flexibility of the sorting operation.

The flexibility gained by not labeling the containers has several aspects. First, different samples require multiple 30-gal containers for different waste categories. Second, many waste categories require a 30-gal container for some samples and only a 5-gal container for others. Third, the need for another empty container arises frequently in an active sorting operation, and grabbing the nearest empty container is quicker than searching for the container with the appropriate label.

Despite the advantages of unlabeled containers, the containers for food waste should be labeled. If individual containers are not designated for food waste, all containers will eventually be coated with food residue. This residue is unpleasant and changes the tare weights of the containers.

The tare weights of the food waste containers should be checked daily. Generally, checking the tare weights of other containers at the beginning of each week of field work is sufficient unless a visible buildup of residue indicates that more frequent checking is required.

SORTING PROCESS

The actual sorting of the sample should be organized in the following basic manner:

Each waste category is assigned a general location around the perimeter of the sorting box. In one effective arrangement, paper categories are sorted to one side of the sorting box, plastic categories are sorted to the other side, other organic categories are sorted to one end, and inorganic categories are sorted to the other end.
Each sorter is assigned a group of categories. With a typical sorting crew of four, each sorter is assigned the categories on one side or at one end of the box.
The sorters place their assigned materials in the appropriate containers and place other materials within reach of the sorters to which they are assigned.
Toward the end of sorting each sample, one of the shallow containers is placed in the middle of the sorting box, and all sorters place other paper in this container (see Table 8.2.1). This process can be repeated for food waste.
When only scattered or mixed bits of waste remain, sorting is suspended.
The material remaining above the screen in the sorting box, or on the bottom of a box without a screen, is scraped or brushed together and either (1) distributed among the categories represented in it in proportion to their abundance, (2) set aside as a separate category, or (3) set aside to be combined with the fine material from below the screen. ASTM D 5231 specifies the first alternative, but it should not be selected if the waste categories are to be subsampled for laboratory testing.
If the sorting box has a screen, the box is upended to allow the fine material from below the screen to fall through the slot at one end of the box. The material that falls out is swept together and shoveled into a container—preferably a wide, shallow container—for weighing.

WEIGHING SAMPLES

ASTM D 5231 specifies the use of a mechanical or electronic scale with a capacity of at least 200 lb (91 kg) and precision of 0.1 lb (0.045 kg) or better. When 30-gal containers are used in sorting samples of 200 to 300 lb, gross weights greater than 100 lb are unusual. Even if larger containers or sample sizes are used, sorting personnel should avoid creating containers with gross weights greater than 100 lb because they are difficult and dangerous to handle. For most sorting operations, a scale capacity of 100 lb is adequate. An electronic scale with a range of 0–100 lb is generally easier to read to within 0.1 lb than a mechanical scale with a range of 0–100 lb.

A platform-type scale is preferred. The platform should be 1 ft square or larger.

The digital displays on electronic scales make data recording easier and minimize recording errors by displaying the actual number to be recorded on the data form. When recording weights from a mechanical scale, interpolation between two values marked on the dial is often required. The advantages of mechanical scales are lower cost, reliability, and durability.

Ideally, one worker places containers on the scale, the supervisor checks the containers for accuracy of sorting and records the weights and container types, and two or more workers dump the weighed containers. If the containers are subsampled for laboratory analysis prior to being dumped, the process is much slower and fewer workers are required.

DUMPING SAMPLES

On landfills, the sorting containers are dumped near the sorting area for removal or in-place burial by facility personnel. In transfer stations and waste-to-energy facilities, the containers can be dumped on the edge of the tipping floor.

When the sorting area is separated from the disposal area, use of the sampling vehicle for disposal is difficult. Loads of waste that should be sampled can be missed, and sorting delays occur because the sampling vehicle is not available for dumping full containers from the previous sample. A better procedure is to dump the sorted waste in a rolloff container provided by the disposal facility. Facility personnel transport the rolloff container to the disposal area approximately once per day. The density of sorted waste is often as low as 150 lb/cu yd, so the rolloff tends to be filled more rapidly than expected. To facilitate dumping sorted waste over the sides, the rolloff container should not be larger than 20 cu yd (15.3 cu m).

Processing the Results of Sorting

After a sample is weighed and the gross weights and container types are recorded on the data form, the net weights are calculated and recorded on the data form. Total net weights are calculated for waste categories sorted into more than one container. Field personnel should calculate net category weights and total net sample weights after each day of sorting to monitor the size of the samples. Undersize samples decrease the accuracy and statistical precision of the results and can violate the contract under which the study is conducted. Oversize samples make sorting the required number of samples more difficult.

The net weights for each waste category in each sample are usually entered into a computer spreadsheet. For each waste category in each group of samples to be analyzed (for example, residential samples and commercial samples), the following should be calculated from the data in the spreadsheet:

- The percentage by weight in each sample
- The mean percentage within the group of samples
- The standard deviation of the percentages within the group of samples
- The confidence interval around the mean

Calculating the overall composition usually involves dividing the total weight of each waste category by the total weight of the samples rather than calculating the composition of each sample and averaging the compositions. If the samples have different weights, which is usually the case, these two methods yield different results. Calculating overall composition based on total weight creates a bias in favor of dense materials, which are more abundant in the heavier samples. Averaging the compositions of the individual samples is preferable because it gives each pound of waste an equal opportunity to influence the results. ASTM D 5231 specifies averaging of sample compositions.

Table 8.2.3 shows mean percentages, standard deviations, uncertainty values, precision levels, and confidence intervals for a group of 200 MSW samples with the characteristics shown in Table 8.2.1. The confidence intervals are based on the uncertainty values (sometimes called precision values). The uncertainty values are typically calculated with the following formula:

$$U_c = t^* s / n^{1/2} \qquad 8.2(6)$$

where:

U_c = uncertainty value at a given level of confidence, typically 90%
t^* = student t statistic corresponding to the given level of confidence
s = sample standard deviation
n = number of samples

This equation is equivalent to the equation for calculating the precision level, Equation 8.2(5), with both sides multiplied by the mean, x. Dividing the uncertainty value by the mean yields the precision level. Adding the uncertainty values for all waste categories yields the weighted average precision level, weighted by the means for the individual waste categories.

Equation 8.2(6), like Equations 8.2(4) and 8.2(5), assumes that the percentage data are normally distributed. As previously discussed, this is not actually the case, and no reliable and reasonably simple method exists for estimating the effect of lack of normality on the statistical precision of the results.

Precision analysis can only be applied to groups of samples that are representative of the waste stream to be analyzed. For example, if 40% of the municipal waste stream is commercial waste but 60% of the samples sorted during a study are collected from commercial loads, statistical precision analysis of the entire body of composition data generated during the study is meaningless. Assuming that the commercial and residential samples represent the

TABLE 8.2.3   ILLUSTRATION OF WEIGHTED-AVERAGE PRECISION LEVEL AND CONFIDENCE INTERVALS[a]

Waste Category	Mean (%) (x)	Standard Deviation (%) (s)	Student t Statistic (t*) for 200 Samples (n) and 90% Confidence	Uncertainty Value (%) ($U_{90} = t^*s/n^{1/2}$)	Precision Level (%) (U_{90}/x)	90% Confidence Interval (%)
Newspaper	6.8	5.4	1.653	0.6	9.4	6.8 ± 0.6
Corrugated & kraft	10.1	8.6	1.653	1.0	9.9	10.1 ± 1.0
Other paper	22.9	9.2	1.653	1.1	4.7	22.9 ± 1.1
Yard waste	9.7	15.5	1.653	1.8	18.7	9.7 ± 1.8
Food waste	12.0	8.4	1.653	1.0	8.2	12.0 ± 1.0
PET bottles	0.4	0.4	1.653	0.05	11.7	0.4 ± 0.05
HDPE bottles	0.7	0.7	1.653	0.1	11.1	0.7 ± 0.1
Other plastic	8.3	4.1	1.653	0.5	5.8	8.3 ± 0.5
Wood	4.0	6.8	1.653	0.8	19.9	4.0 ± 0.8
Textiles/rubber/leather	4.5	5.0	1.653	0.6	12.9	4.5 ± 0.6
Fines	3.3	2.3	1.653	0.3	8.2	3.3 ± 0.3
Disposable diapers	2.5	2.8	1.653	0.3	12.9	2.5 ± 0.3
Other organics	1.4	2.2	1.653	0.3	18.7	1.4 ± 0.3
Aluminum	1.0	0.7	1.653	0.1	8.2	1.0 ± 0.1
Tin & bimetal cans	1.5	1.1	1.653	0.1	8.2	1.5 ± 0.1
Other metal	3.3	4.3	1.653	0.5	15.2	3.3 ± 0.5
Food & beverage containers	4.3	3.7	1.653	0.4	9.9	4.3 ± 0.4
Other inorganics	3.3	5.3	1.653	0.6	18.7	3.3 ± 0.6
Total or weighted average	100.0			10.1	10.1	100.0 ± 10.1

[a]Based on 200 samples, 90% confidence, and the eighteen waste categories listed in the table. Means and standard deviations are based on Table 8.2.1.

respective fractions of the waste stream from which they were collected, separate precision analysis of the commercial and residential results is valid. Representativeness is achieved by either random selection of loads to sample or systematic selection of loads based on preexisting data.

Visual Characterization of Bulky Waste

The composition of bulky waste is typically estimated by observation rather than by sorting samples. Visual characterization of bulky waste is feasible for several reasons: (1) most bulky waste is not hidden in bags, (2) most loads of bulky waste contain few categories of waste, and (3) the categories of waste present are usually not thoroughly dispersed within the load, as they are in loads of MSW. Conversely, sorting samples of bulky waste is problematic for several reasons: (1) because the variation among loads of bulky waste is large, a large number of trucks must be sampled, (2) because the waste categories are not thoroughly dispersed within the loads, the samples must be large, (3) sorting and weighing bulky waste is difficult and dangerous if not done with specialized mechanical equipment.

Estimating the composition of bulky waste based on observation has three phases. First, field personnel prepare field notes describing each load as the load is dumped, as the load sits on the tipping floor or landfill after dumping,

and as the heavy equipment operators move the load around the tipping floor or the working face of the landfill. Second, they determine or estimate the weight of each load. Third, they combine the field notes and load weights to develop an estimate of the composition of each load and of the bulky waste as a whole.

In general, the field notes should include the following elements for each load:

The date and exact time of day

The type of vehicle and its volumetric capacity (e.g., 30-cu-yd rolloff, 40-cu-yd trailer)

Any identifying markings that help match the field notes with the corresponding entry in the facility log for that day. Identifying markings that can be useful include the name of the hauler, the license plate number, and identifying numbers issued by regulatory agencies.

Either (1) a direct estimate of the by-weight composition of the load or (2) an estimate of the by-volume composition of the load combined with an indication of the amount of air space in each component.

If the facility does not have a scale, the facility log generally contains a volume for each load but no weight. If the volume of each load can be determined in the field, as it can when each truck or container is marked with its volumetric capacity, field notes do not have to be matched with log entries. Regardless of whether the facility log is used, the field notes should contain any information that can be helpful in estimating the weight of each load, including its total volume if different from the capacity of the vehicle in which it arrived.

Field personnel should visually characterize most if not all of the loads of bulky waste arriving at the solid waste facility during the period of field work. Because the composition of bulky waste varies from load to load, a large number of loads must be characterized.

Characterized loads of bulky waste should not be regarded as samples because they contain vastly different quantities of waste. The overall composition of bulky waste is not the mean of the results for individual loads, as with MSW. Rather, the overall composition is weighted in accordance with the weights of the individual loads. An estimate of the overall percentage of each component involves calculating the total quantity of the component in all observed loads and dividing it by the total weight of all observed loads, as illustrated by the following equation:

$$p_o = (p_1w_1 + p_2w_2 \cdots + p_nw_n)/w_o \qquad 8.2(7)$$

where:

p_o = the overall percentage of the component in the observed loads
p_1 = the percentage of the component in the first observed load
w_1 = the weight of the first observed load
p_2 = the percentage of the component in the second observed load
w_2 = the weight of the second observed load
p_n = the percentage of the component in the last observed load
w_n = the weight of the last observed load
w_o = the total weight of all observed loads

Before the overall composition can be calculated in this way, the weight of each load must be estimated. If the facility has a scale, environmental engineers can determine the actual weight of the observed loads by matching the field notes for each load with the corresponding entry in the facility log, based on the time of arrival and information about the truck and the load. The time of arrival recorded in the facility log is the time when the truck was logged in rather than the time when the load was discharged. Field personnel must determine the difference between the two times.

If the facility does not have a scale, environmental engineers must estimate the weight of each component and the total weight of the load by converting from cubic yards to tons. The following procedure is suggested:

The total volume of the load is distributed among the components of the load based on the field notes.
The weight of each component is estimated based on its volume and density. Table 8.1.3 shows density ranges for certain waste components.
The estimated component weights are added yielding the estimated total weight of the load.

The cost of a study can be reduced if the same person collects MSW samples for sorting and performs visual characterization of bulky waste during the same period of field work. This technique is feasible if loads of MSW and bulky waste are dumped in the same part of the facility and if a quick method is used for collecting MSW samples.

Sampling MSW for Laboratory Analysis

Obtaining meaningful laboratory results for MSW is difficult. The primary sources of difficulty are (1) the presence of many different types of objects in MSW and (2) the large size of these objects. Collecting small but representative samples from a homogeneous pile of small objects (e.g., a pile of rice) is easier than from a heterogeneous pile of large objects. Secondary sources of difficulty in sampling MSW include the uneven distribution of moisture and inconsistent laboratory procedures.

MIXED SAMPLE VERSUS COMPONENT SAMPLE TESTING

An initial choice to be made is whether to test mixed samples or individual waste components. Testing mixed samples is preferable when:

- The only purpose of the laboratory testing is to determine the characteristics of the mixed waste stream, such as heat value.
- The statistical precision of the laboratory results must be demonstrated.
- The study does not include sorting waste samples.
- No significant changes in the composition of the waste stream are anticipated.

Testing of individual waste components is necessary, of course, when the characteristics of individual waste components must be determined. In addition, component testing makes projecting the impact of changes in the component composition of the waste, such as changes caused by recycling and composting programs, possible. Component testing also enhances quality control because laboratory errors are easier to detect in the results for individual components than in those for mixed samples.

The procedures for collecting mixed samples for laboratory testing are essentially the same as those for collecting mixed samples for sorting. The preceding evaluation of these procedures also applies to the collection of mixed samples for laboratory testing, except for the comments concerning the impacts of various sampling procedures on the sorting process.

Laboratory samples of individual waste components are usually composite subsamples of samples sorted to estimate composition. In general, each component laboratory subsample includes material from each sorted sample. Material for the laboratory subsamples is collected from the sorting containers after the sorting and weighing are complete.

LABORATORY PROCEDURES

A fundamental question is how large should the samples sent to the laboratory be. The answer to this question depends on the procedures used by the laboratory. A state-of-the-art commercial laboratory procedure includes the following steps:

A portion of the sample material sent to the laboratory is weighed, dried, and reweighed to determine the moisture content. The limiting factor at this stage of the procedure is usually the size of the laboratory's drying oven.

A portion of the dried material is ground into particles of $\frac{1}{8}$ to $\frac{1}{4}$ in.

A portion of the $\frac{1}{8}$-to-$\frac{1}{4}$-in material is finely ground into as close to a powder as possible. For flexible plastic, dry ice must be added prior to fine grinding to make it more brittle.

The actual laboratory test is generally performed on 0.5 to 3 g of the finely ground material, depending on the type of test and the specific equipment and procedures.

Variations on this procedure include the following:

Most laboratories do not have equipment for grinding inorganic materials such as glass and metal. In combustion testing, this material is removed from the sample prior to grinding, then weighed and reported as ash. For metals testing, metal objects can be cut up by hand or drilled to create small pieces for testing. Glass and ceramics are typically crushed.

Many laboratories do not have fine grinding equipment, so they perform tests on relatively coarse material.

In addition to using different methods for preparing waste for testing, laboratories use different test methods.

The more sample material the laboratory receives, the more material they must exclude from the small quantity of material that is tested. The real question is not how large the samples should be but how field and laboratory personnel should share the task of reducing samples to a gram or two. For practical purposes, the maximum quantity sent to the laboratory should be the quantity the laboratory is prepared to spread out and mix in preparation for selecting the material to be dried. The minimum quantity should be the quantity the laboratory is prepared to dry and grind up.

Composite laboratory samples are typically accumulated in plastic trash bags, then boxed for shipment. An alternative is to accumulate the samples in 5-gal plastic buckets with lids. Plastic buckets are more expensive than plastic bags but have several advantages:

Plastic buckets (and their lids) are easier to label, and the labels are easier to read.

Adding material to plastic buckets is easier.

The lids, which are lifted only when material is added to the buckets, prevent moisture loss during the active sampling period.

Sample material can be compacted in plastic buckets if it is pushed down around the inside edge.

The buckets can be used as shipping containers.

The buckets can be reused if the laboratory ships them back.

COLLECTING MATERIAL FOR LABORATORY SUBSAMPLES

Three general methods for collecting material for laboratory subsamples from containers of sorted waste are blind grab sampling, cutting (or tearing) representative pieces from large objects, and selecting representative whole objects for inclusion in the sampling. Blind grab sampling is the preferred approach for waste that mainly consists of small objects. Cutting representative pieces is appropriate for waste consisting of large objects with potentially different characteristics. Selecting representative whole objects is appropriate for waste containing only a few different types of objects.

Blind grab samples should be collected by hand or with an analogous grasping tool. The objective is to extract the material from a randomly selected but defined volume within the container of sorted material. When scoops and shovels are used in sampling heterogeneous materials, they tend to create bias by capturing dense, small objects while pushing light, large objects away.

In collecting subsamples from containers of sorted waste, samplers must realize that because sorting progresses from larger objects to smaller, the objects at the top of the container tend to be smaller than those at the bottom. Objects of different sizes can have different characteristics, even within the same waste category. Therefore, the sampler must ensure that the objects at different levels of the containers are represented in the samples. Emptying the container onto a dry and reasonably clean surface prior to collecting the subsample may be necessary.

If the laboratory samples are tested for metals, objects with known metals content should not be represented in the samples. Instead, such objects should be weighed, and the laboratory results should be adjusted to reflect the quantities of metals they contain. For example, if 8 oz of lead weights are found in 10 tn of sorted waste, the weights represent 25 ppm of lead. The weights should be withheld from the laboratory sample, and 25 ppm should be added to the overall lead concentration indicated by the laboratory results. This procedure is more accurate than laboratory testing alone.

Review and Use of Laboratory Results

Laboratory procedures are imperfect, and errors in using the procedures and in calculating and reporting the results are common. Reviewing the results received from a laboratory to see if they make sense is important. This exercise is relatively straightforward for combustion characteristics because much is known about the combustion characteristics of solid waste and its component materials (see Section 8.1). Identification of erroneous laboratory results is more difficult for metals and toxic organic substances.

The following guidelines apply in an evaluation of reasonableness of laboratory results for combustion characteristics on a dry basis:

Dry-basis results for the paper, yard waste, plastics, wood, and disposable diapers categories should be close to those shown in Tables 8.1.4 and 8.1.5.

Greater variability must be accepted in individual results for food waste, textiles/rubber/leather, fines, and other combustibles because of the chemical variety of these categories.

The result for carbon must always be at least six times the result for hydrogen.

No oxygen result should be significantly higher than 50%.

For plant-based materials and mixed food waste, oxygen results should not be significantly less than 30% on an ash-free basis.

Among the paper categories, only those with high proportions of glossy paper, such as magazines and advertising mail, should have ash values significantly greater than 10%.

Nitrogen should be below 1% for all categories except grass clippings, other yard waste, food waste, textiles/rubber/leather, fines, and other organics (see Table 8.1.4).

Chlorine should be below 1% for all categories except for PVC bottles, other plastic, textiles/rubber/leather, and other organics.

Sulfur should be below 1% for all categories except other organics.

The laboratory should be willing to check its calculations and repeat the test if the calculations are not the source of the problem.

Estimating Combustion Characteristics Based on Limited Laboratory Testing

The combustion characteristics of individual waste categories on a dry basis are well documented and fairly con-

TABLE 8.2.4 HEAT VALUE ESTIMATES BASED ON BOIE, CHANG, AND DULONG EQUATIONS

Equation	Dry-Basis HHV (Btu/lb)	As-Received HHV (Btu/lb)
Boie	7395	5310
Chang	7479	5370
DuLong	7510	5392
Average	7461	5357
Laboratory values	7446	5348

sistent within categories. Moisture and component composition are more variable. One option, therefore, is to sort samples to estimate component composition and have subsamples tested for moisture only. Then, with the use of the documented values for the proximate and ultimate composition and heat value of each waste component, the overall combustion characteristics of the waste stream can be estimated.

Another potential cost-saving measure is to estimate heat value based on ultimate composition. Several equations have been proposed for this purpose (Niessen 1995):

BOIE EQUATION

$$HHV = 14{,}976C + 49{,}374H - 4644O + 2700N + 4500S + 1692Cl + 11{,}700P \quad 8.2(8)$$

CHANG EQUATION

$$HHV = 15{,}410 + 32{,}350H - 11{,}500S - 20{,}010O - 16{,}200Cl - 12{,}050N \quad 8.2(9)$$

DULONG EQUATION

$$HHV = 14{,}095.8C + 64{,}678(H - O/8) + 3982S + 2136.6O + 1040.4N \quad 8.2(10)$$

where:

HHV = higher heating value in Btu/lb

Percentages for each element must be converted to decimals for use in these equations (i.e., 35% must be converted to 0.35). Using the values in Table 8.1.4 in the three equations yields the results shown in Table 8.2.4.

These values are close to the overall values in Table 8.1.5, which are based on laboratory testing of the same samples on which the ultimate composition in Table 8.1.4 is based. The laboratory-based values are closer to the average results for the three equations than to the results for any individual equation.

—*F. Mack Rugg*

References

American Society for Testing and Materials. 1992. *Standard test method for determination of the composition of unprocessed municipal solid waste.* ASTM Method D 5231-92 (September).

Britton, P.W. 1971. *Improving manual solid waste separation studies.* U.S. EPA (March).

Franklin Associates, Ltd. 1992. *Characterization of municipal solid waste in the United States: 1992 update.* U.S. EPA, EPA/530-R-92-019, NTIS no. PB92-207 166 (July).

Gay, A.E., T.G. Beam, and B.W. Mar. 1993. Cost-effective solid-waste characterization methodology. *J. of Envir. Eng.* (ASCE) 119, no. 4 (Jul/Aug).

Hilton, D., H.G. Rigo, and A.J. Chandler. 1992. Composition and size distribution of a blue-box separated waste stream. Presented at SWANA's Waste-to-Energy Symposium, Minneapolis, MN, January 1992.

Klee, A.J. 1991. *Protocol: A computerized solid waste quantity and composition estimation system.* Cincinnati: U.S. EPA Risk Reduction Engineering Laboratory.

———. 1993. New approaches to estimation of solid-waste quantity and composition. *J. of Envir. Eng.* (ASCE) 119, no. 2 (Mar/Apr).

Klee, A.J. and D. Carruth. 1970. Sample weights in solid waste composition studies. *J. of the Sanit. Eng. Div., Proc. of the ASCE* 96, no. SA4 (August).

McCamic, F.W. (Ferrand and Scheinberg Associates). 1985. *Waste composition studies: Literature review and protocol.* Mass. Dept. of Envir. Mgt. (October).

Niessen, W.R. 1995. *Combustion and incineration processes: Applications in environmental engineering.* 2d ed. New York: Marcel Dekker, Inc.

SCS Engineers. 1979. *Municipal solid waste survey protocol.* Cincinnati: U.S. EPA.

8.3
IMPLICATIONS FOR SOLID WASTE MANAGEMENT

This section addresses several aspects of the relationship between the characteristics of solid waste and the methods used to manage it. Implications for waste reduction, recycling, composting, incineration, and landfilling are included, as well as general implications for solid waste management as a whole.

MSW is abundant, unsightly, and potentially odorous; contains numerous potential pollutants; and supports both disease-causing organisms and disease-carrying organisms. Like MSW, bulky solid waste is abundant, unsightly and potentially polluting. In addition, the dry, combustible nature of some bulky waste components can pose a fire hazard. Because of these characteristics of MSW and bulky waste, a prompt, effective, and reliable system is required to isolate solid waste from people and the environment.

A beneficial use of solid waste is relatively difficult because it contains many different types of materials in a range of sizes. The only established use for unprocessed MSW is as fuel in mass-burn incinerators (see Section 10.1). Even mass-burn incinerators cannot handle unprocessed bulky waste. In the past, unprocessed bulky waste was used as fill material, but this practice is restricted today. In general, processing is required to recover useful materials from both MSW and bulky waste.

Implications for Waste Reduction

Waste reduction refers to reducing the quantity of material entering the solid waste management system. Waste reduction is distinguished from recycling, which reduces the quantity of waste requiring disposal but does not reduce the quantity of material to be managed.

Based on the composition of MSW (see Section 8.1), each of the following measures would have a significant impact on the quantity of MSW entering the solid waste management system:

- Leaving grass clippings on the lawn
- Increasing backyard composting and mulching of leaves and other yard wastes
- Selling products in bulk rather than in packages, with the consumer providing the containers
- Buying no more food than is eaten
- Substituting reusable glass containers for paper, plastic, and single-use glass containers
- Reusing shopping bags
- Placing refuse directly in refuse containers instead of using trash bags
- Using sponges and cloth hand towels in place of paper towels
- Continuing to use clothing and other products until they are worn out, rather than discarding them when they no longer look new
- Prohibiting distribution of unsolicited printed advertising

Leaving grass clippings on the lawn is becoming increasingly common because of disposal bans in some states and the development of mulching lawn mowers that cut the clippings into smaller pieces. Implementation of the other waste reduction measures on the list is unlikely in the United States because they do not conform to the prevailing standards of convenience, comfort, appearance, sanitation, and free enterprise.

Implications for Waste Processing

Fluctuations in waste generation must be considered when waste processing facilities are planned. If a facility must process the entire waste stream throughout the year, it must be sized to handle the peak generation rate. Storage of MSW for later processing is limited by concerns about odor and sanitation. Limitations on the storage of bulky waste are generally less severe, but long-term storage of combustible materials is usually restricted.

Processing systems for mixed solid waste must be capable of handling a variety of materials in a range of sizes.

Because solid waste does not flow, it must be hauled or moved by conveyor. Because objects in MSW do not readily stratify by size, screening of MSW generally requires a mixing action such as that produced by trommel screens. Abrasive materials in solid waste cause abrasive wear to handling and processing equipment. Heavy, resistant items can damage size reduction equipment. Size reduction is often required, however, because bulky items in solid waste tend to jam conveyors and other waste handling equipment.

Implications for Recovery of Useful Materials

Almost all solid waste materials can be recycled in some way if people are willing to devote enough time and money to the recycling effort. Because time and money are always limited, distinctions must be drawn between materials that are more and less difficult to recycle. Table 8.3.1 shows the compostable, combustible, and recyclable fractions of MSW. The materials listed as recyclable are those for which large-scale markets exist if the local recycling industry is well developed. The list of recyclable materials is different in different areas.

Approximately 75% of the MSW discarded in the United States is compostable or recyclable. No solid waste district of substantial size in the United States has documented a 75% rate of MSW recovery and reuse, however. Reasons for this include the following:

Some recyclable material becomes unmarketable through contamination during use.

A significant fraction of recyclable material cannot be recovered from the consumer.

A portion of both recyclable and compostable material is lost during processing (sorting recyclable materials or removing nonrecyclable and noncompostable materials from the waste stream).

Some compostable material does not decompose enough to be included in the finished compost product and is discarded with the process residue.

TABLE 8.3.1 COMBUSTIBLE, COMPOSTABLE, AND RECYCLABLE COMPONENTS OF MSW[a]

Waste Category	Percentage of Total[b]
Combustible, compostable, and recyclable	**22.6**
Newspaper	6.8
Corrugated cardboard	8.6
Kraft paper	1.5
High-grade paper	1.7
Magazines & mail	4.0
Recyclable and combustible but not compostable	**2.1**
PET bottles	0.4
HDPE bottles	0.7
Polyethylene film other than trash bags	1.0
Recyclable but not compostable or combustible	**7.9**
Aluminum cans	0.6
Tin & bimetal food & beverage cans	1.5
Other metal[c]	1.5
Glass food and beverage containers	4.3
Compostable and combustible but not recyclable	**44.7**
Other paper	17.2
Yard waste	9.7
Food waste	12.0
Disposable diapers	2.5
Fines	3.3
Combustible but not compostable or recyclable	**17.2**
Other plastic	7.3
Wood	4.0
Textiles/rubber/leather	4.5
Other organics	1.4
Not combustible or compostable or recyclable	**5.5**
Other aluminum	0.4
Other metal[c]	1.8
Batteries	0.1
Other inorganics	3.2
Total recyclable[a]	32.6
Total compostable	67.3
Total combustible	86.6

[a]Materials listed as recyclable are those for which large-scale markets exist in areas where the recycling industry is well developed.

[b]Derived from Table 8.1.1. Currently recycled materials are not included.

[c]A substantial portion of this category is readily recyclable, and a substantial portion is not. Some of the material listed here as nonrecyclable can be recovered in recyclable condition by an efficient ferrous recovery system at a combustion facility.

A portion of finished MSW compost cannot be marketed and must be landfilled.

In MSW discharged from compactor trucks, most glass containers are still in one piece, and most metal cans are uncrushed. Most glass and aluminum beverage containers are in recyclable condition. Many glass food containers and steel cans are heavily contaminated with food waste, however. Some of the recyclable paper in MSW received at disposal facilities is contaminated with other materials, but 50% or more is typically in recyclable condition.

The ratio of carbon to nitrogen (C/N ratio) is an indicator of the compostability of materials. To maximize the composting rate while minimizing odor generation, a C/N ratio of 25/1 to 30/1 is considered optimum. Higher ratios reduce the composting rate, while lower ratios invite odor problems.

Table 8.3.2 shows representative C/N ratios of compostable components of MSW. Controlled composting of food waste, with a C/N ratio of 14/1, is difficult unless large quantities of another material such as yard waste (other than grass clippings) are mixed in to raise the ratio. The C/N ratio moves above the optimum level as quantities of paper are added to the mixture, however.

Paper, leaves, and woody yard waste serve as effective *bulking agents* in composting MSW, so the addition of a bulking agent such as wood chips is generally unnecessary.

The metals content of MSW is a major concern in composting because repeated application of compost to land can raise the metals concentrations in the soil to harmful levels. Compost regulations usually set maximum metals concentrations for MSW compost applied to land. Most regulations do not distinguish between different forms of a metal. For example, the lead in printing ink on a plastic bag is treated the same as the lead in glass crystal even though the lead in printing ink is more likely to be released into the environment. Similarly, the hexavalent form of chromium found in lead chromate is treated the same as the elemental chromium used to plate steel even though the hexavalent form is more toxic than the elemental form.

Two extensive, recent studies of metals in individual components of MSW yielded contradictory results. A study in Cape May County, New Jersey found toxic metals concentrated in the noncompostable components of MSW (Camp Dresser & McKee Inc. 1991; Rugg and Hanna 1992). A study in Burnaby, British Columbia, however, found higher metals concentrations in the compostable components of MSW than were found in Cape May (see Table 8.1.6) (Rigo, Chandler, and Sawell 1993).

Disposable diapers are listed as compostable in Table 8.3.1 despite their plastic covers. The majority of the weight of disposable diapers is from the urine, feces, and treated cellulose inside the cover, all of which is compostable. Note, however, that most people wrap used diapers into a ball with the plastic cover on the outside, using the waist tapes to keep the ball from unraveling. Vigorous size reduction is required to prepare these diaper balls for composting.

Wood is biodegradable but does not degrade rapidly enough to be considered compostable. The same is true of cotton and wool fabrics, included in the textiles/rubber/leather category in Table 8.3.1.

Implications for Incineration and Energy Recovery

The heat value of MSW (4800–5400 Btu/lb) is lower than that of traditional fuels such as wood (5400–7200 Btu/lb), coal (7000–15,000 Btu/lb), and liquid or gaseous petroleum products (18,000–24,000 Btu/lb) (Camp Dresser & McKee 1991, 1992a,b; Niessen 1995). The heat value of MSW is sufficient, however, to sustain combustion without the use of supplementary fuel.

Heat value is an important parameter in the design or procurement of solid waste combustion facilities because each facility has the capacity to process heat at a certain rate. The greater the heat value of a unit mass of waste, the smaller the total mass of waste the facility can process.

The ash and moisture content of MSW is high compared to that of other fuels. Most of the ash is contained in relatively large objects that do not become suspended in the flue gas (Niessen 1995). Ash handling is a major consideration at MSW combustion facilities.

Because of its high ash and moisture content and low density, MSW has low *energy density* (heat content per unit volume) (Niessen 1995). Therefore, MSW combustion facilities must be designed to process large volumes of material.

The effect of recycling programs on the heat value of MSW is not well documented. Numerous attempts have been made to project the impact of recycling based on the

TABLE 8.3.2   REPRESENTATIVE C/N RATIOS OF COMPOSTABLE COMPONENTS OF MSW

Waste Category	C/N Ratio
Yard waste	29/1
Grass clippings	17/1
Leaves	61/1
Other yard waste	31/1
Food waste	14/1
Paper	119/1
Newspaper	149/1
Corrugated & kraft	165/1
High-grade paper	248/1
Magazines & mail	131/1
Other paper	85/1
Disposable diapers	95/1
Fines	23/1

Note: Derived from Table 8.1.4.

measured heat values of individual MSW components (for example, see Camp Dresser & McKee [1992a]). Little reliable data exist, however, that document the effect of known levels of recycling on the waste received at operating combustion facilities.

A reasonable assumption is that recycling materials with below-average heat values raises the heat value of the remaining waste, while recycling materials with above-average heat values reduces the heat value of the remaining waste. The removal of recyclable metal and glass containers increases heat value (and reduces ash content), while the recovery of plastics for recycling reduces heat value. The removal of paper for recycling also reduces heat value. Because recycled paper has a low moisture content, its heat value is 30% to 40% higher than that of MSW as a whole.

The increase in heat value caused by recycling glass and metal is probably greater than the reduction caused by recycling paper. Because plastics are generally recycled in small quantities, the reduction in heat value caused by their removal is relatively small. The most likely overall effect of recycling is a small increase in heat value and a decrease in ash content.

Sulfur in MSW is significant because sulfur oxides (SO_x) have negative effects and corrode natural and manmade materials. SO_x combines with oxygen and water to form sulfuric acid. A solid waste combustion facility must maintain stack temperatures above the dew point of sulfuric acid to prevent corrosion of the stack. Niessen (1995) provides additional information.

Like sulfur, chlorine has both health effects and corrosive effects. Combustion converts organic (insoluble) chlorine to hydrochloric acid (HCl). Because HCl is highly soluble in water, it contributes to corrosion of metal surfaces both inside and outside the facility (Niessen 1995).

Chlorine is a component of additional regulated compounds including dioxins and furans. Trace concentrations of dioxins and furans can be present in the waste or can be formed during combustion. Niessen (1995) provides additional discussion.

Oxides of nitrogen (NO_x) form during the combustion of solid waste, both from nitrogen in the waste and in the air. NO_x reacts with other substances in the atmosphere to form ozone and other compounds that reduce visibility and irritate the eyes (Niessen 1995).

Emissions of SO_x, NO_x, chlorine compounds, and hydrocarbons are regulated and must be controlled (see Section 10.1 and Niessen [1995]). Emissions of hydrocarbons and chlorine compounds other than HCl can generally be controlled by optimization of the combustion process. Maintaining complete control of the combustion of material as varied as MSW is difficult, however, so small quantities of hydrocarbons and complex chlorine compounds are emitted from time to time.

Combustion cannot destroy metals. Assuming that a combustion facility is designed with no discharge of the water used to quench the combustion ash, the toxic metals in the waste end up in the ash or are emitted into the air. Regulations limit the emission of toxic metals.

The tendency of a metal to be emitted from a combustion facility is a function of many factors such as:

- The volatility of the metal
- The chemical form of the metal
- The degree to which the metal is bound in other materials, especially noncombustible materials
- The degree to which the metal is captured by the air pollution control system

Emissions of a metal from a solid waste combustion facility cannot be predicted based on the abundance of the metal in the waste.

Mercury is the most volatile of the metals of concern, and a substantial portion of the mercury in MSW escapes capture by the air pollution control systems at MSW combustion facilities. The quantity of mercury in MSW has declined rapidly in recent years because battery manufacturers have eliminated most of the mercury in alkaline and carbon–zinc batteries. One cannot assume that a reduction in the quantity of mercury in batteries proportionately reduces the quantity emitted from MSW combustion facilities, however.

All but a small fraction of each metal other than mercury becomes part of the ash residue either because it never enters the facility stack or because it is captured by the air pollution control system. The environmental significance of a metal in combustion ash residue depends primarily on its leachability and the toxicity of its leachable forms. A portion of the ash residue from some MSW combustion facilities is regulated as hazardous waste because of the tendency of a toxic metal (usually lead or cadmium) to leach from the ash under the test conditions specified by the U.S. EPA.

Niessen (1995) and Chandler & Associates, Ltd. et al. (1993) provide additional information on the implications of solid waste characteristics with combustion as a disposal method. Niessen provides a comprehensive treatise on waste combustion from the perspective of an environmental engineer. The final report of Chandler & Associates, Ltd. et al. provides a detailed study of the relationships among metals concentrations in individual components of MSW, metals concentrations in stack emissions, and metals concentrations in various components of ash residue at a single MSW combustion facility.

Implications for Landfilling

The greater the density of the waste in a landfill, the more tons of waste can be disposed in the landfill. The density of waste in a landfill can be increased in a variety of ways, including the following:

- Using compacting equipment specifically designed for the purpose (Surprenant and Lemke 1994)

- Spreading the incoming waste in thinner layers prior to compaction (Surprenant and Lemke 1994)
- Shredding bulky, irregular materials such as lumber prior to landfilling

Because solid waste contains toxic materials (see Section 8.1), landfills must have impermeable liners and systems to collect water that has been in contact with the waste (leachate). The liner must be resistant to damage from any substance in the waste, including solvents. The first lift (layer) of waste placed on the liner must be free of large, sharp objects that could puncture the liner. For this reason, bulky waste is typically excluded from the first lift.

To some extent, the moisture content of waste placed in a landfill influences the quantity of the leachate generated. In most cases, however, a more important factor is the quantity of the precipitation that falls on the waste before an impermeable cap is placed over it.

For additional information, see Section 10.5.

—*F. Mack Rugg*

References

Camp Dresser & McKee Inc. 1991. *Cape May County multi-seasonal solid waste composition study.* Edison, N.J. (August).

———. 1992a. *Atlantic County (NJ) solid waste characterization program.* Edison, N.J. (May).

———. 1992b. *Prince William County (VA) solid waste supply analysis.* Annandale, Va. (October).

Chandler, A.J., & Associates, Ltd. et al. 1993. *Waste analysis, sampling, testing and evaluation (WASTE) program: Effect of waste stream characteristics on MSW incineration: The fate and behaviour of metals. Final Report of the Mass Burn MSW Incineration Study (Burnaby, B.C.).* Toronto (April).

Niessen, W.R. 1995. *Combustion and incineration processes: Applications in environmental engineering,* 2d ed. New York: Marcel Dekker, Inc.

Rigo, H.G., A.J. Chandler, and S.E. Sawell. 1993. Debunking some myths about metals. In *Proceedings of the 1993 International Conference on Municipal Waste Combustion, Williamsburg, VA, March 30–April 2, 1993.*

Rugg, M. and N.K. Hanna. 1992. Metals concentrations in compostable and noncompostable components of municipal solid waste in Cape May County, New Jersey. *Proceedings of the Second United States Conference on Municipal Solid Waste Management, Arlington, VA, June 2–5, 1992.*

Surprenant, G. and J. Lemke. 1994. Landfill compaction: Setting a density standard. *Waste Age* (August).

9
Resource Conservation and Recovery

9.1
REDUCTION, SEPARATION, AND RECYCLING

Municipal Waste Reduction

Waste reduction is the design, manufacture, purchase, or use of materials (such as products and packaging) which reduce the amount and toxicity of trash generated. Source reduction can reduce waste disposal and handling costs because it avoids the cost of recycling, municipal composting, landfilling, and combustion. It conserves resources and reduces pollution.

PRODUCT REUSE

Reusable products are used more than once and compete with disposable, or single-use, products. The waste reduction effect of a reusable product depends on the number of times it is used and thus the number of single-use products that are displaced.

Used household appliances, clothing, and similar durable goods can be reused. They can be donated as used products to charitable organizations. Such goods can also be resold through yard and garage sales, classified ads, and flea markets.

The following lists common source reduction activities in the private sectors (New Jersey Department of Environmental Protection and Energy 1992):

Office paper. Employees are encouraged to make two-sided copies, route memos and documents rather than making multiple copies, make use of the electronic bulletin board for general announcements rather than distributing memos, and limit distribution lists to essential employees.
Routing envelopes. After large routing envelopes are completely filled, employees can reuse them by simply pasting a blank routing form on the envelope face. Even large envelopes received in the mail can be converted to routing envelopes in this manner.
Paper towels. C-fold towels are replaced with roll towels.
Printers. Recharged laser printer toner cartridges are used.
Tableware. Nondisposable tableware (environmental mug program, china for conferences) is used.
Polystyrene containers. Reusable, glass containers are used,

and all Styrofoam coffee cups in all office areas, shops, and the employee cafeteria are eliminated. Styrofoam peanuts are reused in offices or donated to local businesses.
Beverages and detergents. Some items are available in refillable containers. For example, some bottles and jugs for beverages and detergents are made to be refilled and reused by either the consumer or the manufacturer.
Cleaning rags. Reusable rags are used instead of throwaway rags.
Ringed note binder reuse. Employees take binders to one of several collection points at the facility where they are refurbished for reuse.
Laboratory chemicals. "Just-in-time" chemicals are delivered to labs to preclude stockpiling chemicals which eventually go bad. This method reduces hazardous waste disposal costs through source reduction.
Photocopy machines. New photocopying machines with energy-saving controls are used.
Batteries. Use of rechargeable batteries reduces garbage and keeps the toxic metals in batteries out of the waste stream. Using batteries with reduced toxic metals is another alternative.

INCREASED PRODUCT DURABILITY

When a consumer-durable product has a longer useful life, fewer units (such as refrigerators, washing machines, and tires) enter the waste stream. For instance, since 1973, the durability of the passenger tire has almost doubled as radial tires have replaced bias and bias-belted tires. Radial tires have an average life of 40,000 to 60,000 miles; the average life of bias tires is 15,000 miles, and bias-belted tires is 20,000 miles (Peterson 1989).

Other ways of reducing waste through increased product durability include:

Using low-energy fluorescent light bulbs rather than incandescent ones. These bulbs last longer, which means fewer bulbs are thrown out, and cost less to replace over time.

Keeping appliances in good working order by following the manufacturers' service suggestions for proper operation and maintenance

Whenever intended for use over a long period of time, choosing furniture, luggage, sporting goods, tools, and toys that standup to vigorous use

Mending clothes instead of throwing them away, and repairing worn shoes, boots, handbags, and brief cases

Using long-lasting appliances and electronic equipment with good warranties. Reports are available that list products with low breakdown rates and products that are easily repaired.

REDUCED MATERIAL USAGE PER PRODUCT UNIT

Reducing the amount of material used in a product means less waste is generated when the product is discarded. Consumers can apply this waste reduction approach in their shopping habits by purchasing packaged products in large container sizes. For example, the weight-to-volume ratio of a metal can for a sample food product declines from 5.96 with an 8-oz container (single serving size) to 3.17 with a 101-oz (institutional) size.

Other methods for reducing the material per product unit include:

Using wrenches, screwdrivers, nails, and other hardware available in loose bins. Purchasing grocery items, such as tomatoes, garlic, and mushrooms, unpackaged rather than prepackaged containers.

Using large or economy-size items of household products that are used frequently, such as laundry soap, shampoo, baking soda, pet foods, and cat litter. Choosing the largest size of food items that can be used before spoiling.

Using concentrated products. They often require less packaging and less energy to transport to the store, saving money as well as natural resources.

When appropriate, using products that are already on hand to do household chores. Using these products can save on the packaging associated with additional products.

DECREASED CONSUMPTION

Seldom-used items, like certain power tools and party goods, often collect dust and rust, take up valuable storage space, and ultimately end up in the trash. Renting or borrowing these items reduces consumption and waste. Infrequently used items can be shared among neighbors, friends, or family. Borrowing, renting, and sharing items save both money and natural resources.

Other ways to decrease consumption follow.

Renting or borrowing tools such as ladders, chain saws, floor buffers, rug cleaners, and garden tillers. In apartment buildings or co-ops, residents can pool resources and form banks to share tools and other equipment used infrequently. In addition, some communities have tool libraries, where residents can borrow equipment as needed.

Renting or borrowing seldom-used audiovisual equipment

Renting or borrowing party decorations and supplies such as tables, chairs, centerpieces, linens, dishes, and silverware

Sharing newspapers and magazines with others to extend the lives of these items and reduce the generation of waste paper

Before old tools, camera equipment, or other goods are discarded, asking friends, relatives, neighbors, or community groups if they can use them

REDUCING WASTE TOXICITY

In addition to reducing the amount of material in the solid waste stream, reducing waste toxicity is another component of source reduction. Some jobs around the home require the use of products containing hazardous components. Nevertheless, toxicity reduction can be achieved by following some simple guidelines.

Using nonhazardous or less hazardous components. Examples include choosing reduced mercury batteries and planting marigolds in the garden to ward off certain pests rather than using pesticides. In some cases, less toxic chemicals can be used to do a job; in others, some physical methods, such as sandpaper, scouring pads, or more physical exertion, can accomplish the same results as toxic chemicals.

When hazardous components are used, using only the amount needed. Used motor oil can be recycled at a participating service station. Leftover products with hazardous components should not be placed in food or beverage containers.

For products containing hazardous components, following all directions on the product labels. Containers must be labelled properly. For leftover products containing hazardous components, checking with the local environmental agency or chamber of commerce for any designated days for the collection of waste material such as leftover paints, pesticides, solvents, and batteries. Some communities have permanent household hazardous waste collection facilities that accept waste year around.

Separation at the Source

Kitchen designers and suppliers of kitchen equipment will need to become more sensitive to the needs of recycling. Major manufacturers of kitchen equipment should make sorting drawers, lazy Susan sorting bins, and tilt-out bins as standard kitchen equipment. Kitchen designers should keep in mind small convenience items, such as automatic label scrapers, trash chutes, and can flatteners to make recycling more convenient.

The more finely household waste is separated, the greater its contribution to recycling. Figure 9.1.1 shows an approach where household waste is separated into four containers.

Container 1 would receive all organic or putrescible materials, including food-soiled paper and disposable diapers and excluding toxic substances and glass or plastic items. The contents of this container can be taken to a composting plant that also receives yard wastes and possibly sewage sludge and produces soil additives.

Container 2 would receive all clean paper, newspapers, cardboard, and cartons for paper processing, where contents are separated mechanically and sold to commercial markets.

Container 3 would receive clean glass bottles and jars and aluminum and tin cans free of scrap metals and plastics.

Container 4 would receive all other waste, including plastic, metal, ceramic, textile, and rubber items. (Later, a fifth container could be added for recyclable plastics.) The contents of this container can be considered nonrecyclable and sent to a landfill or a recycling plant for further separation. The contents of this container would represent about 12% of the total MSW.

Separate collections are required for trash items that are not generated on a daily basis, such as yard waste, brush and wood, discarded furniture and clothing, "white goods" such as kitchen appliances, toxic materials, car bat-

teries, tires, used oil, and paint.

"BOTTLE BILLS"

In 1981 Suffolk County on Long Island outlawed nonreturnable soda bottles. By 1983 legislation had been passed in eight states requiring a 5-cent deposit on all soda bottles. The annual return rate on beer bottles in New York is nearly 90% and 80% of six billion soft drink and beer bottles. Further improvement was obtained by raising the deposit on nonrefillable containers to 10 cents and allowing the state to use part of the unredeemed deposits (at present kept by bottlers and totaling $64 million a year) to establish recycling stations.

Bottle bills, while having achieved partial success, should be integrated into overall recycling programs, which include office paper and newspaper recycling, cardboard collection from commercial establishments, curbside recycling, establishment of buy-back recycling centers, wood waste and metal recycling, glass and bottle collection from bars and restaurants, and composting programs. Advertising and public education are important elements in the overall recycling strategy. Street signs, door hangers, utility-bill inserts, and phone book, bus, and newspaper advertisements are all useful. The most effective long-range form of public education is to teach school-children the habits of recycling.

Recycling

PLASTIC

Plastics are strong, waterproof, lightweight, durable, microwavable, and more resilient than glass. For these reasons they have replaced wood, paper, and metallic materials in packaging and other applications. Plastics generate toxic by-products when burned and are nonbiodegradable when landfilled; they also take up 30% of landfill space even though their weight percentage is only 7% to 9%. Recent research has found that paper does not degrade in landfills either and because of compaction in the garbage truck and in the landfill, the original volume percentage of 30% in the kitchen waste basket is reduced 12% to 21% in the landfill. In addition, plastics foul the ocean and harm or kill marine mammals. Other problems include the toxic chemicals used in plastics manufacturing, the reliance on nonrenewable petroleum products as their raw material, and the blowing agents used in making polystyrene foam plastics, such as chlorofluorocarbons (CFCs), which cause ozone depletion. CFCs are now being replaced by HCFC-22 or pentane, which does not deplete the ozone layer but does contribute to smog. For these reasons, recycling appears to be the natural solution to the plastic disposal problem.

Unfortunately, recycling and reuse are not easily accomplished because each type of plastic must go through

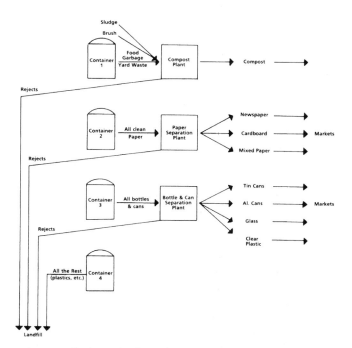

FIG. 9.1.1 Basic separation scheme.

a different process before being reused. There are hundreds of different types of plastics, but 80% of plastic used in consumer products is either high-density polyethylene (milk bottles) or polyethylene terephthalate (large soda bottles). It is not yet possible to separate plastics by types because manufacturers do not indicate the type of plastic used. Plastic parts of automobiles are still uncoded, so salvagers cannot separate them by type. Even if recycled polystyrene were separated and could be used as a raw material for a plastics recycling plant, such plants are just beginning to be built and we do not know if they will be successful. For these reasons, environmentalists would prefer to stop using plastics altogether in certain applications.

TOXIC SUBSTANCES

The careless disposal of products containing toxic or hazardous substances can create health hazards if allowed to decompose and leach into the groundwater from landfills or if vaporized in incinerators. Since hazardous-waste landfills are limited, the available options are either to have manufacturers substitute toxic materials with nontoxic substances or recycle the products that contain toxic materials. Municipalities are just beginning to consider the requirements of toxic-waste recycling. Products that are toxic or contain toxic substances include paint, batteries, tires, some plastics, pesticides, cleaning and drain-cleaning agents, and PCBs found in white goods (appliances). Separate collections are also required for medical wastes.

Batteries play an important role in the recycling of toxic substances. Batteries represent a $2.5 billion-a-year market. At present, practically no batteries are being recycled in the United States. Battery manufacturers feel that recycling is neither practical nor necessary; instead, they feel that all that needs to be done is to lower the quantities of toxic materials in batteries. It is estimated that 28 million car batteries are landfilled or incinerated every year. This number contains 260,000 tons of lead, which can damage human neurological and immunological systems. The billions of household batteries disposed of yearly contain 170 tons of mercury and 200 tons of cadmium. The first can cause neurological and genetic disorders, the second, cancer. Some batteries also contain manganese dioxide, which causes pneumonia. When incinerated, some of these metals evaporate. The excessive emissions of mercury were the reason why Michigan temporarily suspended the operation of the incinerator in Detroit, the nation's largest.

Some states have recently initiated efforts to force manufacturers to collect and recycle or safely dispose of their batteries. The Battery Council International has prompted several states to pass laws requiring recycling of all used car batteries. In many European countries used batteries are returned to the place of purchase for disposal.

The disposal of white goods (appliances such as refrigerators, air conditioners, microwave ovens) is also a prob-

lem. Until 1979 appliance capacitors were allowed to contain PCBs (polychlorinated biphenyls). Even after the ban, some manufacturers were granted an extra year or two to deplete their inventories. When white goods are shredded, the "fluff" remaining after the separation of metals (consisting of rubber, glass, plastics, and dirt) is landfilled. When it was found that the "fluff" contains more than 50 ppm of PCBs, the Institute of Scrap Recycling Industries advised its 1,800 members not to handle white goods. The safe disposal of PCB-containing white goods would require scrap dealers to remove the capacitors before shredding. Similar toxic-waste disposal problems are likely to arise in connection with electronic and computing devices, the printed circuit boards of which contain heavy metals.

A long-range solution to toxic-waste disposal might be to require manufacturers of new products containing toxic substances to arrange for recycling *before* the product is allowed on the market, or at least to provide instruction labels describing the recommended steps in recycling.

PAPER

Paper used to be made of reclaimed materials such as linen rags. Rags were the raw materials used by the first paper mill built in the United States in 1690 in Philadelphia. Only in the nineteenth century did paper mills convert to wood-pulping technology. It takes seventeen trees to make a ton of paper. All Sunday newspapers in the United States, for example, require the equivalent of half a million trees every week. When paper is made from waste paper, it not only saves trees but also saves 4,100 kWh of energy per ton (the equivalent of a few months of electricity used by the average home), 7,000 gallons of water, 60 pounds of air-polluting emissions, and three cubic yards of landfill space and the associated tipping fees. The production of recycled paper also requires fewer chemicals and far less bleaching.

The paper output of the world has increased by 30% in the last decade. In 1990 the United States used more than 72 million tons of paper products, but only 25.5% of that (18.4 million tons) is made from recycled paper. This compares with 35% in Western Europe, almost 50% in Japan, and 70% in the Netherlands. There are some 2,000 waste-paper dealers in the United States who collect nearly 20 million tons of waste paper each year. In 1988, 20% of the collected waste paper was exported, mostly to Japan.

The waste-paper market is very volatile. In some locations the mixed office waste or mixed-paper waste (MPW) has no value at all and tipping fees must be paid to have them picked up. Therefore, what pays for collection and processing is not the prices paid for waste paper, but the savings represented by *not* landfilling them at $70/ton on the East Coast. A ton of old newspapers in California brings $25 to $35 because of the Japanese market demand. In the Northeast an oversupply in 1989 caused the waste-paper price to plummet from $15/ton to about −$10/ton. This oversupply also resulted in increased waste-paper exports to Europe, which in turn caused the collapse of the waste-paper market in Holland, where the value of a kilogram of waste paper dropped from eight cents to one cent.

Waste paper can be classified into "bulk" or "high" grade. The highest-grade papers are manila folders, hard manila cards, and similar computer-related paper products. High-grade waste paper is used as a pulp substitute, whereas bulk grades are used to make paper boards, construction paper, and other recycled paper products. The bulk grade consists of newspapers, corrugated paper, and MPW. MPW consists of unsorted waste from offices, commercial sources, or printing establishments. Heavy black ink used on newspaper reduces its value, however. The value of the paper is also reduced by the presence of other substances that interfere with a single-process conversion into pulp, such as the gum in the binding of telephone directories or the chemical coating of magazines.

The most effective way to create a waste-paper market is to attract a pulp and paper mill to the area. To keep such a plant in operation, however, requires a high-grade waste-paper supply of about 300 tons per day. In addition, facilities are also needed for wastewater treatment.

Newsprint Recycling

A large part of the waste-paper problem has to do with newsprint, which makes up 8% of the total MSW by weight. Some 13 million tons of newsprint are consumed yearly in the United States, 60% imported from Canada.

Connecticut requires the use of 20% recycled paper in the newspapers sold in the state today and 90% by 1998. Suffolk County on Long Island requires 40%. New York State reached a voluntary agreement with its publishers to achieve the 40% goal by the year 2000. Florida applies a ten-cent waste-recovery fee for every ton of virgin newsprint used and grants a ten-cent credit for every ton of recycled newsprint used.

The net effect of such legislation will be an increased and steady demand for waste paper, which is essential for the success of recycling. As the demand for waste paper products rises, paper manufacturers will also increase their capacity to produce recycled paper.

GLASS

About 13 million tons of glass are disposed of in the United States every year, representing more than 7% of the total MSW that is generated. But only about 12% of the total glass production is recycled. In comparison, Japan recycles about 50%.

Salvaged glass has been used in bricks and paving mixtures. "Glasphalt" can be made from a mixture of glass and asphalt or a mix of 20% ground glass, 10% blow sand, 30% gravel, and 40% limestone. In spite of all these other uses, the main purchasers of crushed glass are the glass companies themselves. The use of recycled crushed glass reduces both the energy cost and the pollutant emissions associated with glass making. Crushed glass is easily saleable, with a market almost as good as that for aluminum. Manufacturers use from 20% to as much as 80% of salvaged glass in their glass-making processes.

METALS

In the United States over 15 million tons of metals are discarded every year. This represents almost 9% of MSW by weight. We recycle 14% of our metallic wastes (nearly 64% of aluminum). During the last fifty years, more than half of the raw materials used in steel mills was recycled. At least one-third of the aluminum produced is from recycled sources.

Aluminum recycling is profitable and well established because it requires only 5% of the electric power to remelt aluminum as it does to extract it from bauxite ore. In 1990 the average price paid for crushed, baled aluminum cans was $1,050 per ton and some 55 billion aluminum cans (0.96 million tons) have been recycled. The recycling rate of aluminum increased from 61% in 1989 to 63.5% in 1990. Steel has also been recycled for generations, but the recycling of steel cans is relatively new. It was necessary to reduce the rust-preventing tin layer on the steel cans first, so that they might be added directly to steel furnaces. The recycling of steel cans has increased from 5 billion cans in 1988 to 9 billion in 1990 and its market value varied from $40 to $70 per ton in 1990 depending on location.

The main sources of scrap metals are cans, automobiles, kitchen appliances (white goods), structural steel, and farm equipment. The value of the noncombustibles in incinerator ash varies from area to area.

RUBBER

In the United States some two billion old tires have been discarded, and their number is growing by about 240 million a year. In the past tires were either piled, landfilled, burned, or ground up and mixed with asphalt for road surfacing. These "solutions" were expensive and often caused environmental problems because of the air pollution resulting from massive tire fires. Some newer rubber recycling processes have tried to overcome these limitations. The new processes do not pollute air or water because nothing is burned and no water is used. The tires are shredded and the polyester fibers removed by air classification. The steel from radial tires is removed magnetically. The remaining rubber powder is mixed with chemical agents that restore the ability of the "dead" rubber to bond with other rubber and plastic molecules. The vulcanized or "cured" tire rubber loses its ability to bond during the vulcanizing process.

TABLE 9.1.1   INCINERATOR ASH COMPOSITION

Component	Percentage
Ferrous metal	35
Glass	28
Minerals and Ash	16
Ceramics	8
Combustibles	9
Nonferrous Metal	4

TABLE 9.1.2   COMPOSITION OF INCINERATOR ASH (UNDER-2″ FRACTION)

Component	Percentage
Glass	37
Minerals and ash	21
Ferrous metals	19
Ceramics	9
Combustibles	8
Nonferrous metals	6

Combining old rubber with "virgin" rubber or plastics results in an economically competitive product. The cost of virgin rubber is about 65 cents a pound and polypropylene costs about 68 cents, while the "reactivated" product is about 30 cents a pound ($600/ton).

INCINERATOR ASH

If all the MSW of New York City were incinerated, the residue would amount to 6,000 to 7,000 tons/day, representing a giant disposal problem. About 10% by weight of the incinerator residue is fly ash collected in electrostatic precipitators, scrubbers, or bag filters; the remaining 90% is bottom ash from the primary and secondary combustion chambers. This residue is a soaking-wet complex of metals, glass, slag, charred and unburned paper, and ash containing various mineral oxides. A Bureau of Mines test found that 1,000 pounds of incinerator residue yielded 166 pounds of larger-size ferrous metals, such as wire, iron items, and shredded cans. The total ferrous fraction was found to be 30.5% by weight; glass represented 50% of the total residue by weight.

Common practice in the U.S. is to recover some 75% of the ferrous metals through magnetic separation and to landfill the remaining residue. Incinerator residue has also been used as landfill cover, landfill road base, aggregate in cement and road building applications, and as aggregate substitute in paving materials.

Incinerator residue is processed to recover and reuse some of its constituents and thereby reduce the amount requiring disposal. Processing techniques include the recovery of ferrous materials through magnetic separation, screening the residue to produce aggregate for construction-related uses, stabilization through the addition of lime (which tends to minimize metal leaching), and solidification or encapsulation of the residue into asphaltic mixtures.

An incinerator-residue processing plant might consist of the following operations: (1) fly ash and bottom ash are collected separately, with lime mixed only with the fly ash; (2) ferrous materials are removed from the bottom ash; (3) the ferrous-free residue is screened to separate out the proper particle sizes for use as aggregate; and (4) the remaining oversized items and stabilized fly ash are landfilled. In a more sophisticated ash-processing plant, the ferrous removal and shredding (or oversize removal) are followed by melting of the ash (fusion), resulting in a glassy end-product. This high-tech process has some substantial advantages: It burns all the combustible materials, including dioxins and other trace organics, and encapsulates the metals, thereby preventing their leaching out. The resulting fused product is a glazed, nonabrasive, lightweight black aggregate. The fusion of combined incinerator ash and sewage sludge is currently practiced in Japan.

The first U.S. building to be built from recycled incinerator ash blocks is an 8,000-square-foot boathouse on the campus of the State University of New York at Stony Brook, Long Island. The ash comes from an incinerator in Peekskill and is mixed with sand and cement to form blocks that are as durable as standard cinder blocks. This technology has already been used in Europe. The blocks can be used to build seawalls, highway dividers, and sound barriers, in addition to regular buildings. It is the bottom ash (not the fly ash) portion that is considered safe for such applications.

The ash produced by one New York City incinerator has been extensively sampled and evaluated. Fly ash contains substantial quantities of organic materials. About 20% by weight is larger than 2″ (50.8 mm); the metal content of this fraction is over 80% by weight. The overall composition of all the incinerator residue (Table 9.1.1) differed substantially from the composition of the under-2″ (50.8 mm) fraction (Table 9.1.2). The test also concluded that the New York State Department of Transportation specifications for Type 3 asphalt binder can be met if 10% combined incinerator ash is mixed in with 90% natural aggregate.

—*Béla G. Lipták*

References

New Jersey Department of Environmental Protection and Energy. 1992. *How to reduce waste and save money: Case studies from the private sector.* Division of Solid Waste Management, Office of Recycling and Planning, Bureau of Source Reduction and Market Development, Trenton, N.J. (July).
Peterson, Charles. 1989. What does "waste reduction" mean? *Waste Age* (January).

9.2
MATERIAL RECOVERY

The recycling of postconsumer material found in MSW involves (1) the recovery of material from the waste stream, (2) intermediate processing such as sorting and compacting, (3) transportation, and (4) final processing to provide a raw material for manufacturers. This section emphasizes separation and recovery, and applicable specifications for these materials. It focuses on those materials which are intended for short-term consumer usage, are discarded quickly, and are present in large quantities in the solid waste stream.

Role of MRFs and MRF/TFs

Material Recovery Facilities (MRFs) and Material Recovery/Transfer Facilities (MRF/TFs) are used as centralized facilities for the separation, cleaning, packaging, and shipping of large volumes of material recovered from MSW. These processes include:

Further processing of source-separated wastes from curbside collection programs. The type of source-separated material that is separated includes paper and cardboard from mixed paper and cardboard; aluminum from commingled aluminum and tin cans; plastics by class from commingled plastics; aluminum cans, tin cans, plastics, and glass from a mixture of these materials; and glass by color (clear, amber, and green).

Separating commingled MSW. All types of waste components can be separated from commingled MSW. Waste is typically separated both manually and mechanically. The sophistication of the MRF depends on (1) the number and types of components to be separated, (2) the waste diversion goals for the waste recovery program, and (3) the specifications to which the separated products must conform.

MRFs for Source-Separated Waste

MRFs for source-separated waste further separate paper and cardboard, aluminum and tin cans, and plastic and glass.

PAPER AND CARDBOARD

The principal types of paper recycled are old newspaper (ONP), old corrugated cardboard (OCC), high-grade paper, and mixed paper waste (MPW). These waste papers can be classified into *bulk* or *high-grade*. The highest grade of papers are manila folders, hard manila cards, and similar computer-related paper products. The bulk grade consists of newspapers, corrugated paper, and MPW. MPW consists of unsorted waste from offices, commercial sources, or printing establishments. High-grade waste paper is used as a pulp substitute, whereas bulk grades are used to make paper boards, construction paper, and other recycled paper products. The heavy black ink used on newspaper reduces its value. The value of paper is also reduced by the presence of other substances that interfere with the single-process conversion into pulp, such as gum in the binding of telephone directories or the chemical coating of magazines.

To ensure quality and minimize handling and processing, ONP should be separated from all other waste at or as close as possible to its source of generation. End users can reject an entire shipment of ONP where evidence exists that the paper was commingled with MSW. Care must also be taken to prevent contamination of the paper during collection, loading, transporting, unloading, processing, and storing.

In MRFs, mixed paper and cardboard are unloaded from the collection vehicle onto the tipping floor. There, cardboard and nonrecyclable paper items are removed. The mixed paper is then loaded onto a floor conveyor with a front-end loader. The floor conveyor discharges to an inclined conveyor that discharges into a horizontal conveyor. The horizontal conveyor transports the mixed paper past workers who remove any remaining cardboard from the mixed paper. The paper remaining on the conveyor is discharged to a conveyor located below the picking platform that is used to feed the baler. Once the paper has been baled, the cardboard is baled.

The Paper Stock Institute of America, which represents buyers and processors of waste paper, has listed thirty-three specialty grades whose specifications are agreed upon by buyers and sellers. Table 9.2.1 gives the specifications for the most common grades of postconsumer waste paper.

The four grades from lowest to highest quality are news (grade 6), special news (grade 7), special news de-ink quality (grade 8), and over-issue news (grade 9). Grades 6 and 7 are used primarily in the production of insulation and paperboard as well as in other applications where high quality (absence of contamination) is not of foremost importance. Grade 8 is used to make newspaper again, as is grade 9. Grade 9 is the grade that sellers find provides the most accessible market.

Paper shipped to a paper mill must meet mill specifications on outthrows and prohibited materials. *Outthrows* are defined as all papers that are so manufactured or

TABLE 9.2.1 SPECIFICATIONS FOR RECYCLED PAPER AND CARDBOARD

Grade Number	Class	Description	Prohibitive Materials, %	Total Outthrows, %
1	Mixed Paper	A mixture of various qualities of paper not limited to type of packing or fiber content	2	10
6	News	Baled newspapers containing less than 5% other papers	0.5	2.0
7	Special News	Baled, sorted, fresh, dry newspapers; not sunburned; free from paper other than news; containing not more than the normal percentage of rotogravure and colored sections	None permitted	2.0
8	Special News, De-ink Quality	Baled, sorted, fresh, dry newspapers; not sunburned; free from magazines, white blank, pressroom overissues, and paper other than news; containing not more than the normal percent of rotogravure and colored sections. This packaging must be free from tar.	None permitted	0.25
9	Overissue	Unused, overrun, regular newspaper printed on newsprint; baled or securely tied in bundles; containing not more than the normal percentage of rotogravure and colored sections	None permitted	None permitted
11	Corrugated	Baled, corrugated containers, Containers having liners of test liner, jute, or kraft	1.0	5.0
38	Sorted Colored Ledger	Printed or unprinted sheets, colored shavings, and cuttings of colored or ledger white sulfite or sulfate ledger, bond; and writing and other papers that have a similar fiber and filler content. This grade must be free of treated, coated, padded, or heavily printed stock.	None permitted	2.0
40	Sorted White Ledger	Printed or unprinted sheets, guillotined books, quire waste, and cuttings of white sulfite or sulfate ledger, bond, and writing and other papers that have a similar fiber and filler content. This grade must be free of treated, coated, padded, or heavily printed stock.	None permitted	2.0
42	Computer Printout	White sulfite or sulfate papers in forms manufactured for use in data processing machines. This grade can contain colored stripes and impact or nonimpact (e.g., laser) computer printing, and can contain not more than 5% of groundwood in the packing. All stock must be untreated and uncoated.	None permitted	2.0

Source: Paper Stock Institute, *Guidelines for paper stock* (Washington, D.C.: Institute of Scrap Recycling Inc.).

treated or are in such form to be unsuitable for consumption as the grade specified. *Prohibitive materials* are defined as:

Any material in the packing of paper stock whose presence in excess of the amount allowed makes the packaging unsalable as the grade specified

Any material that may be damaging to equipment

The maximum amount of outthrows in grade specifications is the total of outthrows and prohibitive materials. Examples of prohibitive materials are sunburned newspaper, food containers, plastic or metal foils, waxed or treated paper, tissues or paper towels, bound catalogs or telephone directories, Post-its, and faxes or carbonless carbon paper. Other prohibitive materials are foreign materials such as dirt, metal, glass, food wastes, paper clips, and string.

ALUMINUM AND TIN CANS

Aluminum cans are one of the most common items recovered through municipal and commercial recycling programs because they are easily identified by residents and employers. They also provide higher revenues than other recyclable materials. The recycling of used beverage cans (UBCs) not only saves valuable landfill space but also minimizes energy consumption during the manufacturing of aluminum products. Manufacturing new aluminum cans from UBCs uses 95% less energy than producing them from virgin materials.

A successful aluminum recycling program must have interaction between various entities including those involved with collection, sorting and processing, reclamation, and reuse. Three generator sectors from which aluminum beverage containers can be recovered are residential households, commercial institutions, and manufacturing entities. Curbside collection programs recapture large quantities of recyclables. Aluminum UBCs can be separated as an individual commodity or commingled with other recyclables.

Steel food cans, which make up more than 90% of all food containers, are often called tin cans because of the thin tin coating used to protect the contents from corrosion. Some steel cans, such as tuna cans, are made with tin-free steel, while others have an aluminum lid and a steel body and are commonly called bimetal cans. All these empty cans are completely recyclable by the steel industry and should be included in any recycling program.

At the MRF, the collection vehicle discharges the commingled aluminum and tin cans into a hopper bin, which discharges to a conveyor belt. The conveyor transports the cans past an overhead magnetic separator where the tin cans are removed. The belt continues past a pulley magnetic separator, where any tin cans not removed with the magnet are taken out. The aluminum and tin cans, collected separately, are baled for shipment to markets.

At a reclamation plant, shredded aluminum cans are first heated in a delacquering process to remove coatings and moisture. Then they are charged into a remelting furnace. Molten metal is formed into ingots of 30,000 lb or more that are transferred to another mill and rolled into sheets. The sheets are sent to container manufacturing plants and cut into disks, from which cans are formed.

Aluminum markets have material specifications that regulate the extent of contamination allowed in each delivery as well as the method by which materials are prepared. For example, some markets prohibit aluminum foils and aluminum pans because they are usually contaminated. Noncontainer aluminum products purchased by dealers must simply be dry and free of contaminants.

PLASTIC AND GLASS

The recycling and reuse of plastics are not easily accomplished because each type of plastic must go through a different process before being reused. Hundreds of different types of plastics exist, but 80% of the plastics used in consumer products is either HDPE (milk and detergent bottles) or polyethylene terephthalate (PET) (large soda bottles). The most common items produced from postconsumer HDPE are detergent bottles and motor oil containers. Detergent bottles are usually made of three layers, with the center layer containing the recycled material.

Most plastic container manufacturers code their products. The code is a triangle with a number in the center and letters underneath. The number and letter indicate the resin from which the container is made:

1 = PET   (polyethylene terephthalate)
2 = HDPE (high-density polyethylene)
3 = V   (vinyl)
4 = LDPE (low-density polyethylene)
5 = PP   (polypropylene)
6 = PS   (polystyrene)
7 = Other   (all other resins and multilayered material)

Still, keeping plastics separate is not easy. The most notorious look alikes are PET, the clear, shiny plastic that soda bottles are made from, and PVC, another clear plastic used mainly for packaging cooking oil. Because PVC starts to decompose at the temperature at which PET is just beginning to melt, one stray PVC bottle in a melt of 10,000 PET bottles can ruin the entire batch.

Container glass is the only glass being recycled today. Window panes, light bulbs, mirrors, ceramic dishes and pots, glassware, crystal, ovenware, and fiberglass are not recyclable with container glass and are considered contaminants in container glass recycling.

The consideration in container glass marketing is color separation. Permanent dyes are used to make different colored glass containers. The most common colors are green, brown, and clear (or colorless). In the industry, green glass is called emerald, brown glass is amber, and clear glass is flint. For bottles and jars to meet strict manufacturing specifications, only emerald or amber cullet (crushed glass) can be used for green and brown bottles, respectively.

At the MRF, the collection vehicle discharges the commingled plastic and glass into a hoppered bin, which discharges to a conveyor belt. The material is transported to a sorting area, where the plastic and glass are separated manually from the other materials. The remaining glass is color sorted and sent to a glass crusher. The waste is discharged to vibrating screens where broken glass falls through the openings in the screen. Any residual material is collected at the end of the vibrating screen. The crushed glass is loaded onto large trailers and transported to the vibrating screen. The residual material is disposed of in a landfill. The commingled plastic is separated further by visual inspection or according to the type (PET and HDPE) based on the imprinted code adopted by the plastic industry.

In a glass bottle manufacturing plant, specialized beneficiation equipment performs final cleaning to remove residual metals, plastic, and paper labels. The cullet is then mixed with the raw material used in the production of glass. After the batch is mixed, it is melted in a furnace at temperatures ranging from 2600 to 2800°F. The mix can burn at low temperatures if more cullets are used. The melted glass is dropped into a forming machine where it is blown or pressed into shape. The newly formed glass containers are slowly cooled in an annealing lehr. They are inspected for defects, packed, and shipped to the bottler.

At a reclamation facility, PET bottles and HDPE jugs are transformed into clean flakes. A resin reclamation facility chops and washes the chips to remove labels, adhesives, and dirt and separates the material from their components to produce a clean generic polymer. Clean PET is sold as flakes but most HDPE is made into pellets. The HDPE flakes are fed into an extruder and are compressed as they are carried toward the extrusion die. The combined heat from flow friction and supplemental heating bands causes the resin to melt, and volatile contaminants are vented from the mixture. Immediately before the die, the melted mixture passes through a fine screen that removes any remaining solid impurities. As the melt passes through the orifice, a rotating knife chops the strand into short segments, which fall into a water bath where they are cooled. The pellets are dried to a moisture content of about 0.5% and are packaged for shipment to the end user.

Glass used for new bottles and containers must be sorted by color and must not contain contaminants such as dirt, rocks, ceramics, and high-temperature glass cookware. These materials, known as refractory materials, have higher melting temperatures than container glass and form a solid inclusion in the finished product. Table 9.2.2 gives the material specifications for color-sorted glass. The specifications in the Rotterdam glass processing plant limit the maximum amount of ceramics to 100 g per ton of crushed glass; the same limit for aluminum is only 6 g per ton.

Trade groups representing manufacturers and processors have established specifications for recycled plastics.

These specifications are extensive and beyond the scope of this chapter. In general, buyers require postconsumer plastics to be well sorted, reasonably free of foreign material, and baled within a specified size and weight range.

MSW Processing

A solid waste processing plant in Rhode Island and the Sorain-Cechini MRF plant in Rome, Italy are two examples of MSW processing plants.

MRF PLANT FOR PARTIALLY SEPARATED MSW

In 1989, an 80 tons per day (tpd) MRF was started in Rhode Island (see Figure 9.2.1). Designed and operated by New England CR Inc. in conjunction with Maschinenfabrik Bezner of West Germany, this highly automated plant can sort and recover the recyclables from partially separated MSW containing metallic, glass, and plastic cans, bottles, and other containers but not paper and organics. The partially separated MSW enters the plant on a conveyor belt, which first passes under an electromagnet that attracts the tin-plated steel cans and carries them off to be shredded. As the MSW falls, it encounters a rolling curtain of chains. The lighter objects (aluminum and plastic cans) cannot break through and are diverted toward a magnetized drum. The heavier (mostly glass) bottles pass through the curtain and arrive at a hand-separation belt, where they are separated manually by color.

As the plastic and aluminum containers reach the magnetic drum, the aluminum objects drop into a separate hopper. The plastic objects continue on the conveyor belt and are later sorted according to weight.

The plant design appears to be simple enough to guarantee reliability. The concept of this type of MRF plant is promising because it simplifies the process of source separation by allowing cans and containers of all types to be placed in the same bin.

TABLE 9.2.2 SPECIFICATIONS FOR COLOR-SORTED GLASS

| | *Permissible Color Mix Levels, Percent* | | | |
Color	*Flint*	*Amber*	*Green*	*Other*
Flint (clear)	97 to 100	0 to 3	0 to 1	0 to 3
Amber (brown)	0 to 5	95 to 100	0 to 5	0 to 5
Green	0 to 10	0 to 15	85 to 100	0 to 10

Source: American Society for Testing and Materials (ASTM), 1989, Standard specifications for waste glass as a raw material for the manufacture of glass containers, *1989 Annual book of standards*, Vol. 11.04 (Phila.: ASTM), 299–300.

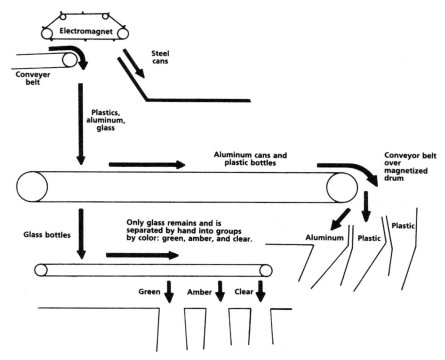

FIG. 9.2.1 The operation of a solid waste processing plant in Rhode Island. (Reprinted, with permission, from New England CRInc., 1989, *New York Times/Bohdan Osyczka*, [2 May], 160.)

MATERIAL RECOVERY PLANT

The Sorain-Cecchini MRF plant in Rome has been in operation for over twenty years. It recovers ferrous metals (6% by weight), aluminum (1%), organics (34%), and film plastics (1%) while generating densified RDF (51%) and rejecting 7% oversized items. The total plant capacity is 1200 tn per day (Cachin and Carrera 1986).

The main processing steps involve magnetic separation for ferrous metal removal, eddy-current separation for aluminum recovery, rotary screens for separation by size, and air classifiers for separation by density. The overall process consists of eighty pieces of equipment, which are flexible and can be used in different combinations as market conditions change.

Figure 9.2.2 shows the resource recovery plant in Rome. The charging conveyor **1** is provided with a pickup device **2** that breaks the bags and removes bulky reject items. A leveling device **3** meters the waste-flow rate and removes rejects. The primary screen **4** separates the large (over 8 in) fraction from the smaller, heavier fraction. The approximately 55% large fraction of paper, wood, and plastic is fed to the 10–20 rpm large breaker **5**, which reduces particle size and breaks plastic bags. The large breaker automatically rejects any items it cannot break (about 2%). The output (53%) is sent to the large air classifier **6**, where the lighter (10%) sheet paper and plastic fraction is separated from the heavier (43%) cardboard, wood, and rags. The reject fraction consists mostly of white goods (appli-

ances) but includes bulky items such as bedsprings. This fraction is hauled away by subcontractors.

The light fraction (10%) passes through a differential shredder **7**, which breaks up only the paper. It is followed by a rotary screen **8**, which separates the lighter 3% of the stream, containing the plastic film. This stream is sent to the small classifier **9**, where the 1% light fraction is taken to plastic recovery, while the remaining paper and rag fragments are included with the densified RDF (DRDF). The recovered plastic film (mostly polyethylene) is shredded into square-inch flakes, cleaned by washing, and air dried. The dry flakes are melted and fed to the extruder, and the pellets are shipped to plastic film manufacturers.

For metal separation, the 40% heavy fraction from the primary screen **4** passes through the primary magnetic separator **10**, which removes 4% and sends that fraction to ferrous recovery. The remaining fraction is further homogenized in the small breaker **11** and separated in the secondary rotary screen **8** into the 15% large fraction (over 4 in), consisting mainly of paper, wood, and plastics, and is sent to DRDF recovery. The 21% small fraction (under 4 in), consisting of organics, glass, ashes, and aluminum, is sent to a conveyor **17**; aluminum (1%) is removed by hand, and the rest (20%) is sent to organic recovery.

The 43% heavy fraction from the large air classifier **6** passes through the secondary magnetic separator **12**, which removes 1% and sends that fraction to ferrous recovery. The remaining fraction travels on the sorting conveyor **13**, where the semiautomatic devices and inspectors

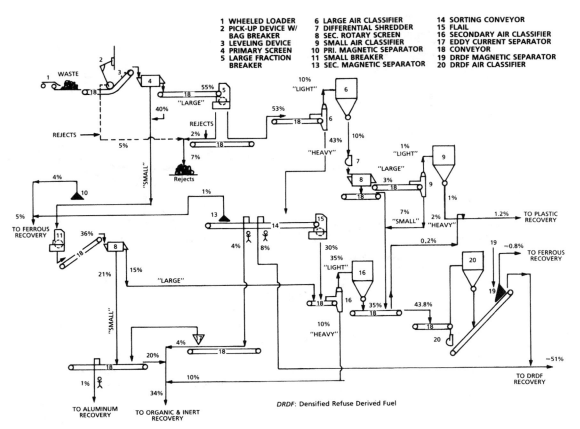

FIG. 9.2.2   Resource recovery plant in Rome, Italy. (Adapted from F.J. Cachin and F. Carrera, 1986, The Sorain-Cecchini system for material recovery, *National Waste Processing Conference, Denver, 1986* [ASME].)

remove the cardboard (8%), which is sent DRDF. Any missed recovery items and rejects (4%) are sent to the eddy-current separator **16** for aluminum recovery and then to organic recovery. The removed aluminum is crushed and densified to a specific gravity of 1.0 (62.4 lb/cu ft) before being placed in storage. The 30% fraction remaining on the sorting conveyor **13** is mostly paper and is sent to the flail **14**, where it is broken down before being sent to the secondary air classifier **15**. The three magnetic separators **10, 12,** and **18** send 5.8% to ferrous recovery, where it is shredded by the abrader hammer mill. The shredding step is followed by cleaning through firing or washing and a final magnetic separation step to remove the nonmetals that were loosened by the abrader.

The organics fraction, left from the plant feed after the removal of metals, plastic film, and paper, is essentially a heavy fraction of small-sized particles containing organics, glass, ceramics, sand, ashes, hard plastics, and small pieces of wood. This fraction is placed into an aerobic digester and broken down into raw compost. After the removal of glass, ceramics, and other inorganic rejects, the raw compost is subcontracted for further processing. This processing splits the organic fraction into a feed fraction (a high-quality compost fraction) and a residue, which is usually landfilled.

The 15% large fraction from the secondary rotary screen **8** and the 30% from the flail **14** are sent to the secondary air classifier **15**, which removes all paper (35%) and sends it to the DRDF air classifier **19** together with the small fraction (7%) from the secondary rotary screen **8** and the heavy fraction (2%) from the small air classifier **9**. After the DRDF magnetic **18** removes the remaining ferrous metals, the DRDF is densified into flakes in the recovery line. The DRDF is stored in a specific gravity of 0.6 (38 lb/ft). The heavy fraction (10%) from the secondary air classifier **15** is sent to organic recovery.

The DRDF obtained is relatively clean, and its sulfur and chlorine content is low as most metals, hard plastics (PVC, PET), and other impurities have been removed from it. The Sorain process can switch from producing DRDF to generating paper pulp depending on market condition.

Adapted from *Municipal Waste Disposal in the 1990s* by Béla G. Lipták (Chilton, 1991).

Reference

Cachin, F.J. and F. Carrera. 1986. The Sorain-Cechini system for material recovery. *National Waste Processing Conference, Denver, 1986.* ASME.

9.3
REFUSE-DERIVED FUEL (RDF)

RDF is the combustible portion of MSW that has been separated from the noncombustible portion through processing such as shredding, screening, and air classifying. The RDF that remains after processing is highly combustible and can be used as is (fluffy material) or in pellet form.

RDF Preparation Plant

Figure 9.3.1 shows the process flow diagram of the RDF preparation plant in Haverhill, Massachusetts. This plant has been operating since 1984, feeding 100% RDF to a 250,000-lb/hr boiler designed for RDF service. In this facility, 1300 tpd of MSW are separated into 983 tpd of RDF fuel, 260 tpd of glassy residue which are landfilled, and 57 tpd of ferrous metals which are sold. The MSW passes through two parallel 70 tph Heil shredders producing an output particle size of 90% under 4 in (101.6 mm). The ferrous metals are removed by dings and head pulley magnets.

The shredded refuse passes through a two-stage, 12.5-ft diameter, 60-ft long (3.8 m × 18.24 m) trommel screen.

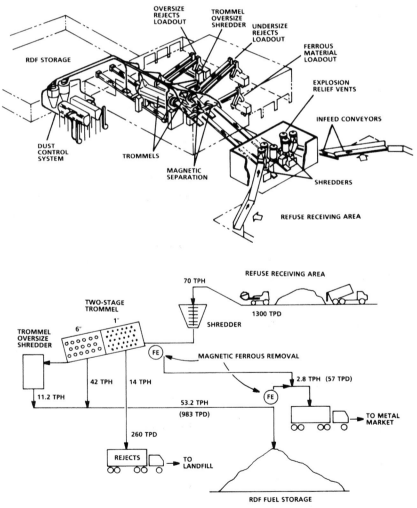

FIG. 9.3.1 RDF preparation plant in Haverhill, Massachusetts. (Reprinted, with permission, from D. Kaminski, 1986, Performance of the RDF delivery and boiler-fuel system at Lawrence, Massachusetts facility, *National Waste Processing Conference, Denver, 1986* [ASME].)

TABLE 9.3.1   ASTM CLASSIFICATION OF RDFS

Class	Form	Description
RDF-1 (MSW)	Raw	MSW with minimal processing to remove oversize bulky waste
RDF-2 (C-RDF)	Coarse	MSW processed to coarse particle size with or without ferrous metal separation such that 95% by weight passes through a 6-in square mesh screen
RDF-3 (f-RDF)	Fluff	Shredded fuel derived from MSW processed for the removal of metal, glass, and other entrained inorganics; particle size of this material is such that 95% by weight passes through a 2-in square mesh screen
RDF-4 (p-RDF)	Powder	Combustible waste fraction processed into powdered form such that 95% by weight passes through a 10 mesh screen (0.035 in. square)
RDF-5 (d-RDF)	Densified	Combustible waste fraction densified (compressed) into pellets, slugs, cubettes, briquettes, or similar forms
RDF-6	Liquid	Combustible waste fraction processed into a liquid fuel
RDF-7	Gas	Combustible waste fraction processed into a gaseous fuel

Source: R.E. Sommerland et al., 1988, Environmental characterization of refuse-derived-fuel incinerator technology, *National Waste Processing Conference, Philadelphia, 1988* (New York: ASME).

The first stage has 1-in holes to remove the glassy residue. The second stage has 6-in (152.4 mm) holes that separate the oversized material for further shredding and send the under-6-in fraction to RDF storage. The RDF produced has a heating value of over 6000 Btu; the ash content is less than 15%, and its particle size is 97% under 4 in (101.2 mm).

Grades of RDF

Different grades of RDF can be produced from MSW. Generally the higher the fuel quality, the lower the fuel yield. For example, an RDF plant in Albany, New York, simply shreds the incoming waste and passes the shredded material across a magnetic separator to remove the ferrous component. The fuel yield is roughly 95%, while the average Btu value of this fuel is similar to raw MSW. Conversely, producing a pellet fuel requires much preprocessing. A field yield of about 50%, based on the total incoming waste, can be achieved and has a heating value which approximates 6500 to 7000 Btu/lb.

Industry-wide specifications for RDF do not exist, but RDF has been classified according to the type and degree of processing and the form of fuel produced (see Table 9.3.1). The properties of RDF to consider and incorporate into supply contracts include the proximate analysis (moisture content, ash content, volatiles, and fixed carbon); ultimate analysis (C, H, N, O, S, and ash percentage); higher heating value (HHV); and content of chlorine, fluorine, lead, cadmium, and mercury.

Modeling RDF Performance

As community recycling increases, the feed to an RDF plant changes. Studies have determined the heating value and composition of RDF from different degrees of recy-

TABLE 9.3.2   COMPOSITION OF WASTE FOR BASE CASE

Component	Percent As-Received
Ferrous	5.5
Aluminum	0.9
Glass	9.5
Mixed paper	22.6
Newsprint	11.8
Corrugated	12.2
NonPVC plastic	2.9
PVC plastic	0.3
Yard waste	12.5
Food waste	2.5
Other noncombustible	9.5
Other combustible	9.8

Source: G.M. Savage and L.F. Diaz, 1986, Key issues concerning waste processing design, *National Waste Processing Conference, Denver, 1986* (ASME).

TABLE 9.3.3 CALCULATED MSW AND RDF PROPERTIES AND COMPOSITIONS RESULTING FROM DIFFERENT DEGREES OF RECYCLING

Scenario	Heating Value (Btu/lb Wet)	Percent Ash (Dry)	Ultimate Analysis (Percent)						Heavy Metal Analysis (mg/kg)									
			C	H	O	N	S	Cl	Sb	As	Ba	Cd	Cr	Cu	Pb	Hg	Ni	Zn
Baseline Case																		
MSW	3970	36.6	32.1	4.3	25.8	0.58	0.17	0.33	53	4.9	2160	14.4	210	720	630	18	220	290
RDF	5670	11.0	44.3	5.9	37.7	0.44	0.16	0.49	68	5.4	2620	14.0	200	170	500	23	40	160
30% Fe, Al, and Glass																		
MSW	4200	32.0	34.0	4.6	27.7	0.62	0.18	0.36	55	5.1	2220	15.0	200	570	600	18	160	270
RDF	5740	9.7	44.9	6.0	38.3	0.45	0.16	0.51	65	5.2	2510	13.4	190	140	470	22	30	130
30% Fe, Al, Glass, Newsprint, and Corrugated																		
MSW	4070	35.0	33.3	4.4	26.1	0.70	0.19	0.37	62	4.8	2500	16.5	210	600	550	21	170	300
RDF	5710	11.0	44.5	6.0	37.3	0.52	0.17	0.56	82	5.3	3200	16.2	210	160	440	28	30	160
30% Newsprint																		
MSW	3905	37.9	31.6	4.2	25.2	0.61	0.17	0.34	55	5.1	2250	15.0	210	750	590	18	230	300
RDF	5635	11.6	44.0	5.9	37.4	0.47	0.16	0.52	74	5.9	2860	15.1	200	180	450	25	40	170
50% PVC																		
MSW	3950	36.8	32.1	4.3	25.8	0.59	0.17	0.24	52	4.8	2190	14.4	210	730	640	18	220	290
RDF	5655	11.1	44.3	5.9	37.8	0.45	0.16	0.34	67	5.3	2680	14.0	200	170	500	23	40	160
50% Yard Waste																		
MSW	4055	37.6	31.7	4.2	25.5	0.50	0.16	0.33	55	5.1	2230	15.0	220	750	660	18	230	300
RDF	5720	11.0	44.2	5.9	37.8	0.41	0.16	0.49	69	5.5	2670	14.1	210	170	500	23	40	160
50% Food Waste																		
MSW	3990	36.7	32.2	4.3	25.8	0.57	0.17	0.32	53	5.0	2170	14.2	210	700	630	18	220	270
RDF	5680	11.0	44.3	5.9	37.8	0.44	0.16	0.48	68	5.4	2620	13.9	200	160	490	22	40	150

Source: Savage and Diaz, 1986.

cling. Table 9.3.2 shows the MSW composition assumed for such a study. In this model, the pretreatment steps include size reduction (to about 5 cm), screening, magnetic separation, and air classification. Table 9.3.3 gives the properties of the MSW and the recovered RDF after various degrees of recycling.

The model shows that the ash content drops and the heating value rises as the MSW is processed into RDF, and the type and degree of recycling has only a limited effect on the ash content or heating value (Savage and Diaz 1986). The nitrogen content of the RDF is consistently lower than that of the MSW, and the sulfur content is relatively unaffected by processing; while PVC recycling has a substantial effect on the chlorine content of the RDF. The calculated heavy metal analysis shows that because of the magnetic separation of ferrous metals, the concentration of lead (Pb) and zinc (Zn) is lower in the RDF than in the MSW.

Modeling is a useful tool in the evaluation of RDF processes. One can estimate the effect of the degree of size reduction, the influence of the opening sizes in screening equipment, and the effect of placing shredders up or downstream of the screening or air-separation equipment. Some modeling calculations can also estimate the base/acid ratio, slagging index, and fouling index values, which can indicate likely maintenance and operating problems associated with a particular process.

Adapted from *Municipal Waste Disposal in the 1990s* by Béla G. Lipták (Chilton, 1991).

Reference

Savage, G.M. and L.F. Diaz. 1986. Key issues concerning waste processing design. *National Waste Processing Conference, Denver, 1986.* ASME.

Treatment and Disposal

10.1
WASTE-TO-ENERGY INCINERATORS

Incineration is the second oldest method for the disposal of waste—the oldest being landfill. By definition, incineration is the conversion of waste material to gas products and solid residues by the process of combustion. Combustion under optimal conditions can cut MSW 90% by volume and 75% by weight. Hot gases generated as a result of combustion exit the furnace and pass through boilers which recover energy in the form of steam. This steam can be sold directly or converted to electricity in a turbine. With dwindling landfill space, incineration reduces volume, but some scientists caution that incinerator residue is more dangerous and should not be disposed of in regular landfills.

The combustion process carries the risk of releasing air pollutants. Emissions from incinerators can include toxic metals and toxic organics. The primary goals of waste-to-energy incineration are to maximize combustion and minimize pollution. Two other goals are high plant availability and low facility maintenance cost.

Mass-Burn and RDF Incinerators

Two main types of waste-to-energy incinerators are mass-burn incinerators and RDF incinerators. Figure 10.1.1 shows the typical structure of a waste-to-energy facility.

The more common *mass-burn incinerators* burn MSW as received with minimal onsite effort to separate objects that do not burn well or do not burn at all. (For example, bulky, oversized items such as tires, bedframes, fences, and logs are often separated by hand to avoid problems, but glass bottles and metals usually are not.)

RDF incinerators burn MSW that has been preprocessed and sorted (either on the site of the incinerator

1. Tipping Hall
2. Refuse Bunker
3. Grapple and Refuse Crane
4. Crane Operator Control
5. Charging Hopper
6. Overfire Air Fan
7. Ram Feed
8. Ignition Burner Fan
9. Underfire Fan
10. Roller Grate
11. Ash Conveyors to Materials Recovery
12. Boiler
13. Overfire Air Intake
14. Turbine Generator
15. Precipitator
16. Stack
17. Control Room
18. Deaerator Storage Tank and Heater
19. Motor Control Center
20. Maintenance Shop
21. Heaters
22. Condenser
23. Switchgear
24. Id Fan
25. Turbine Crane

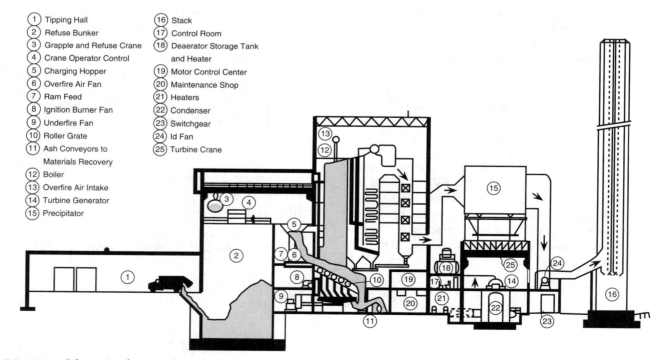

FIG. 10.1.1   Schematic of a typical waste-to-energy resource recovery facility.

or at separate processing facilities). Noncombustible and recyclable material such as ferrous metals, aluminum, and glass are separated mechanically and collected for processing and future sale or disposal. The combustible portion is converted to a more uniform, pellet fuel through particle reduction (usually 4- to 6-in pellets).

RDF technology is preferred by recycling-oriented users partly for economic reasons (e.g., income from the sale of aluminum), and partly because it cuts the incinerator residues to less than half and thereby reduces the amount of leftover material that must be landfilled. RDF-fired boilers can respond faster to load variations, require less excess air, and can operate at higher efficiencies. Comparisons of mass-burner performance on both raw MSW to simple prepared fuels show that prepared fuel plants have many advantages over the mass-burning technology (Sommer and Kenny 1984). However, RDF technology is still in the development stage. The majority of incinerators under construction are mass-burn. Part of the reason for this lack of development is the complexity of the RDF process, which remains an expensive and maintenance-intensive alternative to mass-burning.

Plant Design

The plant design for a waste-to-energy plant should consider state-of-the-art concepts as well as other design criteria.

CONCEPT OF STATE-OF-THE-ART

The term *state-of-the-art* for waste-to-energy plants refers to (1) the best technologies and operating practices for reducing the environmental impacts of these plants and (2) the best regularly attainable emission levels from them for certain air pollutants. The state-of-the-art in waste-to-energy plant design has been improving over time. Over the last decade, as landfill space has become scarce, interest in incineration has been renewed, environmental concerns have increased, and regulations have become more stringent.

The EPA's New Source Performance Standards (NSPS), proposed in 1989 and promulgated in February 1991, were the first regulations to broadly and specifically address the performance of MSW incinerators. The new regulations set standards in four basic areas: good combustion practice, emission levels for six pollutants, monitoring requirements, and operator training and certification. Table 10.1.1 summarizes these regulations.

DESIGN BASIS

In the design of waste-to-energy incinerators, the size of the plant is a critical factor. Planners need accurate information about the amount and type of waste the plant is to burn (see Sections 7.1 to 8.3) as well as projections for future solid waste management practices in the community.

Next, planners must determine what to burn. In keeping with the hierarchy of the Pollution Prevention Act (PPA), a state-of-the-art strategy provides for the maximum amount of source reduction and recycling, including composting, before incineration. Furthermore, materials that are not recyclable and are unsuitable for burning because they are noncombustible, explosive, or contain toxic substances or pollutant precursors, should be separated from the waste to be burned. These activities preserve natural resources, improve incinerator efficiency, and minimize pollutant emissions and ash quantity and toxicity.

A general, overriding principle in the design of a solid waste incinerator is to use the correct size incinerator for the amount of anticipated waste. Combustion is most efficient when an incinerator consistently burns the quantity and quality of MSW it was constructed to burn, as follows:

If the plant is oversized (i.e., if the amount of MSW available for burning is less than the plant was designed to take), it may operate less than full time. Each start up and shutdown causes unsteady burning conditions, resulting in reduced overall efficiency. Such unsteady state conditions increase the generation of incomplete combustion and particulates. More importantly, a plant that is oversized for the amount of waste available to burn has higher per ton disposal costs.

If an incinerator is undersized (that is, more MSW is available to be burned than originally planned), too much MSW may be loaded into the furnace. Overloading an incinerator can result in increased generation of incomplete combustion as well as an increased volume of unburned matter and ash. Also, an undersized incinerator that is not overloaded requires additional expenditures of alternative methods of waste disposal and recycling.

In determining the amount of MSW being generated, planners should collect actual waste data just prior to design and sizing. Waste composition studies should ideally sample waste from different neighborhoods at different times of the week and year, as shown in Figure 8.1.1. Some communities use average waste composition from other towns or cities to estimate their own waste composition. However, this method can be misleading since the composition of MSW changes not only from place to place but also over time.

Information about projected population growth and future trends in the volume and composition of waste is just as critical as current waste data, especially since waste management methods are changing. Incinerators are typically designed for at least a twenty-year lifetime, and incinerator arrangements often include long-term (fifteen- to thirty-

TABLE 10.1.1   KEY FEATURES OF NEW FEDERAL MSW INCINERATOR REGULATIONS (NSPS), COMPARED TO INFORM STATE-OF-THE-ART STANDARDS

New Source Performance Standards	INFORM State-of-the-Art Standard
Materials Separation	
None	Recyclables, noncombustibles, and wastes containing toxic materials or pollutant precursors removed
Good Combustion Practices	
Carbon monoxide emissions: 50–150 ppm (depending on furnace type) Plant-specific maximum load level Plant-specific maximum flue gas temperature at inlet to final particulate control device	50 parts per million
Pollutant Emissions Levels (7% O_2, dry basis)	
PARTICULATES	
0.015 g per dry standard cu ft	0.010 grains per dry standard cubic foot
DIOXINS/FURANS	
30 nanograms per dry standard cu m—total dioxins and furans	0.10 nanograms per dry normal cubic meter—Eadon toxic equivalents
SULFUR DIOXIDE	
80% reduction, or 30 ppm (whichever is less stringent)	30 parts per million
HYDROGEN CHLORIDE	
95% reduction, or 25 ppm (whichever is less stringent)	25 parts per million
NITROGEN OXIDES	
180 ppm	100 parts per million
HEAVY METALS	
No individual standards; particulate emissions as surrogate	Not defined; further research needed to identify lowest regularly attainable emissions levels
Monitoring Requirements	
CONTINUOUS MONITORING	
Carbon monoxide, opacity, sulfur dioxide, nitrogen oxides	Furnace and flue gas temperature, steam pressure and flow, oxygen, carbon monoxide, opacity, sulfur dioxide, oxides of nitrogen
ANNUAL STACK TESTS	
Particulates, dioxins, furans, hydrogen chloride	Particulates, dioxins/furans, hydrogen chloride, metals
Operator Training and Certification	
American Society of Mechanical Engineers certification standards for chief facility operators and shift supervisors	Formal academic and practical education; supervised on-the-job training; formal testing; periodic reevaluation

Source: U.S. Environmental Protection Agency, 1991, Burning of hazardous waste in boilers and industrial furnaces, final ruling, *Federal Register 56*, no. 35 (21 February), 7134–7240.

year) contracts for the quantity of MSW to be delivered to the plant and for the quantity of energy to be sold. Knowing what potentially recyclable material is in the waste stream and in what quantities is essential (see Section 8.1).

Finally, plant designers need information about the composition of the waste stream to determine the optimal physical design of the plant. For instance, different materials generate different amounts of heat energy when burned, and knowing the anticipated overall Btu value is critical to planning boiler capacity and furnace structure. The variability of MSW (specifically density due to changes in composition) is another design consideration for volumetric material handling equipment for RDF incinerators.

Process Design

A typical incinerator system contains basic elements: a feed system, a combustion chamber, an exhaust system, and a residue disposal system. Ancillary equipment includes shredders and a material sorter in the front end and air pollution control devices and a heat recovery device at the back end of the incinerator. Modern incinerators in the United States use continuous-feed systems and moving grates in primary combustion chambers which are lined with refractory material (heat-resistant silica-based material). Secondary combustion chambers burn the gas or solids not burned in the primary combustion chambers before discharging to the air pollution control devices.

WASTE RECEIVING AND STORAGE

A state-of-the-art solid waste management system specifies exactly what waste can be burned (based on combustibility and content of the toxic materials and pollutant precursors). It ensures that prohibited materials are detected and removed from the waste. Table 10.1.2 compiles the materials that are prohibited at several MSW incinerators. In addition, stringy wire items, such as fencing and trolling wire, can become entangled in conveyors and should be removed from the MSW feed. Such specifications are stated in contracts between operators and municipalities. Plant operators should prevent prohibited materials from entering the plant or the furnace.

A preliminary view of the waste is recommended when incoming MSW trucks are weighed. Scales, preferably integrated into an automated recording system, should be provided to record the weight of the MSW entering the plant. Tipping floors, which resemble large warehouse floors, are better suited for visual inspection and the removal of unwanted items. State-of-the-art screening includes opening garbage bags on the tipping floor to identify unwanted items inside the bags. Radioactivity sensors are used as screening devices for hospital waste. The MSW is discharged from the tipping floor into the storage pit or directly into the furnace.

The storage provided depends on variations in the rate of truck delivery of MSW to the plant and the planned burning schedule. Storage permits MSW to be retained during peak loads and thus allows the combustion chambers to be sized for a lower average capacity. Large storage areas are generally required for MSW since it is quite bulky, with a bulk density between 250 and 350 lb/cu yd (180 and 240 kg/cu m). Provisions are often made for as much as one week's MSW at small incinerators to allow for downtime and other operating problems; two to three days of MSW storage is more common at larger plants (less than 500 tn/d). Planners should consider seasonal and cyclic variations and unplanned shutdowns in establishing plant storage requirements. The pit size is usually

TABLE 10.1.2 MATERIALS ROUTINELY PROHIBITED BY MASS-BURNING PLANTS

Bulky wastes (e.g., furniture) (may be acceptable if reduced in size)
Noncombustible wastes (not including glass bottles, cans, etc.)
Explosives
Tree stumps and large branches (may be acceptable if reduced in size)
Large household appliances (e.g., stoves, refrigerators, washing machines)
Vehicles and major parts (e.g., transmissions, rear ends, springs, fenders)
Marine vessels and major parts
Large machinery or equipment
Construction/demolition debris
Tires
Lead acid and other batteries
Ashes
Foundry sand
Cesspool and sewage sludge
Tannery waste
Water treatment residues
Cleaning fluids
Crank case and other mechanical oils
Automotive waste oil
Paints
Acids
Caustics
Poisons
Drugs
Regulated hospital and medical wastes
Infectious waste
Dead animals
Radioactive waste
Stringy wire (e.g., fencing and trolling wire)

Source: M.J. Clark, M. Kadt, and D. Saphire, 1991, *Burning garbage in the US*, edited by Sibyl R. Golden (New York: INFORM, Inc.).

calculated based on an MSW density of 350 lb/cu yd of pit volume.

Refuse tends to flow poorly and can maintain an angle of repose greater than 90°. Thus, plants commonly stack refuse in storage facilities to maximize storage capacity.

Storage pits are usually long, deep, and narrow. A pit can be located in front of the furnace or a pit can be situated on each side of the furnace. If the storage pit is over 25 ft in width, the refuse dumped from the trucks must generally be rehandled. The floor of storage pits is pitched to the facilities' drainage. Storage pits are constructed of reinforced concrete with steel plates or rails along the sides, which protect them against damage from the crane bucket. The pit is usually enclosed in the MSW storage building, in which combustion air for the furnace is drawn. This arrangement creates a slight vacuum inside the building which draws in atmospheric air and prevents the escape of odors and dust.

FEEDING SYSTEMS

The waste feed system introduces refuse into the incinerator from the tipping floor or pit (or, in case of an RDF
fuel plant, from the preprocessing facilities). Of the two
main types of refuse feed systems, a continuous loading
system contributes to more efficient combustion than batch
loading because it allows a more even flow of fuel.

In batch loading, the waste is introduced by a front-end
loader that shoves the garbage, in discrete batches, into
the furnace. The batch method adversely affects combustion since each load pushed into the incinerator causes a
temporary overload, depleting available oxygen and creating poor combustion conditions. Variations in temperature due to air leaks into the furnace have an adverse impact on refractory material and increase air emissions. In
small plants with floor dumps and stored MSW, feeding
is accomplished on a semibatch basis by rams which push
MSW directly to the furnace at 6- to 10-min cycles.

With continuous loading, a traveling bridge crane
equipped with a grapple deposits waste, a few tons at a
time, into the top of an inclined chute. The garbage moves
down the chute onto the drying zone of a moving grate
allowing for continuous introduction of waste into the furnace. RDF is typically continuously fed into the furnace.

A basic requirement of the continuous loading system
is to keep the charging hopper to the furnace fired at all
times and to protect against burnbacks of fire from the
combustion area through the chute to the storage pit area.

Charging Cranes

Two types of cranes are widely used for handling refuse
for municipal incinerators. The most versatile is the bridge
crane with a clam-shell bucket. The bridge itself travels
across the length of the storage pit while a trolley moves
the bucket over the length of the bridge. With the bridge
crane, the storage pit can be as wide as 30 ft. If the storage pit is wide, the crane has to travel to the far side of
the pit to keep refuse from accumulating there. The time
required to traverse the pit affects the carrying capacity of
the system and wide pits with long bridges are not economical. Figure 10.1.2 shows a layout using a bridge crane.
In large furnaces of more than 300 tpd capacity, bridge
cranes are used.

The second type of crane, the monorail, can move in
one direction only, along the rail at the center line of the
pit. The range of the monorail is limited in regard to the
pit. The pit width is limited to about 1 m wider than the
width of the open bucket. If the storage pit is too wide,
the bucket cannot move to the sides of the pit, and the
material accumulates because of its tendency to cling together. The monorail system is normally designed to follow a straight path with the pit at one end and to lift MSW
into charging hoppers at the other. In a medium 100- to
300-tpd plant, the monorail is often used.

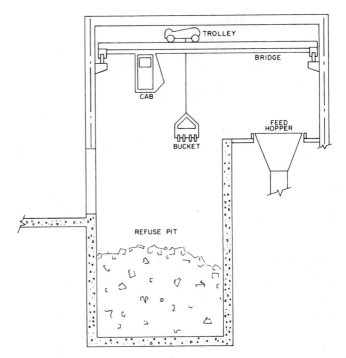

FIG. 10.1.2 Bridge crane installation.

Both crane types have either a clam-shell grapple or an
abrasion-resistant steel bucket with a capacity between 30
and 150 cu ft (1–4 cu m). An automatic lubrication system for the crane is recommended, and a good preventive
maintenance program is essential. Spare buckets are also
recommended. Bridges, trolleys, and hoists travel at speeds
of 6 ft per sec (100 m per min).

The traveling bridge is also used to mix the MSW.
Mixing MSW facilitates combustion particularly if a large
amount of one type of waste is discharged into one part
of the storage pit. In the past, the crane was operated from
an air-conditioned cab mounted on the bridge. With increasing frequency, crane operation is being centralized in
a control room, usually located at the charging floor elevation and either over the tipping positions opposite the
charging hoppers or close to the charging hoppers.

Charging Hoppers and Gates

Charging hoppers hold up some volume of refuse to guarantee a reasonably uniform waste flow into the incinerator. MSW enters charging hoppers in the following ways:

In larger plants where the hoppers are located above the
   storage pits, MSW is lifted by cranes.
In larger plants where the hoppers and storage area are at
   the same elevation, MSW is transferred into storage
   hoppers by ram feeders or by front-end loaders.
In plants under 100 tpd capacity, MSW is loaded directly
   from the trucks into the charging hoppers.
In multicell furnaces, each furnace cell usually has one
   charge hopper.

In a continuous-charging hopper, the outlet gate is kept open, and the air seal is maintained by the MSW and the movement of the mechanical grate charging the furnace. Most hoppers have an angle-of-slide surface of 30 to 60° from the vertical to prevent bridging. The feed chute is normally 4 ft (1.2 m) wide, to pass large objects with minimum bridging, and 12 to 14 ft (3.6 to 4.2 m) long from the hopper to the front end of the furnace. Because of its proximity to the combustion zone, the continuous-charging hopper is usually water cooled.

The continuous-charging hopper allows better furnace temperature control and thereby reduces the need for refractory maintenance. It also spreads the MSW more evenly across the grate, in a relatively uniform and thin layer, while sealing the furnace from cold air.

Lessons learned on RFD facilities suggest (1) using simple RDF floor storage, not bins that can become plugged; (2) using simple RDF transfer via a conveyor rather than pneumatic systems; (3) maintaining a uniform flow of RDF to boiler feeders and avoiding slug feeding, which results in unstable boiler control; and (4) using a proven RDF feeder, which maintains even grate distribution and is responsive to load change (Gibbs and Kreidler 1989).

THE FURNACE

The combustion zones in a refuse incinerator are commonly referred to as furnaces. Several common designs are currently in use: single-chamber furnaces, dual-chamber furnaces, multiple-chambered furnaces, rotary combustors, and fluidized combustors. The most common configuration includes the rectangular furnace, the multicell furnace, the vertical circular furnace, the combined rectangular furnace, and the rotary kiln. Furnaces can also be distinguished according to the type of grates used.

Because all large modern incinerators are continuous, this section discusses only continuous systems. Two classes of continuously feed furnaces are used today: refractory-lined and waterwall furnaces. Waterwall furnaces recover waste heat as well as reduce waste volume, while refractory furnaces are usually designed for volume reduction. Waterwall furnaces have water-filled tubes instead of refractory material lining the combustion chambers. As burning refuse transfers heat through the walls to the water in the tubes, these tubes form a cool wall which is in contact with the flame and hot gas. These cooler walls prevent the accumulation of slag on the side of the combustion chamber and produce steam.

Combustion Process

Efficient and even combustion is a key factor in minimizing the environmental impact of waste-to-energy incinerators, reducing both the amount of unburned material in the ash produced and the amount of air emissions. This reduction depends largely on the design of the furnace and the operating practices.

The following general guidelines foster good combustion (Licata 1986):

The grate should be covered with fuel (a uniform depth of garbage and trash) across its width. The depth at any location on the grate should be consistent with the air that can be delivered for combustion at that point.

The incinerator must include an air distribution system that apportions air according to the burning rate of waste along the entire length and width of the grate.

Underfire air should be introduced carefully. Depending on the technology, it can be concentrated in a small area or spread over a large area. Zones of high-pressure air and blowtorch effects should be eliminated. Bursts of air in one section of the fuel bed prevent even mixing of air in the burning refuse in other areas.

Air must be introduced into burning refuse both above and below the burning bed. Oxygen provided through the overfire system helps complete the combustion of any hydrocarbons (and particulates) not oxidized near the fuel bed.

Steps must be taken to prevent the buildup of slag within the furnace. Slag can damage the boiler system and also results in poor combustion by preventing proper air mixing in the fuel bed.

Gases generated in the incineration process should experience maximum mixing to facilitate oxygen reaching any unburned particles and to provide a maximum dwelling time for the gases before being released into the atmosphere.

The flue gas temperature should be at or above 1600°F for approximately 1 sec after leaving the fire bed. Figure 10.1.3 shows that these combustion conditions destroy

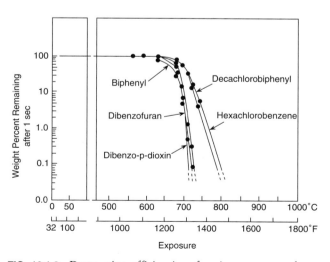

FIG. 10.1.3 Destruction efficiencies of various compounds as a function of temperature. (Reprinted, with permission, from *Air pollution control at resource recovery facilities*, 1984, California Air Resources Board [24 May].)

more than 99.9% of many effluent compounds, including dioxins and furans. Excessively high temperatures and extreme variations cause cracking and spalling, with rapid deterioration of refractories. The minimum burning temperature for carbonaceous waste to avoid the release of smoke is 1500°F (816°C). A temperature less than 1500°F permits the release of dioxins and furans.

Auxiliary burners can be added to maintain combustion efficiency. Reductions in combustion efficiency are usually due to one or more factors: start up and shutdown; large changes in moisture content, heat content, or the quantity of incoming refuse; and maladjustment of the air adjustment system. Auxiliary burners burn another, more uniform fuel (such as natural gas or oil). These burners are used when furnace temperature values fall below 1600°F, thereby stabilizing combustion by maintaining a minimum furnace temperature. Operators can increase residence time by reducing the amount of combustion air.

The design of the furnace interior affects combustion efficiency. Carefully placed protrusions from the furnace wall, called arches or bullnoses, can redirect the flow of air from the grate, guiding it into turbulent eddies within the furnace. Eddy currents maximize turbulence during the combustion of gases.

Grate Designs

The grate (stoker) serves dual functions:

1. Transports the solid waste and residue through the furnace to the point of residue discharge. The grate should be covered with a uniform depth of MSW across its width.
2. Promotes combustion by providing proper waste agitation and by permitting the passage of underfire air through the fuel bed. However, the agitation should not be so violent that it contributes to excessive particulate emissions.

The design of the grate system in the furnace is a critical element in the operation of a RDF facility. Eberhardt (1966) proposes ten elements to consider in the choice of a grate system:

The adaptability of the combustion process to handle wide variations in radiation effects
The adaptability of the refractory to handle wide variations in radiation effects
Provisions for controlling air quantity and temperature
Provisions for an adjustable retention time based on the material being burned
An adjustable height of the waste layer to be burned
A controllable, stabilizing heat supply (auxiliary fuel)
A controlled cooling of residue (by quenching)
A controlled flue gas temperature prior to impinging on the radiation heating surface
The capability of observing the fire layer and the fire gases

Technical design including:
—Prevention of reignition
—Positive conveyance of the refuse mass
—Serviceability and replaceability of worn-out parts
—Proper measuring and control systems

Grate Systems

The key factors for hot, uniform combustion are the constant mixing of air into the material being burned and the use of partially combusted material to heat and ignite new material introduced into the combustion chamber. Three major European grate designs have world-wide application:

Martin process (see Figure 10.1.4). In this design, the grate has a reverse reciprocal action; it moves alternately down and back to provide continuous motion of the refuse. The net motion of the refuse is downward toward the bottom of the furnace, but the agitation caused by oscillation of the grate causes considerable mixing of the burning refuse with the newly introduced material leading to rapid ignition and uniform burning.

Van Roll process (see Figure 10.1.5). This design has three sections: the first dries the newly introduced refuse and ignites it, the second serves as a primary combustion grate, and the last reduces the refuse to ash. Grate elements move so that at a given time for any pair of elements, one is moving and one is stationary. This process results in the refuse moving toward the bottom of the furnace, but the shuffling action of the grates agitates the fuel bed enhancing the combustion process.

VKW or Dusseldorf process (see Figure 10.1.6). This grate is comprised of several horizontal drums with a diameter of 1.5 m (5 ft). The shafts of the drums are parallel one after the other at a 30° slope. The drums are placed on 1.75-m (about 7-ft) centers. Each drum is built of bars (cast iron) in the form of arched segments

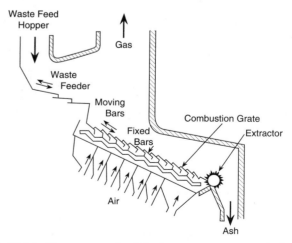

FIG. 10.1.4 Grate system for Martin resource recovery incinerator.

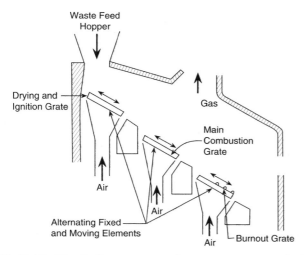

FIG. 10.1.5 Van Roll grate system for resource recovery incinerator.

which are keyed to a central element below. Each drum rests over a separate chamber to control underfire air. The unit rotates in the discharge direction at an adjustable peripheral speed which varies according to the constituents of the waste being burned. The drum shafts lie in the bearings placed in the outside walls of the unit, and each roller is fitted with a driving gear and can be regulated independently of the others. Ignition grates at the front end of the incinerator generally rotate at up to 15 m/hr (50 ft/hr). The burnout grates normally rotate at 5 m/hr (about 16 ft/hr) since they have little waste material to move. The room under the grate is divided into a zone for each roller, to which preheated or cooled flue gas (about 200 to 256°C) can be brought. A special feeding arrangement carries the refuse from the feeding chute to the grate.

Another aspect of grate design is the percentage of air openings provided in the grate. These air openings vary

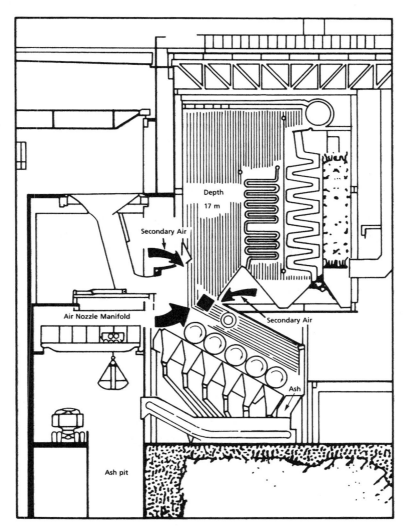

FIG. 10.1.6 VKM or Dusseldorf process (with installation of transverse manifold for 260 secondary air nozzles over roller grate at Wuppertal, Germany).

from 2 to over 30% of the grate area (Velzy 1968). Proponents of the larger openings feel that the siftings (the ash from the fuel bed) should be allowed to fall below the grate as soon as possible and large amounts of air should be permitted to pass through the bed to meet the combustion requirements of varying fuel characteristics. Proponents of the smaller openings cite advantages such as the small volume of siftings, the small amount of underfire air that is required, and the resulting shorter combustion flames, all of which reduce particle entrainment in the escaping gas (Velzy 1968).

In a technique employed at some U.S. facilities, waste is pneumatically injected into the furnace system and burned while suspended in the furnace chamber, rather than being burned on a grate (see Figure 10.1.7). With the removal of ferrous metals and other noncombustibles in typical RDF systems, a boiler system has evolved and has been in commercial operation at Biddeford, Maine, since 1987. The controlled combustion zone (CCZ) boiler design is a state-of-the-art boiler design for both wood and RDF boilers (Gibbs and Kreidler 1989).

Grate Sizing

The hourly burning rate (F_a) varies from 60 to 90 lb of MSW per sq ft of grate area (Velzy and Hechlinger 1987). An hourly rate of 60 lb/sq ft reduces refractory maintenance and provides a safety margin. In coal burning furnaces, the grates are usually covered to a depth of 6 in, which corresponds to an hourly coal load of 30 to 40 lb/sq ft. The heating values and densities of uncompacted MSW are less than half of that. Thus, the same firing densities (on a Btu basis) produced by coal can be produced by MSW when the MSW is supplied at an hourly rate of 60 lb/sq ft and covers the grate to a depth of 3 to 4 ft. The required grate area in square feet is directly proportional to the maximum charging rate F (lb/hr) and inversely proportional to F_a, the grate area A, as follows:

$$A = F/F_a \qquad \text{10.1(1)}$$

The grate design must also be based on the manufacturer's design criteria. Basically, the only consistent design criteria used by manufacturers is the specified kilogram (pounds) of waste that can be loaded per square meter (square foot) of the grate area. Planners need more empirical data for proper design and must develop a more rational approach to select the proper grate.

Furnace Sizing

The firing furnace capacity is a function of its grate area and volume. The furnace volume is usually determined on the basis of an hourly heat release of 20,000 Btu/cu ft. If the hourly release rate is 20,000 Btu/cu ft and the heating value of the MSW is 5000 Btu/lb, the hourly firing rate is 4 lb/cu ft of furnace volume. A typical design basis is to provide 30 to 35 cu ft of furnace volume for each tpd of incinerator capacity (Velzy and Hechlinger 1987).

Air Requirements

The basic requirement of any combustion system is a sufficient supply of air to completely oxidize the feed material. The following chemical and thermodynamic properties must be considered in incinerator design: the elemental composition, the net heating value, and any special properties of the waste that can interfere with incinerator operation. The stoichiometric, or theoretical, air requirement is calculated from the chemical composition of the feed material. Planners must know the percentages of carbon, hydrogen, nitrogen, sulfur, and halogens in the waste as well as its moisture content to calculate the stoichiometric

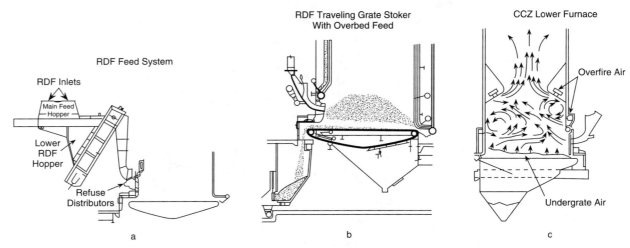

FIG. 10.1.7 RDF furnace. (Reprinted, with permission, from D.R. Gibbs and L.A. Kreidler, 1989, What RDF has evolved into, *Waste Age* [April].)

combustion air requirements and predict combustion air flow and flue gas composition.

Table 10.1.3 shows the stoichiometric oxygen requirements and combustion product yield for each waste component. The stoichiometric air requirement is determined directly from the stoichiometric oxygen requirement with use of the weight fraction of oxygen in air. Given temperature and pressure, the required volume of air can be calculated based on gas laws.

If perfect mixing could be obtained and waste burnout occurred instantaneously, only the stoichiometric requirement of air would be needed. However, neither of these phenomena occurs in real-world applications. Therefore, some excess air is required to ensure adequate waste–air contact. Excess air is usually expressed as a percentage of the stoichiometric air requirement. For example, 50% excess air implies that the total air supply to the incinerator is 50% higher than the stoichiometric requirement.

In general, the minimum excess air requirement for an incinerator depends on the degree of mixing achieved and waste-specific factors. Most incinerators require 80 to 100% excess air to burn all organics in the MSW (Wheless and Selna 1986). Incinerator operation is optimized when sufficient oxygen is provided to achieve complete combustion, but no more. Additional oxygen reduces thermal efficiency and increases nitrogen oxide generation.

The cold air volume required for proper combustion in the incinerator per unit weight of MSW can be calculated as follows (Essenhigh 1974):

Total Cold Air Volume (cu ft/lb) = B (1 − a − M)(S)(1 + e)

10.1(2)

where:

B = the dry and inert-free (DIF) heating value, in Btu/lb of MSW

TABLE 10.1.3 STOICHIOMETRIC OXYGEN REQUIREMENTS AND COMBUSTION PRODUCT YIELDS

Elemental Waste Component	Stoichiometric Oxygen Requirement	Combustion Product Yield
C	2.67 lb/lb C	3.67 lb CO_2/lb C
H_2	8.0 lb/lb H_2	9.0 lb H_2O/lb H_2
O_2	−1.0 lb/lb O_2	—
Cl_2	−0.23 lb/lb Cl_2	1.03 lb HCl/lb Cl_2 −0.25 lb H_2O/lb Cl_2
F_2	−0.42 lb/lb F_2	1.05 lb/HF/lb F_2 −0.47 lb H_2O/lb F_2
Br_2	—	1.0 lb Br_2/lb Br_2
I_2	—	1.0 lb I_2/lb I_2
S	1.0 lb/lb S	2.0 lb SO_2/lb S
P	1.29 lb/lb P	2.29 lb P_2O_2/lb P
Air N_2	—	3.31 lb N_2/lb $(O_2)_{stoich}$
Stoichiometric air requirement		4.31 lb Air/lb $O_{2(stoich)}$

a = the inert and ash fraction of the MSW
M = the moisture fraction of the MSW
S = the cubic feet of stoichiometric cold air required per Btu of heat release
e = the excess air fraction

In cases where metals are not burned and combustibles are predominantly organic, the value of S is approximately 0.01 (i.e., 1 cu ft of cold air per 100 Btu). This approximation is valid, generally to within 10 to 20%, for a wide range of organic fuel. Consequently, variations between wastes depend largely on their noncombustible content (a + M), particularly as the DIF calorific values lie within the narrow range of 8000 to 10,000 Btu/lb. The stoichiometric air requirements of most DIF waste are therefore 80 to 100 cu ft/lb or 6.4 to 8 lb of air per pound of waste. If the waste contains 50% ash and moisture, the calorific values drop to 4000 to 5000 Btu/lb. If this waste is fired at 150% excess, the air requirements are 8 to 10 lb of air per pound of waste as fired.

In modern, mechanical, grate furnace chambers, the underfire and overfire air are usually provided by separate blower systems. Underfire air is admitted to the furnace under the grates and through the fuel bed. It supplies primary air for the combustion process and also cools the grates. Underfire air is usually more than half of the total air (50 to 70%). Particulate emissions from incinerators tend to increase with heat release and underfire air flow, while they tend to decrease with increasing fuel particle size (see Figure 10.1.8).

Overfire air can be introduced at two levels:

Immediately above the fuel bed to promote turbulence and mixing and to complete the combustion of volatile gases driven off the bed of burning solid waste.
From rows of nozzles placed high on the furnace wall. These nozzles allow secondary overfire air to be introduced into the furnace to promote additional turbulence of gases and control temperature. The number, size, and location of the overfire inlet ports determine the amount of turbulence and backmixing in the stirred reaction region above the burning waste. See Figure 10.1.6. For good combustion, the overfire air system must have broad flexibility to accommodate changes in fuel moisture, ash content, and Btu value.

Operators control flue temperature and smoking by modulating the total air flow and the underfire-to-overfire air ratio. For most U.S. grate designs, the required underfire air pressure is about 3 in of water. The overfire air pressure is adjusted so that entrance velocities at the nozzle are high enough to guarantee high turbulence without impinging on the opposite wall and residence times are long enough to assure complete combustion.

Influent air is usually at an ambient temperature, normally 27°C (80°F). It can get as high as 1650°C (2100 to 2500°F) in the immediate proximity of the flame. When the gas leaves the combustion chamber, the temperature

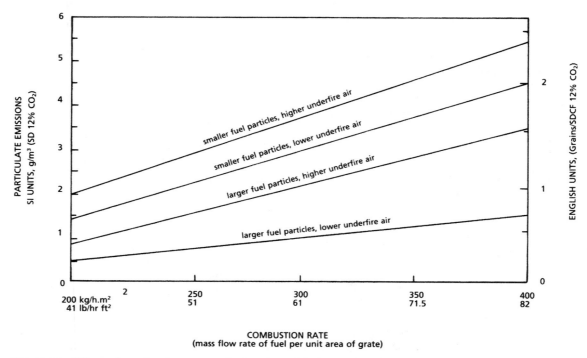

FIG. 10.1.8 Effects of combustion rate, underfire air, and fuel particle size on particulate emissions generated by combustion of wood waste. (Reprinted, with permission, from K.L. Tuttle, 1986, Combustion generated particulate emissions, *National Waste Processing Conference, Denver, 1986* [ASME].)

should be reduced to between 760 and 1000°C (1400 to 1800°F). If air pollution control devices are installed, induced draft fans must be installed, and the temperature should probably not exceed 260 to 370°C (500 to 700°F).

The mathematical modeling of the incinerator presented by Essenhigh (1974) provides a better understanding of the combustion processes taking place in incinerators. Figure 10.1.9 describes the gas-phase (II) and solid-phase (I) zones in a top-charged incinerator (overbed feed).

Calculating Heat Generation

Calculating the amount of heat generated through the incineration of MSW is necessary to determine how much auxiliary fuel is needed for combustion. The moisture content of MSW ranges from 20 to 50% by weight, and the combustible content is 25 to 70% by weight. The heating value of MSW depends on its composition.

Assuming that the average heating value of the combustible is 8500 Btu/lb and the moisture and inert concentration of the MSW is known, environmental engineers can estimate the heat content of MSW using Figure 10.1.10. If the heating value of the combustibles is less than 8500 Btu/lb, the number in Figure 10.1.10 must be multiplied by the ratio of the actual heating value divided by 8500.

Table 10.1.4 gives a material balance of burning 100 lb of MSW. The table assumes the MSW to have a heat content of 5000 Btu/lb, a moisture content of 22.4%, and a noncombustible content of 19% and that it contains 28

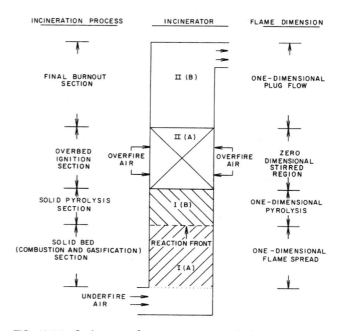

FIG. 10.1.9 Incinerator for continuous overbed feed of waste. Schematic represents solid bed, zone I, with overbed, zone II. For overbed feed, zone I has subzones, including I(A), the combustion and gasification section on the grate, and I(B), the drying and pyrolysis above I(A). Zone II (overbed combustion) has a backmix (stirred) region called subzone II(A), followed by a plug flow burnout region, subzone II(B). With underfeed, zone I is inverted, with drying and pyrolysis below the combustion subzone and the reaction front moving down instead of up as shown.

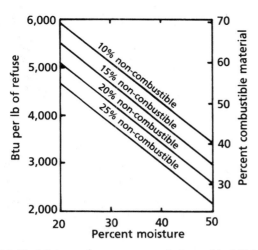

FIG. 10.1.10 Moisture–heat content relation with 8500 Btu/lb combustible material. (Reprinted, with permission, from Velzy and Hechlinger 1987.)

lb of carbon and 0.6 lb of hydrogen. It also assumes that 1–3 lb of combustibles escape unburned, and 140% excess air is needed to cool the refractories. Therefore, the total air required is 2.4 times the stoichiometric requirement, or 8.24 lb of air per pound of MSW.

TABLE 10.1.4   MATERIAL BALANCE FOR FURNACE (IN LB/100 LB OF REFUSE)

Input:			
Refuse			
Combustible material			
Cellulose	52.74		
Oils, fats, etc.	5.86	58.6	
Moisture		22.4	
Noncombustible		19.0	100.0
Total air, at 140% excess air			
Oxygen	191.0		
Nitrogen	633.0	824.0	
Moisture in air		11.0	
Residue quench water		5.0	
Total		940.0	
Output:			
CO$_2$ (28 × 3.667)		102.7	
Air—Oxygen (191–80)	111.0		
Nitrogen	633.0	744.0	
Moisture			
In refuse		22.4	
From burning cellulose		29.3	
From burning hydrogen		5.4	
In air		11.0	
In residue quench water		5.0	73.1
Noncombustible material			19.0
Unaccounted for			1.2
Total			940.0

Source: C.O. Velzy and R.S. Hechlinger, 1987, Incineration, Section 7.4 in *Mark's standard handbook for mechanical engineers*, 9th ed., edited by T. Baumeister and E.A. Avallone (New York: McGraw-Hill).

An enthalpy-balance calculation for this example shows that the enthalpy of each pound of existing gas is 455 Btu higher than the enthalpy of the 80°F air that enters (Velzy and Hechlinger 1987). Based on this enthalpy rise, the expected flue gas temperature of 1680°F can be read from Figure 10.1.11. This flue gas temperature is low enough to protect the refractory.

HEAT RECOVERY INCINERATORS (HRIs)

Three types of HRI designs are used to burn MSW: mass-burning in refractory-walled furnaces, mass-burning in waterwall furnaces, and combustion of RDF in utility boilers.

Mass-Burning in Refractory-Walled Furnaces

In mass-burning in refractory-walled furnaces, the waste-heat boiler, located downstream of the furnace, receives heat from the flue gases. Older HRIs tend to be refractory-walled designs, and their steam production is usually limited to 1.5 to 1.8 lb of steam per pound of MSW burned, assuming that the heating value of MSW is 4400 Btu/lb. In older furnaces, the larger the furnace, the lower the surface-to-volume ratio, because less surface exists to cool the flame. These units need higher quantities of combustion air to prevent overheating the wall, which results in slagging and deterioration. Approximately 50 to 60% of the heat generated in the combustion process can be recovered from such systems.

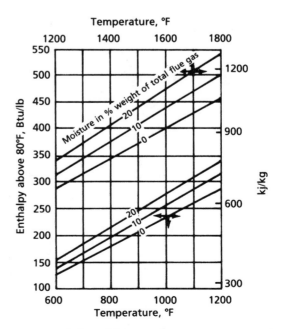

FIG. 10.1.11 Enthalpy of flue gas above 80°F. (Reprinted, with permission, from Velzy and Hechlinger 1987.)

Mass-Burning in Waterwall Furnaces

In mass-burning in waterwall furnaces, most or part of the refractory in the furnace chamber is replaced by water-walls made of closely spaced steel tubes welded together to form a continuous wall. Water is continuously circulated through these tubes. In newer waterwall designs, the steam production is around 3 lb of steam per pound of MSW. The increase in thermal efficiency is mostly due to a reduction in the excess air (from about 150% for refractory-walled furnaces to about 80% for waterwall furnaces).

Coating a substantial height of the primary combustion chamber, which is subject to higher temperatures and flame impingement, with a thin coat of silicon carbide refractory material and limiting the average gas velocities to under 15 ft/sec (4.5 m/sec) is recommended. Gas velocities entering the boiler convection bank should be less than 30 ft/sec (9.0 m/sec) (Velzy 1986). The efficiency of heat recovery in such units ranges from 65 to 70%.

Combustion of RDF in Utility Boilers

In combustion of RDF in utility boilers, the RDF is often burned in a partially suspended state, where some of the RFD stays on the grate. Steam production is about 3 lb of steam per pound of RDF. The efficiency of RDF boiler units ranges from 65 to 75%.

Minimizing Superheater Corrosion

To generate electricity from steam efficiently, HRIs must heat the steam to at least 700°F (371°C). This temperature results in more fireside corrosion in MSW-fired boilers than in regular boilers. Corrosion in refuse boilers is related to the high chlorides in MSW. While an RDF processing system can remove some of the material containing chlorides, removing chloride-containing material in the RDF processing system is not a realistic means to prevent boiler corrosion. High-nickel-alloy superheater tubes (e.g., Inconel 825) minimize superheater corrosion in addition to protecting the furnace from overloading and providing

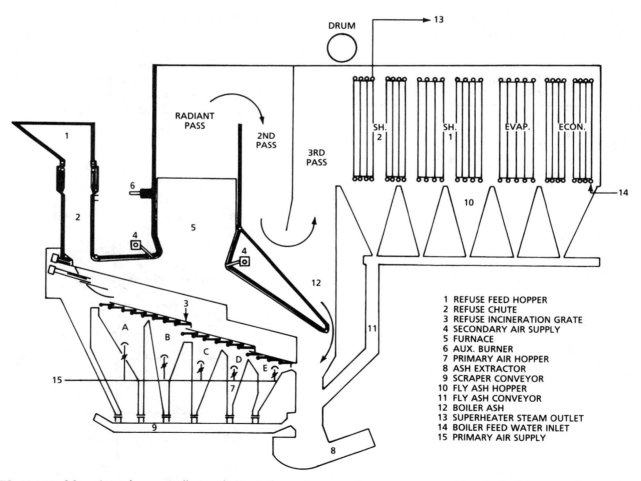

1 REFUSE FEED HOPPER
2 REFUSE CHUTE
3 REFUSE INCINERATION GRATE
4 SECONDARY AIR SUPPLY
5 FURNACE
6 AUX. BURNER
7 PRIMARY AIR HOPPER
8 ASH EXTRACTOR
9 SCRAPER CONVEYOR
10 FLY ASH HOPPER
11 FLY ASH CONVEYOR
12 BOILER ASH
13 SUPERHEATER STEAM OUTLET
14 BOILER FEED WATER INLET
15 PRIMARY AIR SUPPLY

FIG. 10.1.12 Mounting tubes vertically in a horizontal superheater section to prevent particle velocity increases. (Reprinted, with permission, from A.J. Licata, R.W. Herbert, and U. Kaiser, 1988, Design concepts to minimize superheater corrosion in municipal waste combustors, *National Waste Processing Conference, Philadelphia, 1988* [New York: ASME].)

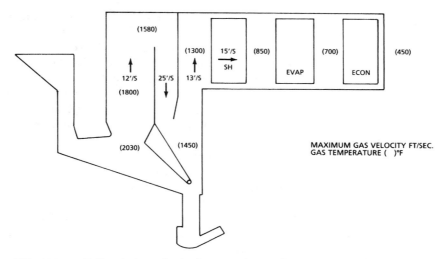

FIG. 10.1.13 Boiler design criteria for corrosion and erosion control. (Reprinted, with permission, from Licata, Herbert, and Kaiser 1988.)

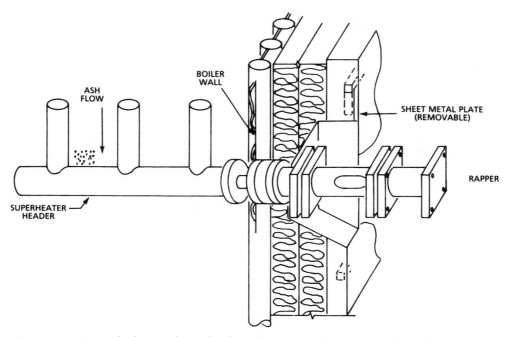

FIG. 10.1.14 Rapper boiler superheater headers. (Reprinted, with permission, from Licata, Herbert, and Kaiser 1988.)

rugged furnace walls. Hydrogen chloride corrosion begins by penetrating a slag layer on the superheater tubes. The tubes must be kept clean by soot blowers or mechanical rapping. Chlorides in hot gases become corrosive and can destroy a superheater.

With improved superheater designs, the operating superheater temperature can be increased from 750 to 825 or 900°F (Licata, Herbert, and Kaiser 1988). This temperature can be achieved when gas velocities are kept between 15 and 18 ft/sec to minimize the erosion caused by the impact of the particles. In addition, tubes should be liberally spaced to mitigate the increase in velocity as ash

buildup occurs. Figures 10.1.12 and 10.1.13 show the recommended superheater design criteria for velocities and temperatures.

Another design improvement is the elimination of the harmful effects of soot-blowing by steam or air which damages the protective oxide film, creates hot spots from nonuniform cleaning, and reentrains ash into the flue gas. Rapping rather than blowing can eliminate these effects (Licata, Herbert, and Kaiser 1988). Figure 10.1.14 shows pneumatically actuated mechanical rappers that allow deposits to slide down the tube surfaces into the hoppers below.

The boiler design should also protect against stratification (which can result in reduced atmosphere quality) by forcing the flue-gas stream to make a 180° turn before entering the superheater (see Figure 10.1.12). When the excess air level is maintained at 80 to 85%, high levels of CO concentration caused by incomplete combustion can be prevented. Another recommended feature is a ceramic lining for the postcombustion zone. This lining provides a 1-sec (minimum) residence time for flue gases at temperatures in excess of 1800°F (980°C) before they enter the superheater section.

An increased soot removal frequency and innovative cleaning techniques can minimize the secondary formation of dioxins and furans. Cleaner tubes have fewer fly ash particles on which dioxins and furans can form and allow more heat to be transferred away from flue gases. This heat transfer further cools the gases below the 450°F (250°C), which is conducive to dioxin and furan formation. Additionally, minimizing the production of precursors in the furnace by maximizing combustion efficiency helps decrease secondary dioxin and furan formation.

RESIDUE HANDLING

In a continuously fed incinerator, the ash or residue is discharged continuously through a chute into a conveyor trough, which is filled with water to cool the residue before it is hauled away for final disposal. The chute is submerged under quench water to seal the furnace outlet and prevent entry of atmospheric air. In newer, mass-burning facilities, full-size discharge chutes minimize hangups with large pieces of residue.

The residue conveyor pulls the settled residue from the bottom of the trough and transports it to an ash hopper, storage bin, roll-off carrier, or dump truck. The trough is constructed of steel or concrete, and the residue–discharge system usually has two conveyor troughs so that a full standby is available. Having a full standby permits switching between systems for even wear and scheduled maintenance.

As the conveyor carries the residue, most of the quench water runs off and returns to the trough. The conveyor should run at velocities not exceeding 5 to 10 ft/min (1.5 to 3 m/min) for good dewatering and minimal wear (Stelian and Greene 1986). The moisture content of the ash is usually 25 to 40% or more by weight. Reducing the water content of the ash minimizes transportation costs and water pollution. By reducing the speed of variable speed conveyors, operators can achieve this reduction by maximizing the residence time of residue on the wet-drag conveyor. Wet-drag conveyors can operate at slopes up to 45°, but some operators prefer lower slopes to protect bulky items from rolling back.

The design of a residue-handling system should minimize the discharge of water pollutants. Ash can be acid or alkaline; therefore, the water pH must be controlled in the range of 6 to 9 pH. The water can also contain high concentrations of BOD, dioxins, heavy metals, and other suspended or dissolved toxic or polluting constituents. For this reason, the ash-handling system must operate in the zero discharge mode (Stelian and Greene 1986). A water circulation and clarification system, including properly designed basins, sumps, and an easily maintained pumping station, is required. To capture the water that might drain off in the ash-transfer process, the system should have catch troughs where the conveyor transfers the ash into the receiver.

In cold regions with freezing winter temperatures, the ash-handling system must be protected against freezing. In cold areas, heated trucks transport the ash, and the fly-ash conveyors are insulated for protection against corrosion and caking. The ash conveyor can unload wet residue into a temporary container or directly into a transport vehicle for removal from the site. In mass-burn systems, directly discharging into dump trucks is best and simplest.

AIR POLLUTION CONTROL (APC)

Although incinerator design, operating practices, and fuel cleaning (waste reduction and separation systems) can significantly reduce the amount of pollutants produced in waste-to-energy plants, some pollutants are inevitably generated. Add-on emission control devices neutralize, condense, or collect these pollutants and prevent them from being emitted into the air. Most of these devices are placed at the back end of the incinerator, treating flue gases after they pass out of the boiler.

Different types of pollutants require different control devices: scrubbers and condensers for acid gases, scrubbers and condensers with electrostatic precipitators (ESPs) or fabric filters (baghouses) for particulates and other heavy metals and chemical neutralization systems for oxides of nitrogen. Variations exist within these basic categories; while some devices are more likely than others to achieve high removal efficiencies, operational factors, such as temperature, play a key role.

Acids, Mercury, Dioxin, and Furan Emissions

Scrubbers, followed by an efficient particulate control device, are the state-of-the-art equipment for controlling emissions of acids such as hydrogen chloride, sulfur dioxide, and sulfuric acid. Scrubbers generally use impaction, condensation, and acid–base reactions to capture acid gases in flue gas. Since greater removal efficiencies usually accompany greater condensation, devices that lower gas temperatures and thus increase condensation can enhance scrubber effectiveness. The lower temperatures also allow mercury, dioxins, and furans to condense so that they can subsequently be captured by a particulate device.

Three types of scrubbers are in use: wet scrubbers, spray-dry scrubbers, and dry injection scrubbers. The first two scrubbers are condensers, while dry injection scrubbers require a separate condenser (either a humidifier or a heat exchanger). In all cases, temperature and, for dry and spray-dry scrubbers, the amount of lime (an alkaline substance that neutralizes acids) are the key factors affecting scrubber effectiveness. In general, to maximize emission control, the scrubber should be adequately sized, operate at temperatures below 270°F, and allow flue gas circulation through the scrubber for at least 10–15 sec.

WET SCRUBBERS

Wet scrubbers capture acid gas molecules onto water droplets; sometimes alkaline agents are added in small amounts to aid in the reaction. New designs report on removing over 99% of the hydrogen chloride and, in some cases, sulfur dioxide and over 80% of the dioxin, the lead, and mercury (Hershkowitz 1986). The disadvantages include the added cost to treat the wastewater produced, corrosion of the metal parts, and incompatibility with the fabric type of the particulate control device. However, wet scrubbers collect gases as well as particulates, especially sticky ones.

SPRAY-DRY OR SEMI-DRY SCRUBBERS

With these scrubbers, acid gases are captured by impaction of the acid gas molecules onto an alkaline slurry, such as lime. Here, the evaporation water from the scrubbing liquid is carefully controlled so that when the material reaches the bottom of the tower, it is a dry powder (a dry fly ash and lime mixture). This method eliminates the scrubber water that must be treated or disposed; additionally, the power requirements and corrosion potential are reduced. Emission tests have demonstrated control efficiencies of 99% or better for hydrogen chloride and sulfur dioxide removal under optimal conditions (temperatures below 300°F, sufficiently high lime/acid ratios, and sufficiently long gas residence time in the scrubber). Dioxins were also considerably reduced (Hershkowitz 1986).

DRY INJECTION SCRUBBERS

Dry injection scrubbers inject dry powdered lime or another agent that reacts with the acid gases in flue gas. In one research test, removal efficiencies of 99% for hydrogen chloride and 96% for sulfur dioxide were measured under optimal temperature conditions (230°F); dioxins were also considerably reduced (Platt et al. 1988).

TRENDS

A report by the German equivalent to the U.S. EPA predicts a trend toward wet scrubbing because of better elimination of sulfur dioxide, heavy metals, and other toxic substances (McIlvaine 1989). In addition, spray drying and other dry processes have the disadvantage of increased residue production. The report does cite potential problems with wet scrubbers; however, water treatment and heavy metal precipitation and evaporation are promising solutions. Typically, a German APC system uses a packed tower to remove hydrogen chloride, sulfur dioxide, and condensed heavy metals and a high-efficiency ESP to remove dust.

Particulate and Heavy Metal Emissions

The emissions of particulates and heavy metals are best reduced by collecting them in one of two basic types of add-on particulate control devices: fabric filters and ESPs. Heavy metals are captured because they are condensed out of flue gas onto the particles. These devices are designed to operate at temperatures lower than 450°F for flue gas leaving the boiler; some operate at temperatures as low as 250°F, which is beneficial for condensing and collecting acids, volatile metals, and organics. The state-of-the-art level for particulate emission is 0.010 g per dry cu ft.

FABRIC FILTERS

Fabric filters (also called baghouses) are a state-of-the-art particulate control technology with a consistent 99% removal efficiency over the range of particulate sizes. Figure 10.1.15 shows a schematic diagram of a scrubber followed by a baghouse for particulate control. Particulates as small as 0.1 microns can be captured. The accumulated particulates or fly ash fall into a hopper when the fabric filters are cleaned, and this ash must be disposed of appropriately. Table 10.1.5 lists the advantages and disadvantages of fabric filter systems.

ESPs

ESPs consist of one or more pairs of electric charge plates or fields. The particulates in flue gases are given an electric charge, forcing them to stick to the oppositely charged plate. ESPs with four or more fields are state-of-the-art. Table 10.1.6 lists the advantages and disadvantages of ESPs.

CYCLONES

A third type of particulate control device is the cyclone, a mechanical device that funnels flue gases into a spiral, creating a centrifugal force that removes large particles. When combined with baghouses and ESPs cyclones improve their efficiency by removing larger particles before they reach these other more efficient devices.

TRENDS

When they are placed after a scrubber, particulate control devices also collect heavy metals and other pollutants that have condensed out of flue gas onto particle surfaces. Placing a scrubber first helps lower the temperature of gases entering a fabric filter. However, wet scrubbers cannot precede fabric filters because the wet particles in flue gases clog the filters. Thus, facilities with wet scrubbers

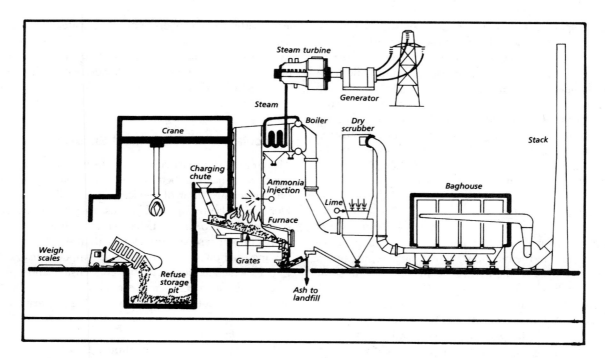

FIG. 10.1.15 Schematic of Commerce waste-to-energy plant in southern California. (Reprinted, with permission, from Commerce Refuse-to-Energy Authority.)

place their scrubbers after the particulate control device.

The smallest particles are the most potentially damaging when inhaled into the lungs, and dioxins, furans, acid gases, and heavy metals are adsorbed in the largest quantities on these smaller particles. Thus, a state-of-the-art particulate control device should achieve even lower emission levels for particulates below 2 microns in diameter.

Since many heavy metals condense at temperatures of 450°F (230°C), both ESPs and fabric filters collect heavy metals that condense onto particulate matter. Effective

TABLE 10.1.5   ADVANTAGES AND DISADVANTAGES OF FABRIC FILTER SYSTEMS

Advantages:

High particulate (coarse to submicron) collection efficiencies

Dry collection and solids disposal

Relatively insensitive to gas stream fluctuations. Efficiency and pressure drop are unaffected by large changes in inlet dust loading for continually cleaned filters

Corrosion and rusting of components usually not a problem

No hazard of high voltage, simplifying maintenance and repair and permitting the collection of flammable dust

Use of selected fibrous or granular filter aids (precoating) which permits the high-efficiency collection of submicron smokes and gaseous contaminants

Filter collectors available in a number of configurations, resulting in a range of dimensions and inlet and outlet flange locations to suit a range of installation requirements

Simple operation

Disadvantages:

Special refractory mineral or metallic fabrics (that are still in the developmental stages and can be expensive) required for temperatures in excess of 550°F

Fabric treatments to reduce dust seeping or to assist in the removal of the collected dust required for certain particulates

A fire or explosion hazard due to concentrations of some dusts in the collector ($\approx$ 50 g/cu m) when a spark or flame is accidently admitted. Fabrics can burn if readily oxidizable dust is being collected.

High maintenance requirements (bag replacements, etc.)

Fabric life shortened at elevated temperatures and in the presence of acid or alkaline particulate or gas components

Crusty caking or plugging of the fabric caused by hydroscopic materials, condensation of moisture (or tarry), and adhesive components which may require special additives

Respiratory protection for maintenance personnel required in replacing the fabric

Medium pressure-drop requirements, typically in the range of 4 to 10 in of water

TABLE 10.1.6   ADVANTAGES AND DISADVANTAGES OF ESPs

Advantages:
High particulate (coarse and fine) collection efficiencies with a relatively low expenditure of energy
Dry collection and solids disposal
Low pressure drop (typically less than 0.5 in of water)
Designed for continuous operation with minimum maintenance requirements
Low operating costs
Capable of operation under high pressure (to 150 psi) or vacuum conditions
Capable of operation at high temperatures (to 1300°F)
Capable of handling large gas flow rates effectively

Disadvantages:
High capital costs
Sensitive to fluctuations in gas stream conditions (flow, temperature, particulate and gas composition, and particulate loading)
Difficulty in collecting certain particulates due to extremely high or low resistivity characteristics
Relatively large space requirements for installation
Explosion hazard when treating combustible gases and collecting combustible particulates
Special precautions required to safeguard personnel from high voltage equipment
Ozone produced by the negatively charged discharge electrodes during gas ionization
Sophisticated maintenance personnel required

mercury emission control technology, while evolving, has not been implemented in MSW incinerators. A volatile metal, mercury vaporizes under the high temperatures of combustion although recent research suggests that mercury can also be present as mercuric chloride, mercuric oxides, and mercury solids. Whereas most vaporized metals return to a solid state when combustion gases cool, mercury remains in the vapor state. Wet scrubbing, activated carbon and sodium sulfide technologies show promising results.

Mercury control requires that the vapor be adsorbed onto particulates or absorbed into a liquid which is evaporated to leave the solids. The mercury-laden solids are collected in traditional collection devices. Some technologies, used in conjunction with other pollution control systems, can simultaneously remove dioxins, furans, mercury, and other metals as well as acid gases and particulates (Seigies and Trichon 1993).

Nitrogen Oxide Emissions

State-of-the-art control of nitrogen oxides requires both minimizing the formation of nitrogen oxides in the furnace and transforming them into nitrogen and water. Strategies for minimizing formation include using appropriate furnace designs (such as flue gas recirculation and dual-chambered furnaces) and operating practices (such as optimal temperatures and amount of excess air). Techniques for destroying nitrogen oxides involve injecting chemicals that neutralize them.

Chemical injection devices use ammonia, urea, or other compounds to react with nitrogen oxides to form nitrogen and water. The technologies for neutralizing and removing nitrogen oxides from flue gases are called selective noncatalytic reduction (SNCR) and selective catalytic reduction (SCR). Both technologies have been successfully demonstrated on MSW incinerators. Wet scrubbing and condensation also have the capacity to control nitrogen oxides.

Emission Control Devices

The arrangement of emission control devices other than the devices for nitrogen oxides is usually standard: a scrubber and condenser, followed by a particulate collector, followed by an induction fan that sucks flue gases up to the stack. Two reasons for this arrangement are:

Fabric filters cannot operate at the high temperatures at which gases exit the boiler without risk of fire. Thus, placing the scrubber between the boiler and the fabric filter or ESP permits cooling and often humidification that prevent fire. Cooling the gases also plays a role in reducing acid gas, mercury, and dioxin emissions.
Dioxins and heavy metals are trapped more effectively by particulate control devices when they are first condensed out of the flue gas and adsorbed onto the surface of particulate matter, as happens in a scrubber–condenser system.

An alternate arrangement, common in European plants, involves an ESP followed by a wet scrubber. The ESP is not damaged by high temperatures, and the wet scrubber cools and condenses gases and captures particulates.

The location of control devices for nitrogen oxides depends on the type of technology used. These devices can be in the furnaces or the boiler as well as at the back end of the plant.

Ash Management

The first priority in state-of-the-art ash management is to reduce both the volume and toxicity of the residue left after burning MSW. Removing noncombustibles and material containing toxic substances from the MSW before incineration followed by efficient combustion accomplishes this reduction. The amount of toxic material in ash has been increasing as more effective air pollution control devices capture more pollutants in the fly ash.

State-of-the-art ash management practices are designed to minimize worker and citizen exposure to potentially toxic substances in ash during handling, treatment, and storage, long-term storage, or reuse. Safe ash management has several components:

The bottom ash or residue (noncombustible and partially burned solids left in the incinerator) and fly ash (material captured by emission control devices) is kept separate for rigorous handling of the potentially more toxic fly ash.

The ash is contained while still in the plant. A closed system of conveyors is preferable to handling ash in the open.

The ash is transported wet in leakproof, covered trucks to disposal sites.

The ash is treated to minimize its potential toxic impact.

The ash is disposed in ash-only monofills because codisposal of ash with MSW increases the leachability of the ash when it is exposed to acid.

Fly ash from APC is fine-grained, not unlike soot from fireplaces. For every ton of MSW burned, approximately $\frac{1}{4}$ tn becomes some form of ash. Fly ash accounts for about 10 to 15% of the total ash residue; the remaining 85 to 90% is bottom ash.

Operational data from resource recovery incineration facilities throughout the world indicate certain heavy metals, such as lead and cadmium, tend to concentrate in the fly ash, scrubber residue, and small particles (less than $\frac{3}{8}$ in) in the bottom ash. Heavy metals, including lead, cadmium, and total soluble salts (including chlorides and sulfates), are potentially leachable components which can impact the environment. Leachable components are those chemical species which dissolve in water and are transported with water. The toxicity characteristic leaching procedure (TCLP) and numerous other techniques exist to estimate the potential environmental impact resulting from ash generation, handling, and disposal.

Two main categories of ash treatment, both recently developed and being improved, are fixation or cementation and vitrification. Both techniques minimize the environmental impact of ash and enable its reuse in situations such as cinderblocks, reefs, and roads. A few incinerators have onsite vitrification facilities.

Another new treatment technology involves washing the toxic materials out of the ash with hot water and then treating the water to remove soluble toxic materials. The system has been used in Europe, particularly in incinerators with wet scrubbers (Clark, Kadt, and Saphire 1991).

Instrumentation

Continuous process monitors (CPMs) and *continuous emission monitors (CEMs)* track the performance of incinerators so that when combustion upsets or high emissions of one or more pollutants occur, timely corrective measures can be implemented. These monitors are usually connected to alarms that warn plant operators of any combustion, emission, or other operating condition that requires attention. Table 10.1.7 lists the typical instrumentation on continuous-feed incinerators for closed-loop control of the temperature and draft controllers (Shah 1974).

State-of-the-art CPMs and CEMs measure nine operating and emission factors: furnace and flue gas temperatures, steam pressure and flow, oxygen, carbon monoxide, sulfur dioxide, nitrogen oxides, and opacity (a crude measure of particulates). Continuous monitoring of hydrogen chloride is possible and may soon be a state-of-the-art requirement. By monitoring parameters that indicate combustion efficiency (carbon monoxide, oxygen, and furnace temperature), plant operators also obtain indications of levels of incomplete combustion. Operators must perform frequent maintenance, including periodic calibration, on continuous monitors to ensure their accuracy.

Adapted from *Municipal Waste Disposal in the 1990s* by Béla G. Lipták (Chilton, 1991).

TABLE 10.1.7   INSTRUMENT LIST FOR CONTINUOUS-FEED INCINERATOR

Temperature Recorders
Furnace temperature at furnace sidewall near outlet, range 38 to 1250°C
Stoker compartment temperature, range 38 to 1250°C
Dust collector inlet temperature, range 38 to 500°C

Temperature Controllers
Furnace outlet temperature controlled by regulating total air from forced draft fan; set point in 800 to 1000°C range
Dust collector inlet temperature controlled by regulating water spray into flue gas; set point in 300 to 400°C range

Draft Gauges
Forced-draft-fan outlet duct
Induced-draft-fan inlet duct
Furnace outlet
Stoker compartments
Differential gauge across dust collector

Draft Controller
Furnace draft control by regulating damper opening

Oxygen Analyzer
Furnace outlet

References

Clark, M.J., M. Kadt, and D. Saphire. 1991. *Burning garbage in the US* Edited by Sibyl R. Golden. New York: INFORM, Inc.

Eberhardt, H. 1966. European practices in refuse and sewage sludge disposal by incineration. *ASME National Incinerator Conference, New York, 1966.*

Essenhigh, R.H. 1974. Incinerators—the incineration process. Vol. 2 in *Environmental engineers' handbook,* edited by B.G. Lipták. Radnor, Pa.: Chilton Book Company.

Gibbs, D.R. and L.A. Kreidler. 1989. What RDF has evolved into. *Waste Age* (April).

Hershkowitz. 1986. Garbage burning: Lessons from Europe: Consensus and controversy in four European states. New York: INFORM, Inc.

Licata, A.J. 1986. Design for good combustion. *24 January, 1986, Municipal Solid Waste Forum, Marine Sciences Research Center, State University of New York, 1986.*

Licata, A.J., R.W. Herbert, and U. Kaiser. 1988. Design concepts to minimize superheater corrosion in municipal waste combustors. *National Waste Processing Conference, Philadelphia, 1988.* New York: ASME.

McIlvaine, R.W. 1989. Incineration and APC trends in Europe. *Waste Age* (January).

Platt, Brenda et al. 1988. *Garbage in Europe technologies, economics, and trends.* Institute for Local Self Reliance (May).

Seigies, J. and M. Trichon. 1993. Waste to burn. *Pollution Engineering* (15 February).

Shah, I.S. 1974. Scrubbers. Sec. 5.12–5.21 in *Environmental engineers handbook,* edited by B.G. Lipták. Radnor, Pa.: Chilton Book Company.

Sommer, Jr., E.J. and G. Kenny. 1984. Effects of materials recovery on waste-to-energy conversion at Gallatin, Tennessee mass fired facility. *Proc. Waste Processing Conf.* New York: ASME.

Stelian, J. and H.L. Greene. 1986. Operating experience and performance of two ash handling systems. *National Waste Processing Conference, Denver, 1986.* ASME.

U.S. Environmental Protection Agency. 1991. Burning of hazardous waste in boilers and industrial furnaces, final ruling. *Federal Register* 56, no. 35 (21 February): 7134–7240.

Velzy, C.O. 1968. The enigma of incinerator design. *ASME Winter Annual Meeting, New York, 1968.*

Velzy, C.O. and R.S. Hechlinger. 1987. Incineration. Sec. 7.4 in *Mark's standard handbook for mechanical engineers,* 9th ed. Edited by T. Baumeister and E.A. Avallone. New York: McGraw-Hill.

Wheless, E. and M. Selna. 1986. Commerce refuse-to-energy facility: An alternative to landfilling. *National Waste Processing Conference, Denver, 1986.* ASME.

10.2
SEWAGE SLUDGE INCINERATION

Sewage sludge, the stabilized and digested solid waste product of the wastewater treatment process, can be disposed of by landfilling, incineration, composting, or ocean dumping. Nature returns organic material to the soil as fertilizer. Organic material becomes waste when it is not returned to the soil but instead is burned, buried, or dumped in the ocean. These unhealthy practices began when chemical fertilizers took the market away from sludge-based compost and when industrial waste began to contaminate sewage sludge with toxic metals (lead and cadmium), making it unusable for agricultural purposes. Until recently, the bulk of the sewage sludge generated by metropolitan areas has been either landfilled or dumped in the ocean. These options are gradually disappearing and as a result municipalities will have to make some hard decisions.

Sludge Incineration Economics

Incinerating sewage sludge has been practiced for the last sixty years. Early designs were either flash-drying or multiple-hearth types, while in recent years fluidized-bed incinerators have also been used. The flash-drying process has a low capital cost and is flexible in that it can produce the amount of dried sludge that markets need; the remainder can be incinerated. Its limitations are the added cost of pay fuel and the associated odor and pollution problems. The multiple-hearth design, the most widely used for sludge incineration, reduces odor and pollution but provides less operating flexibility because it cannot dry the sludge without incinerating it. The most recent and advanced design is the fluidized-bed sludge incinerator, which can operate in either the combustion or pyrolysis mode. The exhaust temperature from a fluidized-bed incinerator is higher than from a multiple-hearth furnace so afterburners are less likely to be needed to control odor.

The auxiliary fuel cost of sludge incineration is higher with fluidized-bed incinerators than with multiple hearths. The cost varies according to the moisture content of the sludge and the degree of heat recovery (Sebastian 1974a). Eliminating the need for auxiliary fuel requires that the dry–solid content exceed 25% for multiple-hearth and 32% for fluidized-bed incinerators (Sebastian 1974a). In some fluidized-bed installations in Japan, operating costs have been cut in half through heat recovery (Henmi, Okazawa, and Sota 1986).

In a multiple-hearth incinerator with a feed containing 10% solids, the ash is about 10% of the feed. Table 10.2.1 gives the composition of incinerator ashes. The ash is either landfilled or marketed as a soil conditioner. Table 10.2.2 gives the composition of Vitalin, the ash from Tokyo's Odai plant. (The Japanese word lin means phos-

TABLE 10.2.1   TYPICAL ANALYSIS OF ASH FROM TERTIARY QUALITY-ADVANCED WASTE TREATMENT SYSTEMS

Content	Percent of Total Sample			
	Lake Tahoe 11/19/69	Lake Tahoe 11/25/69	Minn.–St. Paul 9/30/69	Cleveland 3/2/70
Silica (SiO$_2$)	23.85	23.72	24.87	28.85
Alumina (Al$_2$O$_3$)	16.34	22.10	13.48	10.20
Iron oxide (Fe$_2$O$_3$)	3.44	2.65	10.81	14.37
Magnesium oxide (MgO)	2.12	2.17	2.61	2.13
Total calcium oxide (CaO)	29.76	24.47	33.35	27.37
Available (free) calcium oxide (CaO)	1.16	1.37	1.06	0.29
Sodium (Na)	0.73	0.35	0.26	0.18
Potassium (K)	0.14	0.11	0.12	0.25
Boron (B)	0.02	0.02	0.006	0.01
Phosphorus pentoxide (P$_2$O$_5$)	6.87	15.35	9.88	9.22
Sulfate ion (SO$_4$)	2.79	2.84	2.71	5.04
Loss on ignition	2.59	2.24	1.62	1.94

phorus). The term soil conditioner is used instead of fertilizer because the phosphate content is less than 12%, the nitrogen content is under 6%, and the total NPK content is less than 20%.

Incineration Processes

A description of the flash-dryer, multiple-hearth, fluidized-bed and fluidized-bed with heat recovery incineration processes follows.

FLASH-DRYER INCINERATION

The flash-dryer incinerator process was first introduced in the 1930s as a low-capital-cost, space-saving alternative to air drying sludge on sand beds. This method of drying is advantageous because the resulting heat-dried sludge is virtually free of pathogens and weed seeds and the process is flexible enough to produce only the amount of dried sludge that could be marketed. The disadvantages of this process are dust and odor. These problems, while manageable through the use of dust collectors and afterburners, make the flash-dryer less popular than multiple-hearth and fluidized-bed incinerators.

TABLE 10.2.2   COMPOSITION OF MULTIPLE-HEARTH INCINERATION ASH FROM THE ODAI PLANT IN TOKYO

Silica oxide	30.00	Potassium	1.00
Magnesium oxide	3.30	Nitrogen	0.20
Calcium oxide	30.00	Manganese	0.06
Phosphoric oxide	6.20	Copper	0.61
Ferric oxide	18.20	Boron	200.00 ppm

Note: Ash is marketed under the trade name Vitalin.

In the flash-drying process (see Figure 10.2.1), the wet, dewatered sludge is mixed with dry sludge from the dryer cyclone. This preconditioned mixture contacts a gas stream of 1000 to 1200°F, which moves it at a velocity of several thousand feet per minute. In this turbulent, high-temperature zone, the moisture content of the sludge is reduced to 10% or less in only a few seconds. As the mixture enters the dryer cyclone, the hot gases are separated from the fine, fluffy heat-dried sludge. Depending on the mode of operation, the flash-dried sludge is either sent to the sludge burner and incinerated at 1400°F, or it is sent to the fertilizer cyclone and recovered as a saleable fertilizer product. When the incoming wet sludge contains about 18% solids, about 6500 Btu ($\frac{1}{2}$ lb of coal plus 1 cu ft of natural gas) are required to produce 1 lb of dry sludge (15,000 kJ/kg) (Shell 1979).

MULTIPLE-HEARTH INCINERATION

Multiple-hearth incineration was developed in 1889 and was first applied to sludge incineration in the 1930s. It is the most widely used method of sludge incineration (Sebastian 1974b). The multiple-hearth furnace consists of a steel shell lined with a refractory (see Figure 10.2.2). Horizontal brick arches separate the interior into compartments. The sludge is fed through the roof by a screw feeder or a belt feeder and flapgate at a rate of about 7 to 12 lb per sq ft. Rotating rabble arms push the sludge across the hearth to drop holes, where it falls to the next hearth. As the sludge travels downward through the furnace, it turns into a phosphate-laden ash (see Tables 10.2.1 and 10.2.2).

The sludge is dried in the upper, or first, operating zone of the incinerator. In the second zone, it is incinerated at a temperature of 1400 to 1800°F (760 to 982°C) and deodorized. In the third zone, the ash is cooled by the in-

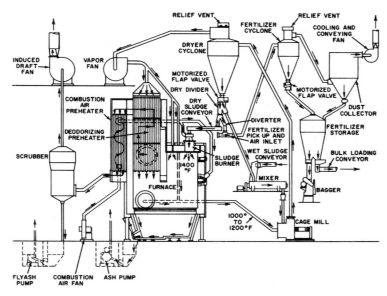

FIG. 10.2.1 Sludge drying and incineration using a deodorized flash-drying process.

coming combustion air. The air, which travels in counterflow with the sludge, is first preheated by the ash, then participates in the combustion, and finally sweeps over the cold incoming sludge drying it until the moisture content is about 48%. At this percentage of moisture content, a phenomenon called *thermal jump* occurs as the sludge enters the combustion zone. The thermal jump allows the sludge to bypass the temperature zone where the odor is distilled. The exhaust gases are 500 to 1100°F (260 to 593°C) and are usually odor-free. The sludge temperature profile across the furnace is shown in Table 10.2.3.

The pollution control equipment usually includes three-stage impingement-type scrubbers for particulate and sulfur dioxide removal and standby after-burners, which de-

stroy malodorous substances such as butyric and caproic acids. The need for afterburners is a function of the exhaust gas temperature. Usually at temperatures above 700 to 800°F (371 to 427°C) in a well-controlled incinerator where the combustion process is complete, afterburners are not necessary for odor-free operation. If combustion is not complete, however, the exhaust gas temperature might have to rise to 1350°F (732°C) before the odor is distilled. In such cases, installing an afterburner is less expensive than using auxiliary fuel to achieve such high exhaust temperatures.

If the incoming sludge contains 75% moisture and if 70% of the sludge solids are volatile, the incineration process produces about 10% ash. The ash can be used as a soil conditioner and as the raw material for bricks, concrete blocks, and road fills, or it can be landfilled. In the United States the supply of phosphates is sufficient for less than a century (Sebastian 1974b), so the phosphate content of sludge ash is important. If the ash also contains

FIG. 10.2.2 Multiple-hearth incineration of sludge.

TABLE 10.2.3   SLUDGE FURNACE TEMPERATURE PROFILE

Hearth No.	Approximately at Half Capacity (°F)	Nominal Design Capacity (°F)
1	670	800
2	1380	1200
3	1560	1650
4	1450	1450
5	1200	1200
6	325	300

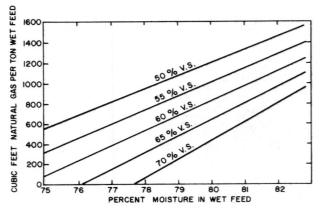

FIG. 10.2.3 Multiple-hearth incinerator fuel consumption as a function of moisture content in the feed and percentage of volatile solids. *Notes*. 1. Curves are not applicable for feed rates below 4 tn per hr. 2. Curves do not include allowance for lime as a filter aid. 3. To correct for lime, downgrade volatile solids according to lime dosage. Assuming lime forms calcium hydroxide, each pound of CaO forms 1.32 lb calcium hydroxide. 4. Natural gas calorific value is assumed to be 100 Btu per cu ft. 5. Heat content of sludge is based on 10,000 Btu per lb of volatile solids. 6. %V.S. represents percentage of volatile solids in the feed.

zinc or chromium, it can damage certain crops although it does not damage cereals or grass (Sebastian 1974b). The harmful effects are more likely to occur in acidic soils and can be offset by the addition of lime to the sludge.

The main advantages of multiple-hearth incinerators include their long life and low operating and maintenance costs (Sebastian 1974b); their ability to handle sludges with a moisture content of up to 75% without requiring auxiliary fuel; their ability to incinerate or pyrolize hard-to-handle substances, such as scum or grease; their ability to reclaim chemical additives, such as lime in combination with or separately from incinerating the sludge; and their flexibility, in that they can be operated intermittently or continuously at varying feed rates and exit-gas temperatures (Sebastian 1974b). The auxiliary fuel requirement varies with the dry–solids content of the sludge and with the percentage of volatiles in the solids (see Figure 10.2.3).

FLUIDIZED-BED INCINERATION

Fluidized-bed incineration can handle sewage sludge containing as much as 35% solids. The sludge is injected into a fluidized bed of heated sand. The incinerator is a vertical cylinder with an air distribution plate near the bottom, which allows the air to enter the sand bed while also sup-

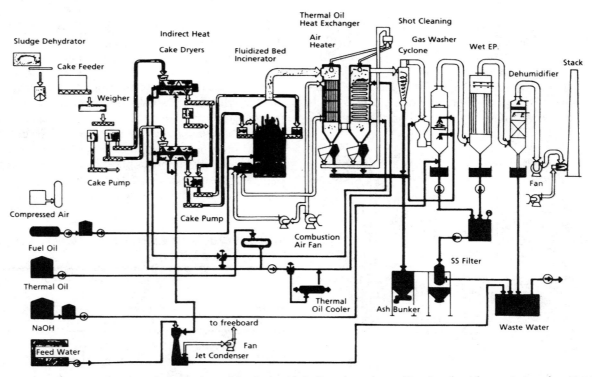

FIG. 10.2.4 Flow diagram of sewage sludge incineration plant with indirect heat dryer. (Reprinted, with permission, from M. Henmi, K. Okazawa, and K. Sota, 1986, Energy saving in sewage sludge incineration with indirect heat drier, *National Waste Processing Conference, Denver, 1986* [ASMER].)

TABLE 10.2.4 OPERATIONAL DATA OF FLUIDIZED-BED SLUDGE INCINERATOR WITH AND WITHOUT INDIRECT HEATING

| | Direct Incineration (Without Drying) | | | Incineration with Dried Cake | | | | |
	Design	Run 1	Run 2	Design	Run 3	Run 4	Run 5	Run 6
Input Cake (tpd)	80	60	80	96	60	80	96	70
Cake Property								
Moisture (%)	78	83–86 (84.7)	← ←	75–80 (78)	83–86 (84.8)	←	←	78–80 (79)
Combustibles (%ds)	56–64 (60)	80.2–80.7 (80.5)	← ←	56–64 (60)	80.1–80.2 (80.2)	79.1–79.4 (79.3)	79.3–80.4 (84.7)	80 (80)
Lower heating value (MJ/kg ds)	12.6	18.5–18.9 (18.7)	←	12.6	18.7	18.8	18.5	18.8
Dried Cake								
Moisture (%)	same	same	same	66–73 (70)	80.1–80.9 (80.6)	79.8–80.6 (80.2)	79.2–80.2 (79.7)	70–72 (71)
Weight (tpd)	80	60	80	70	46.7	61.8	72.4	43
Supporting Fuel								
(l/h)	312	238	265	47	92	94	96	0
(l/tn cake)	94	95	80	12	37	28	24	0
Furnace Temperature								
Sand bed (°C)	800	790	780–820 (800)	800	770–810 (780)	770–810 (780)	760–810 (780)	(780)
Outlet (°C)	850	860	850–950 (900)	800	760–800 (780)	760–800 (780)	760–810 (780)	(780)
Fluidizing air (cu m Normal/h)	7500	6400	7500	5500	5500	5500	5500	5500
Excess air ratio	1.40	1.62	1.52	1.56	2.02	1.61	1.41	(1.8)

Source: M. Henmi, K. Okazawa, and K. Sota, 1986, Energy saving in sewage sludge incineration with indirect heat drier, *National Waste Processing Conference, Denver, 1986* (ASMER).

Note: Values in parentheses are average values. All units are in metric (SI).

porting it. As the air flow increases, the bed expands and becomes fluidized. The solid waste in the sludge can be destructed by either combustion or pyrolysis.

During combustion the organic material is turned into carbon dioxide:

$$\text{Sewage Sludge} + \text{Oxygen} \longrightarrow CO_2 + H_2O + \text{Ash} + \text{Heat} \quad \textbf{10.2(1)}$$

The heat of combustion helps maintain the fluidized bed at a temperature of about 1400°F (760°C).

In the pyrolysis process, the sludge is decomposed in the presence of inert gases at 1400°F (760°C), which yields hydrogen, methane, carbon monoxide, and carbon dioxide. For pyrolysis, auxiliary heat is required to maintain the fluidized bed at the high reaction temperature:

$$\text{Sewage Sludge} + \text{Inert Gas} \longrightarrow$$
$$H_2 + CO + CO_2 + CH_4 + \text{Auxiliary Heat} \quad \textbf{10.2(2)}$$

The operation of the fluidized-bed incinerator is optimized by control of the airflow rate and the bed temperature. Since reaction rates are related to bed mixing and the source of agitation is the fluidizing air, operators can adjust reaction rates by changing the airflow supply. The bed temperature is usually maintained between 1300 and 1500°F (704 to 815°C). For complete combustion (odor-free operation), about 25% excess air is needed (Rabosky 1974).

Organic material can be deposited on the sand particles (agglomerative mode) and removed by continuous or intermittent withdrawal of excess bed material. An alternative mode of operation (nonagglomerative) combines the organic ashes with exhaust gases and collects them downstream with dust collectors.

The main advantages of fluidized-bed incinerators include the uniformity of the bed, the elimination of stratification and hot or cold spots, the high rate of heat transfer for rapid combustion, the elimination of odor and the need for afterburners, and the low maintenance requirements of the process. The disadvantages include the high operating-power requirement, the need for auxiliary fuel

TABLE 10.2.5   OPERATING COSTS OF FLUIDIZED-BED SLUDGE INCINERATION WITH AND WITHOUT HEAT RECOVERY

Items	Utility Unit Cost	Normal Undried Cake Incineration		Newly Developed Indirectly Dried Cake Incineration	
		Amount	Cost	Amount	Cost
Plant Capacity		80 tn/d		96 tn/d	
Operation Cost			US$/d		US$/d
1. Supporting fuel	0.35 US$/l	6360 l/d	2226	2304 l/d	806
2. Electricity	0.1 US$/kWh	10800 kWh/d	1080	10300 kWh/d	1030
3. Chemical (NaOH)	0.35 US$/kg	220 kg/d	77	151 kg/d	53
4. Lubrications	2.6 US$/l	0.8 l/d	2	0.8 l/d	2
Total per Day	—	—	3385 US$/d	—	1891 US$/d
Unit Treatment Cost per Input Cake Vol.	—	—	42.3 US$/tn	—	19.7 US$/tn

Source: Henmi, Okazawa, and Sota 1986.
Note: All units are in metric (SI).

if the dry solid concentration is less than 32% and the need for dust-collection devices.

FLUIDIZED-BED INCINERATION WITH HEAT RECOVERY

The addition of heat-recovery equipment can increase the capacity of fluidized-bed incinerators by about 20%

(Henmi, Okazawa, and Sota 1986). The plant shown in Figure 10.2.4 has been in operation in Tokyo since 1984 (Henmi, Okazawa, and Sota 1986). Heat recovery involves inserting a heat exchanger into the stream of the hot gases, which generates a supply of hot thermal oil. The oil is then used as the energy source to heat the sludge cake dryers.

The hot oil passes through the hollow inside of the motor-driven screws, while the sludge cake is both moved and

TABLE 10.2.6   CONCENTRATION OF POLLUTANTS IN ATMOSPHERIC EMISSIONS AND IN ASH PRODUCED BY A FLUIDIZED-BED TYPE SLUDGE INCINERATOR

Items	Regulations	Run 1	Run 4
Exhaust Gas			
Sulfur oxides (SO_x)	292 ppm	4 ppm	25 ppm
Nitrogen oxides (NO_x)	250 ppm	—	41 ppm
Hydrogen chloride (HCl)	93 ppm	3 ppm	—
Dust density	0.05 g/m³N	0.006 g/m³N	0.002 g/m³N
Residual Ash			
Amount	7.0 tn/d	2.4 tn/d	2.9 tn/d
Ignition loss	<15%	0.6%	0.6%
Dissolution to water			
Alkyl mercury (Hg)	—	<0.0005	—
Total mercury (Hg)	0.05	<0.0005	—
Cadmium (Cd)	0.1	<0.05	—
Lead (Pb)	1	<0.2	—
Phosphorus (P)	0.2	<0.01	—
Chromium 6+ (Cr[VI])	0.5	<0.05	—
Arsenic (As)	0.5	<0.19	—
Cyanide (CN)	1	<0.05	—
Polychlorinated biphenyls (PCB)	0.003	<0.0005	—

Source: Henmi, Okazawa, and Sota 1986.
Notes: Values marked (—) were not measured. Suffix "N" means the value converted at normal condition of 273 K, 1 atm.

TABLE 10.2.7   CONCENTRATION OF WASTEWATER GENERATED BY FLUIDIZED-BED INCINERATOR

Content (mg/l)	Feed Water	Dryer Condensate	Wet Electrostatic Precipitation Effluent	Dehumidifier Effluent
Tkj N	24.71	29.67	24.97	24.52
NH_4^+ N	21.14	24.40	21.24	20.75
Org N	3.57	5.27	3.73	3.77
NO_2^- N	0.18	0.16	N.D.	0.11
NO_3^- N	0.63	0.60	0.98	0.62
T N	25.52	30.43	25.95	25.25
BOD	6.86	41.5	2.24	3.44
CODmn	12.2	21.4	38.8	12.5
SS	3.3	15.7	84.0	2.3
Cl^-	74.3	74.5	58.3	76.1

Source: Henmi, Okazawa, and Sota 1986.

heated by these screws. As shown in Table 10.2.4, the cake is dried to a substantial degree (some 20% of the inlet flow is evaporated) as the sludge cake passes through the cake dryer. The heating oil circulates in a closed cycle and is maintained at about 480°F (250°C) inside the screw-conveyor dryers by the throttling of two three-way valves. One valve can increase the outlet oil temperature from the cake dryers by blending in warmer inlet oil; the other can lower the outlet temperature by sending some of the oil through an oil cooler.

The operators of the Tokyo incinerator feel that the total capital cost of the plant is unaffected by the addition of the heat-recovery feature because the cost of the heat-transfer equipment is balanced by the reduced capacity requirement. The operating costs, on the other hand, are cut in half with the heat-recovery system (Table 10.2.5).

Another interesting feature of this system is the method of cleaning the accumulation of ash and soot from the heat-transfer surfaces. This automated system uses 3- to 5-mm-diameter steel-shot balls that are dropped every three to six hours from the top of the hot-air heaters. The random movement of the balls removes the dust from the heater tubes. The dust is removed at the bottom, while the balls are collected and returned to the top.

Table 10.2.6 gives the composition of the ash residue and the stack gases (after they have been cleaned by wet electrostatic precipitation); both meet Japanese regulations. Table 10.2.7 gives the composition of the wastewater produced by this process. According to the operators, the process produces almost no odor.

Adapted from *Municipal Waste Disposal in the 1990s*
by Béla G. Lipták (Chilton, 1991).

References

Henmi, M., K. Okazawa, and K. Sota. 1986. Energy saving in sewage sludge incineration with indirect heat drier. *National Waste Processing Conference, Denver, 1986.* ASMER.

Rabosky, J.G. 1974. Incineration—fluidized bed incineration. Vol. 3, Sec. 2.23 in *Environmental engineers' handbook*, edited by B.G. Lipták. Radnor, Pa.: Chilton Book Company.

Sebastian, F. 1974a. Incinerator economics. Vol. 3, Sec. 2.17 in *Environmental engineers' handbook*, edited by B.G. Lipták, Radnor, Pa.: Chilton Book Company.

———. 1974b. Multiple hearth incineration. Vol. 3, Sec. 2.22 in *Environmental engineers' handbook*, edited by B.G. Lipták. Radnor, Pa.: Chilton Book Company.

Snell, J.R. 1974. Flash drying or incineration. Vol. 1, Sec. 8.6 in *Environmental engineers' handbook*, edited by B.G. Lipták. Radnor, Pa.: Chilton Book Company.

10.3
ONSITE INCINERATORS

Air Preheater Co., Inc.; American Schack Co., Inc.; Aqua-Chem, Inc.; BSP Corp., Div. of Envirotech Systems, Inc.; Bartlett-Snow; Beloit-Passavant Corp.; Best Combustion Engrg. Co.; Bethlehem Corp.; Brule Pollution Control Systems; C-E Raymond; Carborundum Co.; Pollution Control Div., Carver-Greenfield Corp.; Coen Co.; Combustion Equipment Assoc. Inc.; Copeland Systems, Inc.; Dally Engineering-Valve Co.; Dorr-Oliver Inc.; Dravo Corp.; Environmental Services Inc.; Envirotech; First Machinery Corp.; Foster Wheeler Corp.; Fuller Co.; Garver-Davis, Inc.; Haveg Industries Inc.; Holden, A. F., Co.; Hubbell, Roth & Clark, Inc.; Intercontek, Inc.; International Pollution Control, Inc.; Ishikawajima-Harima Heavy Industries Co., Ltd.; Kennedy Van Saun Corp.; Klenz-Aire, Inc.; Koch & Sons, Inc.; Koch Engrg. Co., Inc.; Kubota, Ltd., Chuo-Ku, Tokyo, Japan; Lawler Co., Leavesley Industries; Lurgi Gesellschaft fuer Waerm & ChemotecHnik mbH, 6 Frankfurt (Main) Germany; Maxon Premix Burner Co., Inc.; Melsheimer, T., Co., Inc.; Midland-Ross Corp., RPC Div.; Mid-South Mfg. Corp.; Mine & Smelter Supply Co.; MSI Industries; Mitsubishi Heavy Industries, Ltd., Tokyo, Japan; Monsanto Biodize Systems, Inc.; Monsanto Enviro-Chem Systems Inc.; Nichols Engrg. & Research Corp.; North American Mfg. Co.; Oxy-Catalyst, Inc.; P.D. Proces Engrg. Ltd., Hayes, Middlesex, England; Peabody Engrg. Corp.; Picklands Mather & Co., Prenco Div.; Plibrico Co.; Prenco Mfg. Co.; Pyro Industries, Inc.; Recon Systems Inc.; Renneburg & Sons Co.; Rollins-Purle, Inc.; Ross Engrg. Div., Midland-Ross Corp.; Rotodyne Mfg. Corp.; Rust Engrg. Co., Div. of Litton Industries; Sargent, Inc.; Surface Combustion Div., Midland-Ross Corp.; Swenson, Div. of Whiting Corp.; Tailor & Co., Inc.; Takuma Boiler Mfg. Co., Ltd., Osaka, Japan; Thermal Research & Engrg. Co.; Torrax Systems, Inc.; Vulcan Iron Works, Inc.; Walker Process Equip., Div. of Chicago Bridge & Iron Co.; Westinghouse Water Quality Control Div., Infilco; Zink, John, Co.; Zurn Industries, Inc.

This section describes some of the smaller incinerator units used onsite in domestic, commercial, and industrial applications. Onsite incineration is a simple and convenient means of handling the waste transportation problem since it reduces the volume of disposable waste. Onsite incinerators are smaller and their fuel more predictable in composition than the MSW burned in municipal incinerators. Therefore, this section discusses separately considerations that affect the location, selection, and operating practices of these onsite units.

Location

The onsite incinerator should be located close to larger sources of waste and expected waste collection routes. Onsite incinerators are constructed of 12-gauge steel casing with high-temperature (over 1000° F) insulation and high-quality refractory lining. Indoor installations are preferred, but even when the incinerator is situated outdoors, the charging and cleanout operations should be shielded from the weather. Incinerator rooms should be designed for two-hour fire resistance and should comply with the National Fire Protection Association (NFPA) recommendations contained in bulletin NFPA No. 82.

The Incinerator Institute of America (IIA) separates incinerators into nine classes according to their use and size (see Table 10.3.1) and provides minimum construction and performance standards for each class. The NFPA has also

TABLE 10.3.1   CLASSIFICATION OF INCINERATORS

Class I—Portable, packaged, completely assembled, direct-feed incinerators having not over 5 cu ft storage capacity or 25 lb/hr burning rate, suitable for type 2 waste.

Class IA—Portable, packaged or job assembled, direct-feed incinerators having a primary chamber volume of 5 to 15 cu ft or a burning rate of 25 lb/hr up to, but not including, 100 lb/hr of type 0, 1, or 2 waste; or a burning rate of 25 lb/hr up to, but not including, 75 lb/hr of type 3 waste.

Class II—Flue-fed, single chamber incinerators with more than 2 cu ft burning area for type 2 waste. This incinerator type is served by one vertical flue functioning as a chute for both charging waste and carrying the products of combustion to the atmosphere. This incinerator type is installed in apartment or multiple dwellings.

Class IIA—Chute-fed, multiple chamber incinerators for apartment buildings with more than 2 cu ft burning area, suitable for type 1 or 2 waste. (Not recommended for industrial installations). This incinerator type is served by a vertical chute for charging waste from two or more floors above the incinerator and a separate flue for carrying the products of combustion to atmosphere.

Class III—Direct-feed incinerators with a burning rate of 100 lb/hr and more suitable for burning type 0, 1, or 2 waste.

Class IV—Direct-feed incinerators with a burning rate of 75 lb/hr or more suitable for burning a type 3 waste.

Class V—Municipal incinerators suitable for type 0, 1, 2, or 3 waste or a combination of all four wastes and are rated in tons per hour or tons per twenty-four hours.

Class VI—Crematory and pathological incinerators suitable for burning type 4 waste.

Class VII—Incinerators designed for specific by-product waste, type 5 or 6.

Note: For waste type numbers, see Tables 10.3.2 and 10.3.3.

TABLE 10.3.2 CLASSIFICATION OF WASTES TO BE INCINERATED

Classification of Waste Type and Description	Principal Components	Approximate Composition, % by Weight	Moisture Content, %	Incombustible Solids, %	Btu Value/lb of Refuse as Fired	Requirement for Auxiliary Fuel Btu per lb of Waste	Recommended Minimum Btu Burner Input per lb Waste	Density lb/cu ft
0 Trash	Highly combustible waste, paper, wood, cardboard cartons, and up to 10% treated papers, plastic or rubber scraps; commercial and industrial sources	Trash 100	10	5	8500	0	0	8–10
1 Rubbish	Combustible waste, paper, cartons, rags, wood scraps, combustible floor sweepings; domestic, commercial, and industrial sources	Rubbish 80 Garbage 20	25	10	6500	0	0	8–10
2 Refuse	Rubbish and garbage; residential sources	Rubbish 50 Garbage 50	50	7	4300	0	1500	15–20
3 Garbage	Animal and vegetable wastes, restaurants, hotels, markets; institutional commercial, and club sources	Garbage 65 Rubbish 3	70	5	2500	1500	3000	30–35
4 Animal solids and organics	Carcasses, organs, solid organic wastes; hospital, laboratory, abattoirs, animal pounds, and similar sources	Animal and human tissue 100	85	5	1000	3000	8000 (5000 primary) (3000 secondary)	45–55
5 Gaseous, liquid, or semi-liquid	Industrial process wastes	Variable	Dependent on major components	Variable	Variable	Variable	Variable	Variable
6 Semisolid and solid	Combustibles requiring hearth, retort, or grate equipment	Variable	Dependent on major components	Variable	Variable	Variable	Variable	Variable

TABLE 10.3.3   INDUSTRIAL WASTE TYPES

Type of Wastes	Description
1	Mixed solid combustible materials, such as paper and wood
2	Pumpable, high heating value, moderately low ash, such as heavy ends, tank bottoms
3	Wet, semisolids, such as refuse and water treatment sludge
4	Uniform, solid burnables, such as off-spec or waste polymers
5	Pumpable, high ash, low heating value materials, such as acid or caustic sludges, or sulfonates
6	Difficult or hazardous materials, such as explosive, pyrophoric, toxic, radioactive, or pesticide residues
7	Other materials to be described in detail

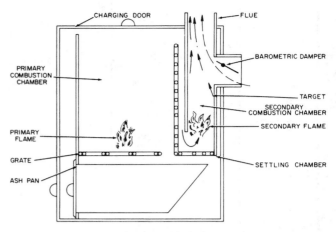

FIG. 10.3.1   Features of a domestic incinerator.

instituted similar classifications and construction standards in its standard *Incinerators and rubbish handling*. The IIA also classifies waste into seven types (see Tables 10.3.2 and 10.3.3). Planners must also comply with local and state codes when selecting an incinerator.

Selection

The first step in incinerator selection is to record the volume, weight, and classes of waste collected for a period of at least two weeks. The survey should be checked against typical waste-production rates. The maximum daily operation can be estimated as three hours for apartment buildings; four hours for schools; six hours for commerical buildings, hotels, and other institutions; and seven hours per shift for industrial installations.

The results of the waste survey help to determine whether a continuous or batch-type incinerator should be installed. Batch-type units consist of a single combustion chamber (see Figure 10.3.1). If the batch furnace has no grate, the ash accumulation reduces the rate of burning. The batch incinerator is sized according to the weight of each type of waste per batch at the number of batches per day. The continuous incinerator consists of two chambers:

one for charge storage and the other an evacuated chamber for combustion. The charge chamber can be loaded at any time. Sizing is based on the pounds-per-hour burning rate required.

The nature and characteristics of the waste are usually summarized in a form such as that in Table 10.3.4. Most incinerator manufacturers offer standard, pre-engineered packages for waste types 0, 1, 2, 3, and 4 (see Tables 10.3.2 and 10.3.3). Waste types 5 and 6 usually require unique designs because the physical, chemical, and thermal characteristics of these wastes are variable. Type 6 waste tends to have low heating values but contains material that can cause intense combustion. Plastics and synthetic rubber decompose at high temperatures and form complex organic molecules that require auxiliary heat and high turbulence before they are fully oxidized. In extreme cases, three combustion chambers are necessary; operators must recycle flue gases from the secondary combustion chamber back into the primary chamber to complete the combustion process.

Charging

Incinerators can be charged manually or automatically; they can also be charged directly (see Figure 10.3.2) or

TABLE 10.3.4   WASTE ANALYSIS SHEET

% Ash _____ % Sediment _____ % Water _____
Waste material soluble in water? _____ Water content well mixed, emulsified? _____
If there are solids in the liquid, what is their size range? _____
Conradson carbon _____ Corrosion (copper strip) _____
Is the material corrosive to carbon steel? _____ Corrosive to brass? _____
What alloy is recommended for carrying the fluid? _____
Distillation data (if applicable) 10% at _____ °F; 90% at _____ °F; end point _____ °F.
Flash point _____ °F; fire point _____ °F; pour point _____ °F.
Viscosity _____ SSF at 122°F or _____ SSU at 100°F.
pH _____ ; acid number _____ ; base number _____
Heating value _____ Btu/gal   Specific gravity (H_2O = 1.0) _____
Will the material burn readily? _____
Toxic? (explain) _____

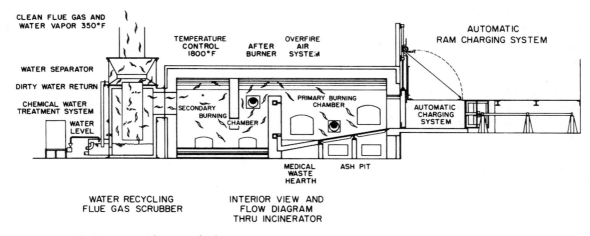

FIG. 10.3.2 Incinerator with a ram-feed system.

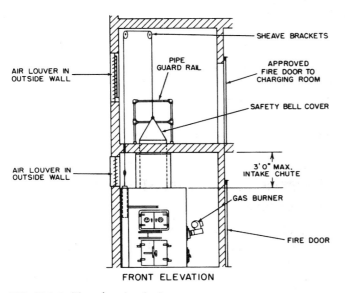

FIG. 10.3.3 Top-charging incinerator.

Hydraulic plungers or rams offer a more controlled method of automatic charging. The movement of the reciprocating plunger forces the refuse from the bottom of the charge hopper into the furnace (see Figure 10.3.2). This method is the most common for automatic charging for capacities exceeding 500 lb per hour.

Incinerators that burn sawdust or shredded waste are frequently charged by screw feeders or pneumatic conveyors. Screw feeders are at least 6 in (15 cm) in diameter and are designed with variable pitch to minimize the compression (and therefore blocking) of the shredded waste. Container charging, which is being used in a few isolated cases, has the advantage of protecting against exposure to flashback caused by aerosol cans or the sudden combustion of highly flammable substances.

Accessories

For smaller incinerators, chimneys provide sufficient draft to discharge flue gases at a high enough point where no nuisance is caused by the emissions. A fully loaded chemistry should provide at least 0.25 in of water draft (-62 Pa). Table 10.3.5 lists the diameters and heights of chimneys according to the weight rate of waste burned in a continuously charged multiple-chamber incinerator. The table assumes that the incinerator uses no dilution air and

from charging rooms (see Figure 10.3.3). Direct incinerators are the least expensive but are limited in their hourly capacities to 500 lb, while indirect incinerators operate at capacities up to 1000 lb/hr. A manually charged incinerator (see Figure 10.3.3) is fed through a bell-covered chute from the floor above the furnace. This labor-saving design also guarantees good combustion efficiency and protection against flashbacks. The separate charging room is also convenient for sorting waste for recycling. Incinerators can also be fed from the same floor where the furnace is located. This arrangement also permits sorting and is labor-efficient although the radiant heat can be uncomfortable for the operator.

In high-rise buildings, the installation of a waste chute eliminates the labor involved in charging the incinerator (see Figure 10.3.4). The chute automatically directs the solid waste into a top-charged, mechanical incinerator. The charging rate can be regulated by rotary star feeders or by charging gates that open at 15- to 30-min intervals. Both offer protection against momentary overloading.

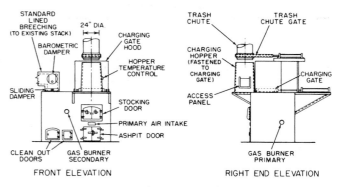

FIG. 10.3.4 Incinerator with automatic charging system.

that the breechings between the furnace and chimney are of minimum length. The lining thicknesses shown are for outdoor chimneys; chimneys inside buildings need additional insulation.

For proper incinerator operation, the cold air supply to the furnace should not be restricted. In most designs, the furnace receives its air supply from the incinerator room. The air supply should be sized for about 15 lb of air per lb of MSW burned. If the air supply is insufficient, the mechanical ventilation system of the building can cause smoking due to downdrafts. When a chimney's natural draft is insufficient, fans are installed to generate the required draft. In onsite incinerators, the forced-draft air is usually introduced underfire. Introducing overfire air to improve combustion efficiency is not widely used in onsite units.

When the waste is wet or its heating value is low, auxiliary fuels are needed to support combustion. In continuously charged incinerators, the primary burner is sized for 1500 Btu per lb of type 3 waste or for 3000 Btu per lb of type 4 waste (see Table 10.3.2). The heat capacity of the secondary burner is also 3000 Btu per lb of waste. When the incinerator is fully loaded, the secondary burner runs for only short periods at a time.

Controls

Onsite incinerators are frequently operated automatically from ignition to burndown (see Figure 10.3.5). The cycle is started by the microswitch on the charging door, which automatically starts the secondary burner, the water flow to the scrubber, and the induced-draft fan. When the door is closed, the primary burner is started and stays on for an adjustable time of up to an hour or until the door is reopened. Unless interrupted by a high-temperature switch, the secondary burner stays on for up to five hours. To provide each charge with the same preset burndown protection, the secondary burner timer is reset every time the charging door opens. Both burners have overtemperature and flame-failure safety controls. Also, a separate cycle timer controls the induced-draft fan and scrubber water flow to guarantee airflow and scrubbing action during

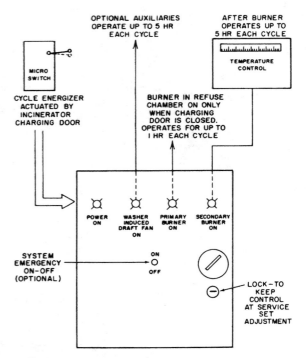

FIG. 10.3.5   Incinerator control system.

burning. In some installations, the charging sequence is also automated.

Domestic and Multiple-Dwelling Incinerators

Domestic incinerators are sized to handle a few pounds of solid waste per person per day. In single dwellings, a typical incinerator has about 40,000 Btu/hr of auxiliary heat capacity. Because domestic incinerators are less efficient than their municipal counterparts, the amount of auxiliary fuel used is high. The domestic incinerator in Figure 10.3.1 has two combustion chambers. The main purpose of the secondary chamber is to eliminate smoke and odor. As a result, the pollutant emissions from domestic incinerators are not excessive (see Table 10.3.6).

In multiple dwellings, the main purpose of incineration is to reduce the volume of the MSW prior to disposal. The refuse from a dwelling of 500 residents producing 2000 lb/day of MSW at a density of 4 lb/cu ft fills 100 trash cans. If incinerated onsite, the residue fits into 10 trash cans.

Incinerators in multiple dwellings can either be chute fed or flue fed. In the chute-fed design, waste is discharged into the chute and then into the incinerator feed hopper in the basement (see Figure 10.3.4). In the flue-fed design (see Figure 10.3.6), the chimney also serves as the charging chute for the waste, which falls onto grates above an ash pit inside a boxlike furnace. The main purpose of the charging door is to ignite the waste, while the purpose of the underfire and overfire air ports is to manually set the airflow for smokeless burning. The walls of the incinera-

TABLE 10.3.5   CHIMNEY SELECTION AND SPECIFICATION CHART

Incinerator Capacity All Types Waste, lb/hr	Chimney Size		Lining Thickness	Steel Casing Thickness
	Inside Diameter	Height above Grate		
100–150	12″	26′	2″	10 ga.
175–250	15″	26′	2″	10 ga.
275–350	18″	32′	2″	10 ga.
400–550	21″	37′	2½″	10 ga.
600–750	24″	39′	3″	¼″
800–1400	30″	44′	3″	¼″
1500–2000	36″	49′	3½″	¼″

TABLE 10.3.6 INCINERATOR EMISSIONS—TYPICAL VALUES

Pollutant	New Domestic Wastes Incinerator	Municipal Incinerator with Scrubber
Particulates, grain/SCF	0.01–0.20	0.03–0.40
Carbon monoxide, ppm	200–1000	<1000
Ammonia, ppm	<5	—
Nitrogen oxides, ppm	2–5	24–58
Aldehydes, ppm	25–40	1–9

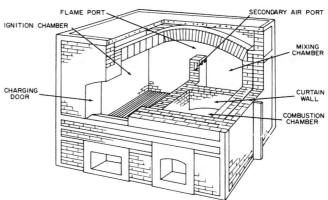

FIG. 10.3.7 Retort-type, multiple-chamber incinerator.

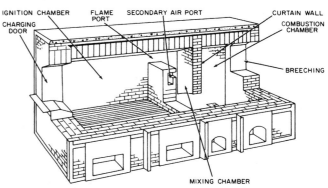

FIG. 10.3.8 Inline multiple-chamber incinerator.

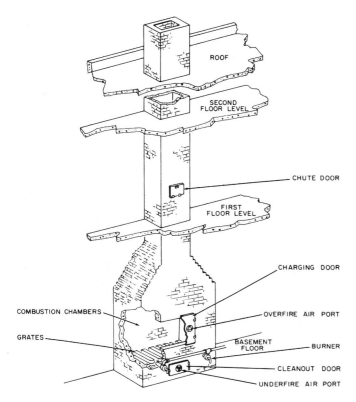

FIG. 10.3.6 Flue-fed incinerator.

tor consist of two brick layers with an air space between them. The inner layer is made of 4.5 in of firebrick and the 9-in outer layer is made of regular brick.

Flue-fed apartment incinerators have a draft-control damper in the stack, right above the furnace. This damper is pivoted and counterweighted to close when a chute door opens to charge refuse into the furnace. As a result, draft at the furnace remains relatively constant. To withstand flame impingement, the draft-control damper should be made of 20-gauge 302 stainless steel.

Miscellaneous Onsite Incinerators

Some incinerator designs are specially developed for on-site industrial applications. The outstanding design fea-

ture of the retort incinerator (see Figure 10.3.7) is the multiple chambers connected by lateral and vertical breechings; the combustion gases must pass through several U-turns for maximum mixing. The inline design (see Figure 10.3.8) also emphasizes good flue-gas mixing. Here, the combustion gases are mixed by passing through 90° turns in the vertical plane only. Both designs are available in mobile styles for use in temporary applications such as land clearance or housing construction. The retort design is for smaller waste-burning capacities (under 800 lb/hr), while the inline design is for higher burning rates.

Rotary incinerators for burning solid or liquid wastes can be continuous or batch and can be charged manually or by automatic rams. Their capacities range from 100 to 4000 lb/hr. For burning waste that contains chlorinated organics, the incinerator chamber must be lined with acid-resistant brick, and the combustion gases must be sent through absorption towers to remove the acidic gases from the flue gas.

Adapted from *Municipal Waste Disposal in the 1990s* by Béla G. Lipták (Chilton, 1991).

10.4
PYROLYSIS OF SOLID WASTE

Pyrolysis is an alternate to incineration for volume reduction and partial disposal of solid waste. A large portion of MSW is composed of long-chain hydrocarbonaceous material such as cellulose, rubber, and plastic. This organic material represents a storehouse of organic building blocks that could be retained as organic carbon. Pyrolysis is a process that is less regressive than incineration and recovers much of the chemical energy.

Long-chain organic material disintegrates when exposed to a high-temperature thermal flux according to the following equation:

$$\text{Polymeric material} + \text{Heat flux} \longrightarrow aA(gas) + bB(liquid) + cC(solid) \quad \text{10.4(1)}$$

The resulting gas includes CO_2, CO, H_2, CH_4, and various C_2 and C_3 saturated and unsaturated hydrocarbons. The liquid contains a variety of chemical compounds, and the liquid ranges from a tar substance to a light water-soluble distillate. The solid is primarily a solid char.

The relative yield of each of these groups of pyrolysis products depends on the chemical structure of the solid to be pyrolyzed, the temperature for decomposition, the heating rate, and the size and shape of the material.

If the products of pyrolysis react with oxygen, they react according to the following equations:

$$A(gas) + O_2 \longrightarrow CO_2 + H_2O + heat \quad \text{10.4(2)}$$

$$B(liquid) + O_2 \longrightarrow CO_2 + H_2O + heat \quad \text{10.4(3)}$$

$$C(solid) + O_2 \longrightarrow CO_2 + H_2O + heat \quad \text{10.4(4)}$$

Pure pyrolysis involves only the reaction in Equation 10.4(1), the destructive distillation in an oxygen-free atmosphere. This definition can be expanded to include systems in which a limited amount of oxygen is made available to the process to release enough chemical energy for the pyrolysis reaction.

Comparing the results of various experimental investigations on pyrolysis is difficult because of the many variables influencing the results. No reliable design methods have been developed that allow for the scale-up of the experimental results. However, certain guiding principles underlying all pyrolysis systems can help in the selection of a process that most likely satisfy a particular need.

The process and operating conditions vary depending upon the relative demand for the char, liquid, and gas from the process.

Pyrolysis Principles

An understanding of the energy relationships, the effect of thermal flux, solid size, and the types of equipment is requisite to an understanding of pyrolysis principles.

ENERGY RELATIONSHIPS

No single value exists for the total energy required to pyrolyze any material. It depends upon the products formed which depend on the temperature, rate of heating, and sample size. Therefore, the reported values for the heat of pyrolysis conflict among various experimenters.

Figure 10.4.1 expresses the general energy requirements to pyrolyze a material as the amount of oxygen varies. The lower solid line represents the amount of heat added or removed from the system. The upper solid line represents the chemical energy of the pyrolysis products. For pure pyrolysis, no oxygen is available, and all energy for the pyrolysis reaction is supplied from indirect heating. The heat required is given by q, which represents the heat necessary to pyrolyze the solid feed and heat the products to the pyrolysis temperature. The value ΔH_1 represents the chemical energy of the gas. As oxygen is made available, energy is released within the system, and less indirect energy is supplied.

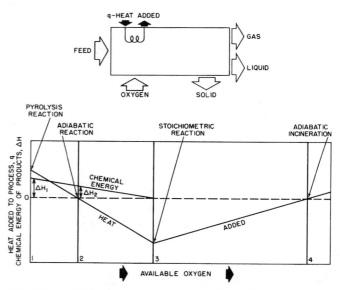

FIG. 10.4.1 General energy requirements for solid–gas reactions as a function of oxygen availability.

At point 2, an adiabatic condition is reached where the heat released from the oxidation of a portion of the pyrolysis products can furnish the energy required for the pyrolysis reaction as well as the energy necessary to heat the pyrolysis products, oxidation products, and nitrogen to the pyrolysis temperature. The value ΔH_2 represents the total chemical energy of the gas under these conditions. The larger fraction of the energy goes to sensible heat if nitrogen is present and ΔH_2 is smaller.

As the available oxygen increases, heat must be removed to maintain a constant reaction temperature. At point 3, the stoichiometric oxygen for complete combustion is reached, and the reaction products contain no chemical energy. Additional oxygen acts only as a coolant; therefore, less energy must be removed until point 4 is reached. This point is where the feed is being incinerated adiabatically, and no heat recovery is possible. This figure shows that the combined energy of the pyrolysis products is higher when the available oxygen is reduced. An advantage of the oxygen dependency is that it eliminates the limitation of pyrolysis systems on the rate of external heat demand. When enriched oxygen is used rather than air, the fraction of energy tied up in sensible heat is less, leaving more chemical energy in the pyrolysis products (the greater the fraction of chemical energy).

EFFECT OF THERMAL FLUX

The products resulting from the thermal destruction of hydrocarbonaceous solids depend upon the maximum temperature of pyrolysis and the time needed to bring the feed to this temperature. The products formed during slow heating are far different than the products obtained during rapid heating. At very slow heating rates to low temperatures, the molecule has sufficient time to break at the weakest level and reorganize itself into a more thermally stable solid that becomes increasingly hard to destroy. On rapid heating to a high temperature, the molecule explodes and forms a range of smaller organic molecules.

For the cellulose molecule, slow heating forms high char yields and low gas and liquid yields. The gas is composed primarily of CO, H_2O, and CO_2 and has a low heating value. For rapid heating rates and high temperature, the gas yield increases and the liquid is smaller. The gas is composed primarily of CO_2, CO, H_2, CH_4, C_2H_2, and C_2H_4 and has a reasonable heating value. For intermediate heating rates and temperatures, high liquid yields are obtained. The gas produced is composed of many C_1, C_2, C_3, and C_4 compounds and has a high heating value.

Table 10.4.1 shows some values obtained for pyrolysis gas obtained at 1300 and 1600°F at a slow heating rate in a retort. For comparison the table also shows 1450°F pyrolysis at a high heating rate in a fluidized bed. Both systems pyrolyzed MSW.

SOLID SIZE

For a large retort requiring indirect heating, the time required to pyrolyze a batch often exceeds twenty-four hours. The products change drastically as the reaction proceeds because a long time is needed for the center of the batch to reach the pyrolysis temperature. The mass near the center goes through a much slower heating cycle than the material near the walls. For the produced gas and liq-

TABLE 10.4.1 PYROLYSIS PRODUCTS OF MSW

Product Data	Composition of Pyrolysis Products		
Speed of Pyrolysis	Slow	Slow	Fast
Pyrolysis Temperature, °F	1382	1652	1450
Weight %			
Residue	11.59	7.7	3.0
Gas	23.7	39.5	61.0
Tar	1.2	0.2	26.0
Light Oil	0.9		
Liquor	55.0	47.8	4.0
Gas (Volume %)			
H_2	30.9	51.9	37.16
CO	15.6	18.2	35.50
CH_4	22.6	12.7	11.10
C_2H_6	2.05	0.14	not
C_2H_4	7.56	4.68	measured
CO_2	18.4	11.4	16.3
Btu/ft^3	563	447	366
10^6 Btu/ton	5.42	7.93	6.36
Gas Volume, cu ft/tn	9620	17,300	17,400

uids to be collected, they must pass through thick layers of pyrolysis char, and numerous secondary reactions result. For these reasons, the pyrolysis products of wood from a large retort can contain more than 120 products.

The same conditions are true for individual particles. For a material having the thermal properties of wood, the time required for the center temperature of a sphere to reach the surface temperature can be given by the following equation:

$$t = 0.5r^2 \qquad 10.4(5)$$

where r is the radius in inches and t is in hours. This equation indicates that about one hour is needed for the center of a 3-in particle to approach the surface temperature.

TYPES OF EQUIPMENT

Several types of equipment are available for the pyrolysis of waste. The general types include retorts, rotary kilns, shaft kilns, and fluidized beds. The type of equipment and the manner of contacting have a significant effect on the pyrolysis product yield.

The retort has the longest application history and has been used extensively to make wood charcoal and naval stores. It is a batch system where the retort is charged, sealed, and heated externally. The heating cycle is long (often over twenty-four hours). The products are complex. They are normally solid char and a pyroligneous acid plus the gas produced which is used as the energy source for indirect heating. The process is limited by the rate of heat addition; a typical analysis for demolition lumber shows a yield of 35% char, 30% water, 12% wood tar, 5% acetic acid, 3% methanol, and 15% gas with a heating value of 300 Btu/cu ft.

Rotary Kiln

The rotary kiln is more flexible and provides increased heat transfer. The kiln can be heated indirectly, or the heat can be furnished by partially burning the pyrolysis products. The gas flow can be either parallel or countercurrent to the waste flow. The gas and liquid products do not have to escape through thick layers of char as in the batch retort; therefore, fewer complex solid–gas reactions occur. The heat cycle is much faster than in the retort, and the gas yield is higher and the liquid yield lower. The size of the indirectly fired kiln, because of the high temperatures involved and the need to transfer energy through the walls, is severely limited. The maximum capacity is in the range of 2 tn per hour for wood waste and is similar for solid waste.

If a limited amount of oxygen is used as the energy needed for pyrolysis, refractory lined kilns can be used, and large systems become feasible. If the oxygen and the feed are introduced countercurrently, the oxygen contacts the pyrolysis char first and tends to burn this char to furnish the heat for pyrolysis, which reduces the char yield. If air is introduced parallel to the feed, the oxygen reacts with the raw feed and the pyrolysis gas and gives a lower gas yield.

Rotary equipment, however, is more expensive to build, is more difficult to design with positive seals, and requires more maintenance.

Shaft Kiln

In the shaft kiln, the solids descend through a gas stream. The oxygen enters the bottom countercurrently to the feed and burns the char product reaching the bottom of the shaft. The combustion gases produced flow past the solid feed causing pyrolysis. The char is used to furnish the energy for pyrolysis. This use is undesirable if the char is a valuable product and, in that case, the pyrolysis gas can be used to preheat the air.

Fluidized-Bed Reactor

The fluidized-bed reactor is a system where the heat transfer rate is rapid. This design gives low liquid yields and high gas yields. In the fluidized bed, the feed is injected into a hot bed of agitated solids. To keep the bed in the fluid state, the system passes gas upward through the bed. If air is introduced into the bed, the oxygen contacts and reacts with the pyrolysis gas more readily than the char does, reducing the gas yield. To assure that the produced pyrolysis gas does not react with oxygen, operators can remove the char produced and burn it in a separate unit and return the hot gas used to pyrolyze the feed into the fluidized bed. The capacity is limited by the sensible heat available from this gas. In a fluidized bed, circulating the solids in the bed adds heat. The heat source necessary to pyrolyze the feed can be heated solids which can be easily added and removed in the fluidized bed.

Experimental Data

Little data is published on the pyrolysis of MSW, and no data is published for full-scale operating units. The data in Tables 10.4.1 and 10.4.2 are based on solid waste from a small pilot plant scale or bench scale lab experiments.

Table 10.4.2 presents the yield of various materials exposed to a 1500°F temperature. The yields of gas, liquid, and char vary widely between materials. The feed in all cases is newspaper, and the final pyrolysis temperature is 1500°F. An increase in the gas yield and a decrease in liquid organics due to an increase of the heating rate is evident from these data. The total amount of energy available from the gas also increases with the heating rate.

Table 10.4.1 gives the products for a slow pyrolysis

TABLE 10.4.2 PYROLYSIS YIELDS, IN WEIGHT PERCENT OF REFUSE FEED

Type of Waste Feed / *Pyrolysis Products*	*Gas*	*Water*	*CnHmOx*	*Char C + S*	*Ash*
Ford Hardwood	17.30	31.93	20.80	29.54	0.43
Rubber	17.29	3.91	42.45	27.50	8.85
White Pine Sawdust	20.41	32.78	24.50	22.17	0.14
Balsam Spruce	29.98	21.03	28.61	17.31	3.07
Hardwood Leaf Mixture	22.29	31.87	12.27	29.75	3.82
Newspaper I	25.82	33.92	10.15	28.68	1.43
II	29.30	31.36	10.80	27.11	1.43
Corrugated Box Paper	26.32	35.93	5.79	26.90	5.06
Brown Paper	20.89	43.10	2.88	32.12	1.01
Magazine Paper I	19.53	25.94	10.84	21.22	22.47
II	21.96	25.91	10.17	19.49	22.47
Lawn Grass	26.15	24.73	11.46	31.47	6.19
Citrus Fruit Waste	31.21	29.99	17.50	18.12	3.18
Vegetable Food Waste	27.55	27.15	20.24	20.17	4.89

process at 1382 and 1652°F along with data for fast pyrolysis at 1450°F. The increase in gas yield and decrease in organic liquid with an increase in reactor temperature are evident. The hydrocarbon fraction of the gas decreases from 32.2 to 17.5% with an increase in temperature from 1382 to 1652°F, while the H_2 and CO portion increases from 46.5 to 70.1%. This data shows that higher temperature pyrolysis gives significantly higher yields to lower Btu gas. The higher temperature apparently results in the destruction of hydrocarbons in the gas. Comparing values for solid waste is difficult because of the variability between feed stocks of MSW. Table 10.4.1 shows the data for a run at 1450°F where pyrolysis is rapid along with data for slow pyrolysis. These data indicate that the fast pyrolysis at 1450°F gives results which are closer to those of the slow, high-temperature pyrolysis process. The total CO and H_2 is 72.6%, the total hydrocarbon 10.1%+ (undoubtedly higher but only CH_4 is evaluated), and the gas volume 17,400 cu ft/tn. Unfortunately, the data for fast pyrolysis is not complete, and a full comparison on the yield is not possible.

Table 10.4.3 presents data for a process to convert MSW to fuel oil. The temperature for pyrolysis is low (932°F), and the reaction rate rapid. This low-temperature pyrolysis gives higher yields of organic liquids, and the gas has significant quantities of C_2–C_7 hydrocarbons not present at higher temperatures. Reducing the temperature for this same process by several hundred degrees results in an increase in the gas yield of 80%.

The data available substantiate the guiding principles previously outlined and explain the composition and quantities of the products and how they are affected by changes in composition, temperature, and heating rate. They do not accurately predict the products from a particular

TABLE 10.4.3 FUEL OIL PRODUCTION FROM MSW

Char fraction, 35 wt%, heating value 9000 Btu/lb
CO	48.8 wt%
H	3.9
N	1.1
S	0.03
Ash	31.8
Cl	0.2
O (by difference)	12.7

Oil fraction, 40 wt%, heating value 12,000 Btu/lb
C	60%
H	8
N	1
S	0.2
Ash	0.4
Cl	0.3
O_2 (by difference)	20.0

Gas fraction, 10 wt%, heating value 600 Btu/cu ft
H_2O	0.1 mol%
CO	42.0
CO_2	27.0
H_2	10.5
CH_3Cl	<0.1
CH_4	5.9
C_2H_6	4.5
C_3–C_7	8.9

Water fraction contains:
  Acetaldehyde
  Acetone
  Formic acid
  Furfural
  Methanal
  Methylfurfural
  Phenol

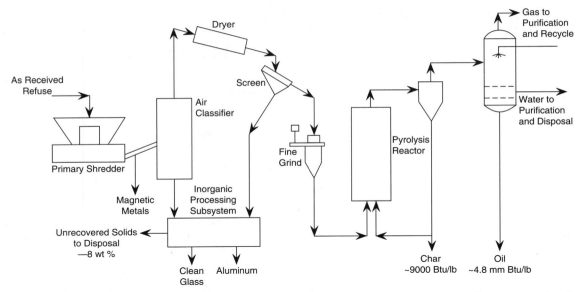

FIG. 10.4.2 Schematic diagram of Occidental Flash Pyrolysis System for the organic portion of MSW. (Reprinted, with permission, from G.T. Preston, 1976, Resource recovery and flash pyrolysis of municipal refuse, presented at Inst. Gas Technol. Symp., Orlando, FL, January.)

process or waste. However, the data do furnish sufficient evidence for environmental engineers to suggest the type of process for a given application.

Status of Pyrolysis

Pyrolysis is widely used as an industrial process to produce charcoal from wood, coke and coke gas from wood, coke and coke gas from coal, and fuel gas and pitch from heavy petroleum fractions. In spite of these industrial uses, the pyrolysis of MSW has not been as successful. No large-scale pyrolysis units are used for MSW operation in the United States as of April 30, 1995.

Only one full-scale MSW pyrolysis system was built in the United States. A simplified flowsheet of the Occidental Flash Pyrolysis System is shown in Figure 10.4.2. The front-end system consists of two stages of shredding, air classification, trommeling, and drying to produce finely divided RDF. Because of the short residence time of RDF in the reactor, this process is described as flash pyrolysis. The heat required for the pyrolysis reaction in the reactor is supplied from recirculation of the hot char. The hot char is removed from the reactor, passed through an external fluidized bed in which some air is added to partially oxidize the char, and recirculated to furnish energy for the endothermic pyrolysis reaction which yields the liquid by-products.

The end products were gases, pyrolytic oil, char, and residues. The liquid product had several noxious qualities making it a poor substitute for Bunker C fuel oil. It was corrosive, requiring special storage and fuel nozzles, and was more difficult to pump and smelled poorly. These

qualities resulted largely from highly oxygenated organics (including acids). Furthermore, the oil produced had a moisture content of 52%, not the 14% predicted from the pilot plant results. The increase in moisture in the oil decreased the energy content to 3600 Btu/lb, compared to the 9100 Btu/lb predicted by the pilot plant tests. The 100 tpd plant was built in El Cajun, California but never ran successfully and was shut down after only two years of operation.

The principal causes for the failure of pyrolysis technology appear to be the inherent complexity of the system and a lack of appreciation by system designers of the difficulties of producing a consistent feedstock from MSW (Tchobanoglous, Theisen, and Vigil 1993).

Although systems such as the Occidental Flash Pyrolysis System were not commercial successes, they produced valuable design and operational data that can be used by future designers. If the economics associated with the production of synthetic fuels change, pyrolysis may again be an economical, viable process for the thermal processing of solid waste. However, if gaseous fuels are required, gasification is a simpler, more cost-effective technology.

—R.C. Bailie (1974)
and David H.F. Liu (1996)

Reference

Tchobanoglous, G., H. Theisen, and S. Vigil. 1993. *Integrated solid waste management.* McGraw-Hill, Inc.

10.5
SANITARY LANDFILLS

The landfill is the most popular disposal option for MSW in the United States. Not only has it traditionally been the least-cost disposal option, it is also a solid waste management necessity because no combination of reduction, recycling, composting, or incineration can currently manage the entire solid waste stream. Barring unforeseen technological advances, landfills will always be needed to handle residual waste material.

Landfill Regulations

Solid waste landfills are federally regulated under Subtitle D of the Resource Conservation and Recovery Act of 1976 (RCRA). In the past, landfill regulation was left to the discretion of the individual states. The Solid Waste Disposal and Facility Criteria, promulgated by the U.S. EPA, specify how MSW landfills are to be designed, constructed, operated, and closed and were implemented in 1993 and 1994. The criteria were developed to ensure that municipal landfills do not endanger human health and are based on the assumption that municipal landfills receive household hazardous waste and hazardous waste from small generators. States are required to adopt regulations at least as strict as the EPA criteria. Although some states had some acceptable regulations in place, many did not. The EPA is currently considering criteria for non-hazardous industrial waste landfills.

In most states, landfills constructed under the new regulations are more expensive to construct and operate than past landfills because of requirements concerning daily cover, liners, leachate collection, gas collection, monitoring, hazardous waste exclusion, closure and postclosure requirements, and financial assurances (to cover anticipated closure and postclosure costs). Some of the major aspects of the landfill criteria are briefly described below next (40 *CFR* Parts 257–258).

LOCATION RESTRICTIONS

Location restrictions exclude landfills from being near or within certain areas to minimize environmental and health impacts. Table 10.5.1 summarizes location restrictions. Other location restrictions not mentioned in the federal disposal criteria but found in other federal state regulations include public water supplies, endangered or threatened species, scenic rivers, recreation or preservation areas, and utility or transmission lines.

EMISSIONS, LEACHATE, AND MONITORING
Gaseous Emissions

Landfills produce gases comprised primarily of methane and carbon dioxide. Emissions are controlled to avoid explosive concentrations of methane or a build-up of landfill gases that can rupture the cover liner or kill cover vegetation. Landfill design and monitoring must ensure that the concentration of CH_4 is less than 25% of the lower explosion limit in structures at or near the landfill and less than the lower explosion limit at the landfill property boundary.

A final rule announced by EPA in March, 1996 requires large landfills that emit volatile organic compounds in excess of 50 megagrams (Mg) per year to control emissions by drilling collection wells into the landfill and routing the gas to a suitable energy recovery or combustion device. It also requires a landfill's surface methane concentration to be monitored on a quarterly basis. If the concentration is greater than 500 parts per million, the control system must be modified or expanded to insure that the landfill gas is collected. The rule is expected to effect only the largest 4% of landfills in the United States.

Leachate

Leachate is water that contacts the waste material. It can contain high concentrations of COD, BOD, nutrients, heavy metals, and trace organics. Regulations require leachate to be collected and treated to avoid ground or surface water contamination. Composite bottom liners are required, consisting of an HDPE geomembrane at least 60 mil over 2 ft of compacted soil with a hydraulic conductivity of less than 1×10^{-7} cm/sec. However, equivalent liner systems can be used, subject to approval. The composite liner is covered with a drainage layer and leachate collection pipes to remove leachate for treatment and maintain a hydraulic head of less than 1 ft. Leachate is generally sent directly to a municipal wastewater treatment plant but can be pretreated, recirculated, or treated on-site.

Surface Water

Leachate generation can be reduced when water is kept from entering the landfill, especially the working face. Surface water control also reduces erosion of the final cover. Regulations require preventing flow onto the active

TABLE 10.5.1   SITING LIMITATIONS CONTAINED IN SUBTITLE D OF THE RCRA AS ADOPTED BY THE EPA

Location	Siting Limitation
Airports	Landfills must be located 10,000 ft from an airport used by turbojet aircraft, 5000 ft from an airport used by piston-type aircraft. Any landfills closer must demonstrate that they do not pose a bird hazard to aircraft.
Flood plains	Landfills located within the 100-year floodplain must be designed to not restrict flood flow, reduce the temporary water storage capacity of the floodplain, or result in washout of solid waste, which would pose a hazard to human health and the environment.
Wetlands	New landfills cannot locate in wetlands unless the following conditions have been demonstrated: (1) no practical alternative with less environmental risk exists, (2) violations of other state and local laws do not exist, (3) the unit does not cause or contribute to significant degradation of the wetland, (4) appropriate and practicable steps have been taken to minimize potential adverse impacts, and (5) sufficient information to make a determination is available.
Fault areas	New landfill units cannot be sited within 200 ft of a fault line that has had a displacement in Holocene time (past 10,000 years).
Seismic impact zone	New landfill units located within a seismic impact zone must demonstrate that all contaminant structures (liners, leachate collection systems, and surface water control structures) are designed to resist the maximum horizontal acceleration in lithified material (liquid or loose material consolidated into solid rock) for the site.
Unstable areas	Landfill units located in unstable areas must demonstrate that the design ensures stability of structural components. The unstable areas include areas that are landslide prone, are in karst geology susceptible to sinkhole formation, and are undermined by subsurface mines. Existing facilities that cannot demonstrate the stability of the structural components must close within five years of the regulation's effective date.

Source: Data from G. Tchobanoglous, H. Theissen, and S. Vigil, 1993, *Integrated solid waste management: Engineering principles and management issues* (New York: McGraw-Hill).

portion of the landfill (i.e., the working face) during peak discharge from the twenty-five-year storm of twenty-four-hour duration. Collection and control of water running off the active area during the twenty-five-year storm of twenty-four-hour duration is also required. Landfills should have no discharges that violate the Clean Water Act.

Daily Cover

Exposed waste must be covered with at least 6 in of soil at the end of each operating day. Alternative covers, such as foam or temporary blankets, can be approved for use.

Hazardous Waste

Hazardous waste should be kept out of the landfill so that the quality of leachate and gaseous emissions is improved. Landfill operations must have a program for detecting and preventing the disposal of regulated hazardous wastes.

Monitoring

Monitoring is done to identify, quantify, and track contaminants and to determine where and when corrective action should take place. At least three wells, one upgradi-

ent and two downgradient, should be maintained, and the well water should be tested at specified intervals. Most sites have more than three wells; the applicable regulations vary by state. Monitoring is conducted before, during, and after the landfill operating period. Remedial action is required when downgradient water quality is significantly worse than upgradient water quality.

Closure and Postclosure

To reduce, control, or retain leachate, gaseous emissions, and surface water, landfill operators must close the landfill properly and maintain it until waste material stabilizes. They must install and maintain a final cover to keep rainwater out of the landfill and establish vegetation to reduce erosion. The postclosure period is thirty years, during which all previously mentioned regulations must be followed, erosion must be controlled, and site security must be maintained.

Financial Assurance

In case of bankruptcy or other circumstances, facilities must be closed in a proper manner. The financial capability to safely close the facility at any time during its operational life must be maintained by the operator in a manner acceptable to the governing agency.

Siting New Landfills

Proper siting of sanitary landfills is crucial to providing economic disposal while protecting human health and the environment. The siting process consists of the following tasks (Walsh and O'Leary 1991a):

- Establishing goals and gathering political support
- Identifying facility design basis and need
- Identifying potential sites within the region
- Selecting and evaluating in detail superior sites
- Selecting the best site
- Obtaining regulatory approval

Goals include delineating the region to be served, facility lifetime, target tipping fees, maximum hauling distance, potential users, and landfill services. Political support is crucial to successful siting. Because opposition to a new landfill is almost always present, strong political support for a new landfill must exist from the start of the siting process. A solid waste advisory council—made up of interested independent citizens and representatives of interested groups—should be formed early in the process, if one does not already exist.

The design basis and needs of a landfill depend on the applicable regulations and the required landfill area (which in turn depend on the amount of waste to be handled and the required lifetime). The amount of waste to be handled depends on the present and future population served by the landfill, the projected per capita waste generation rate, and the projected recycling, composting, and reduction rates.

Developing a new landfill involves finding the most suitable available location. The main criteria involved in siting a new landfill are:

SITE SIZE—The site should have the capacity to handle the service area's MSW for a reasonable period of time.

SITE ACCESS—All-weather access roads with sufficient capacity to handle the number and weight of waste transport vehicles must be available.

HAUL DISTANCE—This distance should be the minimum distance that does not conflict with social impact criteria.

LOCATION RESTRICTIONS—These restrictions are summarized in Table 10.5.1. Additional or stricter constraints can also be imposed.

PHYSICAL PRACTICALITY—Sites with, for example, surface water or steep slopes should be avoided.

LINER AND COVER SOIL AVAILABILITY—This soil should be available onsite; Offsite sources increase construction and operating costs.

SOCIAL IMPACT—Siting landfills far from residences and avoiding significant traffic impacts minimizes this impact.

ENVIRONMENTAL IMPACT—The effect on environmentally sensitive resources, such as groundwater, surface water,

wetlands, and endangered or threatened species should be minimized. Siting landfills on impermeable soils with a deep water table avoids groundwater impacts.

LAND USE—The land around the potential site should be compatible with a landfill.

LAND PRICE AND EASE OF PURCHASE—A potential site is easier to purchase if it is owned by one or a few parties.

ESTIMATING REQUIRED SITE AREA

Before attempting to identify potential landfill sites, planners must estimate the area requirement of the landfill. Landfill sizing is a function of:

- Landfill life (typically five to twenty-five years)
- Population served
- Waste production per person per day
- Extent of waste diversity
- Shape and height of the landfill
- Landfill area used for buffer zone, offices, roads, scalehouse, and optional facilities such as MRF, tire disposal and storage, composting, and convenience center

A number of formulas can help determine the acreage required for waste disposal (Tchobanoglous, Theissen, and Vigil 1993; Noble 1992). The total annual waste produced by the population to be served by the landfill for each year of the expected landfill's life is estimated as:

$$V_{ip} = \frac{(365 \text{ d/yr})PW_g(1-f)}{C_d} \qquad 10.5(1)$$

where:

V_{ip} = annual in-place waste volume (cu yd/yr)
P = population served by landfill in a given year
W_g = waste generation in a given year (lb/person/d)
f = fraction of waste stream diverted in a given year
C_d = specific density of the waste (lb/cu yd)

Population predictions for the years of expected landfill operation can usually be obtained from local government agencies. The total amount of waste generated in a community per person can be developed from waste characterization studies. State or national data can be used if no other data are available. Data on recycling trends should be gathered locally.

As landfilling costs increase, larger and heavier compactors are becoming more common, resulting in higher compaction. Compaction densities achieved in landfills vary from around 800 to as high as 1400 lb/cu yd depending on the type of compaction equipment used. Values of 1000 to 1200 lb/cu yd are often used as estimates.

Cover material adds to the amount of material placed in the landfill, reducing the landfill's effective volume. Typical waste-to-cover-soil-volume ratios are in the range of 4:1 to 10:1. A value of 5:1 or 4:1 is often assumed, indicating that for every 4 or 5 cu yd of waste, 1 cu yd of cover soil is deposited. In a 5:1 ratio, the cover-soil-to-waste ratio is 1 divided by 5, or 0.2. Incorporating waste and cover soil, the annual in-place waste and soil volume is:

$$V_{ap} = V_{ip}(1 + CR) \qquad 10.5(2)$$

where:

V_{ap} = annual in-place waste and soil volume, including waste and cover soil (cu yd/yr)
CR = cover-soil-to-waste ratio

Sometimes planners assume that all of the cover soil will come from the landfill excavation. In this case, all of the soil material excavated from the landfill ends up in the landfill. With this assumption, planners can estimate the area using the assumed shape and height of the landfill above ground level and the sum of the annual in-place waste volume the landfill expects to receive. Height regulations are generally included in state or local landfill regulations and vary from place to place.

The simplest shape that can be assumed is the cube. A more realistic shape is the flat-topped pyramid. Both shapes are shown in Figure 10.5.1. The volume of the cubic landfill is $V = (H)(B^2)$, where H = the height and B = the length of the base. Thus, area = B^2 = V/H. The volume of the flat-topped pyramid landfill, with a square base and 3:1 side slopes, is:

$$V = HB_{3:1}^2 - 6H^2B_{3:1} + 12H^3 \qquad 10.5(3)$$

where a 3:1 side slope means that for every 3 ft horizontal run, the slope rises 1 ft. Solving for B with the quadratic equation gives:

$$B_{3:1} = \frac{6H^2 \pm \sqrt{4HV_{ip} - 12H^4}}{2H} \qquad 10.5(4)$$

Planners can determine the area by squaring $B_{3:1}$. They can take the buffer zone and area needed for roads, facilities, and lagoons into account by increasing B or B^2. If 4:1 side slopes are used:

$$B_{4:1} = \frac{8H^2 + \sqrt{4HV_{ip} - 21.33H^4}}{2H} \qquad 10.5(5)$$

If planners do not assume that all of the excavated soil comes from the landfill excavation, then their solution must incorporate the annual in-place waste and soil volume, the excavated landfill volume, the aboveground landfill volume, and the cover-soil-to-waste ratio. Excavation side slopes are often assumed to be 1:1 but may be more gradual. Excavation bottom slopes are slight and can be assumed level for the purpose of initial size estimates.

EXCLUSIVE AND NONEXCLUSIVE SITING CRITERIA

Landfill siting criteria can be divided into two main groups: exclusive and nonexclusive criteria. If a site fails an exclusive criterion, it is excluded from consideration. Exclusive criteria include federal, state, or local location restrictions or physical restrictions. Exclusive criteria can be applied with maps and transparent overlays. For example, a U.S. Geological Survey (USGS) quadrangle map can be used as the base map. Transparent overlays with darkened restricted areas can be placed over the base map, as shown in Figure 10.5.2. Areas that remain clear are con-

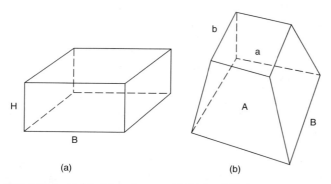

FIG. 10.5.1 Cubic (a) and pyramid (b) landfill shapes.

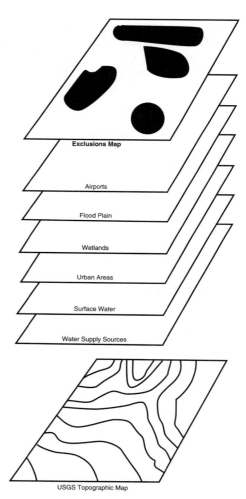

FIG. 10.5.2 Exclusive criteria mapping with overlays.

sidered potential landfill sites. If information is available in digitized format, geographical information systems (GIS) can be used to complete overlay analyses (Siddiqui 1994).

A small number of potential landfill sites are selected from the areas that remain after the exclusive criteria are applied. The final selection process uses nonexclusive criteria, such as hydrogeological conditions, hauling distance, site accessibility, and land use. This process can be done in one or two steps.

If digitized data exist for the entire region under investigation, planners can rank the remaining areas using GIS and an appropriate decision making model, such as the analytical hierarchy process (Siddiqui 1994; Erkut and Moran 1991). For example, USGS soil maps are available in digitized format and include depth to water table, depth to bedrock, soil type, and slope although information is available only down to 5 ft. Planners can also use USGS digitized maps to identify urban areas, rivers and streams, and land use.

If areas are ranked, planners use this information, along with nondigitized information and field inspection, to select a number of sites from the best areas. Otherwise, planners select a number of sites based only on nondigitized information and simple field assessments without the aid of area rankings.

Once a number of sites have been identified, planners should rank the sites in a scientifically justifiable manner using established decision making models such as the analytical hierarchy procedure, interaction matrices, or multiattribute utility models (Camp Dresser & McKee 1984; Morrison 1974; Anandalingham and Westfall 1988–89). This process identifies a small number of sites, usually less than four, to undergo detailed investigations regarding hydrogeologic characteristics such as drainage patterns, geologic formations, groundwater depth, flow directions, and natural quality and construction characteristics of site soils. In addition, detailed information about existing land use, available utilities, access, political jurisdictions, and land cost is gathered (Walsh and O'Leary 1991a). Planners use this information to select the site for which regulatory approval will be sought.

Hydrogeologic information is crucial to the final site selection and has many uses. The main consideration is the proximity of groundwater, groundwater movement, and the potential for attenuation of leachate should it escape from the landfill. The proximity of groundwater is simply the depth to the groundwater table. Measuring the piezometric elevation of the water table in a number of wells on and around the potential site determines groundwater movement. The direction of flow is perpendicular to the lines of constant piezometric elevation. Groundwater movement is important (1) to assess the potential for landfill contamination to impact human health, for example if nearby drinking water wells are downgradient of the landfill site, and (2) to determine the placement of monitoring

wells should a site be selected. Leachate attenuation is a function of mechanical filtration; precipitation and coprecipitation; adsorption, dilution, and dispersion; microbial activity; and volatilization, most of which can be assessed via a subsurface investigation (O'Leary and Walsh 1991c).

The approval process can be demanding. Application writers should work closely with state permitting personnel, who can offer guidance on what is acceptable. To keep costs low, planners should start the siting process with the consideration of large areas based on limited and readily available information and end it with the selection of one site based on detailed information on a small number of sites.

At some point during the selection and permitting of a new landfill, planners hold at least one, and perhaps several, public meetings to ensure that public input is obtained concerning the selection of the landfill site. Generally all landfill sites inconvenience some portion of the local population, and thus most sites generate some public opposition. The siting process must be clear, logical, and equitable. The site selected must be the best available site. However, even if the best site is selected, equity considerations may necessitate offering compensation to residents near the site.

Design

Landfill design is a complex process involving disciplines such as geomechanics, hydrology, hydraulics, wastewater treatment, and microbiology. Design goals can include the following (Walsh and O'Leary 1991c):

- Protection of groundwater quality
- Protection of air quality
- Production of energy
- Minimization of environmental impact
- Minimization of disposal costs
- Minimization of dumping time for site users
- Extension of site lifetime
- Maximum use of land upon site closure

Planners must consider the final use of the landfill site during the design process to ensure that landfill design and operation are compatible with the end use.

Table 10.5.2 summarizes the sanitary landfill design steps. Table 10.5.3 summarizes the landfill design factors. Procurement of the requisite permits can take several years, thus the design process, an integral part of any application, should be started long before the current disposal option is scheduled to close.

The design package includes plans, specifications, a design report, an operator's manual, and cost estimates (Walsh and O'Leary 1991b). An important part of landfill design is the development of maps and plans which describe the landfill's construction and operation, including the location map, base map, site preparation map, devel-

TABLE 10.5.2 SANITARY LANDFILL DESIGN STEPS

1. Determination of solid waste quantities and characteristics
   a. Existing
   b. Projected
2. Design of filling area
   a. Selection of landfilling method based on site topography, bedrock, and groundwater
   b. Specification of design dimensions: cell width, length, and depth; fill depth; liner thickness; interim cover thickness; and final cover thickness
   c. Specification of operational features: method of cover application, need for imported soil for cover or liner, equipment requirements, and personnel requirements
3. Design features
   a. Leachate controls
   b. Gas controls
   c. Surface water controls
   d. Access roads
   e. Special working areas
   f. Special waste handling
   g. Structures
   h. Utilities
   i. Convenience center
   j. Fencing
   k. Lighting
   l. Washracks
   m. Monitoring facilities
   n. Landscaping
4. Preparation of design package
   a. Development of preliminary site plan of fill areas
   b. Development of landfill contour plans: excavation plans; sequential fill plans; completed fill plans; fire, litter, vector, odor, surface water, and noise controls
   c. Computation of solid waste storage volumes, cover soil requirement volumes, and site life
   d. Development of final site plan showing normal fill areas; special working areas (i.e., wet weather areas), leachate controls, gas controls, surface water controls, access roads, structures, utilities, fencing, lighting, washracks, monitoring facilities, and landscaping
   e. Preparation of elevation plans with cross sections of excavated fill, completed fill, and phase development of fill at interim points
   f. Preparation of construction details: leachate controls, gas controls, surface water controls, access roads, structures, and monitoring facilities
   g. Preparation of ultimate landuse plan
   h. Preparation of cost estimate
   i. Preparation of design report
   j. Preparation of environmental impact assessment
   k. Submission of application and obtaining required permits
   l. Preparation of operator's manual

Source: Data from P. Walsh and P. O'Leary, 1991, Landfill site plan preparation, *Waste Age* 22, no. 9: 97–105 and E. Conrad et al., 1981, *Solid waste landfill design and operation practices*, EPA draft report, Contract no. 68-01-3915.

opment plans, cross sections, phase plans, and the completed site map (Walsh and O'Leary 1991c).

The location map is a topographic map which shows the relationship of the landfill to surrounding communities, roads, etc. The base map usually has a scale of 1 in to 200 ft and contour lines at 2 to 5 ft intervals. It includes the property line, easements, right-of-ways, utility corridors, buildings, wells, control structures, roads, drainage ways, neighboring properties, and land use. The site preparation map shows fill and stockpile areas and site facilities. The landfill should be designed so that the excavated material is used quickly as cover. Development plans show the landfill base and top elevations and slopes. Cross sections at various places and times during the landfill lifetime should also be developed. Phase plans show the order in which the landfill is constructed, filled, and closed. The completed site map shows the elements of the proposed end use and includes the final landscaping. Construction details should be available detailing leachate controls, gas controls, surface water controls, access roads, structures, and monitoring facilities.

Equally important as the design maps is the site design report, which describes the development of the landfill in sequence (Walsh and O'Leary 1991b). The four major elements of the design report are:

- Site description
- Design criteria
- Operational procedures
- Environmental safeguards

Landfill Types

Two types of lined landfills are excavated and area. At excavated landfills, soil is excavated from the area where waste is to be deposited and saved for use as daily, intermediate, or final cover. Excavated landfills are constructed on sites where excavation is economical and the water table is sufficiently below the ground surface. Area landfills do not involve soil excavation and are built where excavation is difficult or the water table is near the surface. All cover soil is imported to area landfills. Both types of landfills are lined; the excavated landfill on the bottom of the excavation, the area landfill on the ground surface.

If the entire available area is lined at the beginning of a landfill's life, the large lined area collects rainwater for the life of the landfill, generating a large quantity of unnecessary leachate. For this reason, landfill liners and leachate collection systems are constructed in phases. Each phase consists of constructing a liner and leachate collection system on a portion of the available area, depositing waste in the lined area, and installing intermediate cover. Construction of the next phase begins before the current

TABLE 10.5.3 LANDFILL DESIGN FACTORS

Factors	Remarks
Access	Paved all-weather access roads to landfill site; temporary roads to unloading areas
Land area	Area large enough to hold all community waste for a minimum of five years, but preferably ten to twenty-five years; area for buffer strips or zones also
Landfilling method	Based on terrain and available cover; most common methods are excavated and area landfills
Completed landfill characteristics	Finished slopes of landfill, 3 or 4 to 1; height to bench, if used, 50 to 75 ft; slope of final landfill cover, 3 to 6%
Surface drainage	Drainage ditches installed to divert surface water runoff; 3 to 6% grade maintained on finished landfill cover to prevent ponding; plan to divert storm water from lined but unused portions of landfill
Intermediate cover material	Use of onsite soil material maximized; other materials such as compost produced from yard waste and MSW also used to maximize the landfill capacity; typical waste-to-cover ratios from 5 to 1 to 10 to 1
Final cover	Multilayer design; slope of final landfill cover, 3 to 6%
Landfill liner, leachate collection	Multilayer design incorporating the use of a geomembrane and soil liners. Cross slope for leachate collection systems, 1 to 5%; slope of drainage channels, 0.5 to 1.0%. Size of perforated pipe, 4 in; pipe spacing, 20 ft
Cell design and construction	Each day's waste forms one cell; cover at end of day with 6 in of earth or other suitable material; typical cell width, 10 to 30 ft; typical lift height including intermediate cover, 10 to 14 ft; slope of working faces, 2:1 to 3:1
Groundwater protection	Any underground springs diverted; if required, perimeter drains, well point system, or other control measures installed. If leachate leakage occurs, control with impermeable barriers, pump and treat, or active or passive bioremediation
Landfill gas management	Landfill gas management plan developed including extraction or venting wells, manifold collection system, condensate collection facilities, vacuum blower facilities, flaring facilities, and energy production facilities; operating vacuum located at well head, 10 in of water
Leachate collection	Maximum leachate flow rates determined and leachate collection pipe and trenches sized; leachate pumping facilities sized; collection pipe materials selected to withstand static pressures corresponding to the maximum height of the landfill
Leachate treatment	Pretreatment determined based on expected quantities of leachate and local environmental and political conditions
Environmental requirements	Vadose zone gas and liquid monitoring facilities installed; up- and downgradient groundwater monitoring facilities installed; ambient air monitoring stations located
Equipment requirements	Number and type of equipment based on the type of landfill and the capacity of the landfill
Fire prevention	Onsite water available; if nonpotable, outlets must be marked clearly; proper cell separation prevents continuous burn-through if combustion occurs

Source: Data from G. Tchobanoglous, H. Theissen, and S. Vigil, 1993, *Integrated solid waste management: Engineering principles and management issues* (New York: McGraw-Hill).

phase is filled so that it is ready to receive waste as soon as the current phase is filled. The liners and leachate collection systems of adjacent phases are usually tied together.

The size of landfill phases depends on the rate at which waste is deposited in the landfill, local precipitation rates, state permitting practice, and site topography. At landfills receiving large amounts of waste per day, phase size can be chosen so that phase construction equipment is always in use. As soon as the construction of one phase ends, construction of the next phase begins. Smaller landfills cannot operate this way.

Figure 10.5.3 shows several points in a normal, excavated, landfill lifetime, simplified because the landfill has

only one phase. Part (a) shows the landfill just before waste is deposited. The liner is installed at grades that cause leachate to flow toward leachate collection pipes. Groundwater monitoring wells are also installed. Part (b) shows the second waste lift of an operating landfill cell being created. Each lift consists of a layer of daily waste cells. Each daily cell consists of the waste deposited during a single operating day. Daily cells are separated by the cover soil applied at the end of each day. To keep the daily cover and litter to a minimum, operators should keep the working face as small as possible.

Temporary roads on the landfill allow truck traffic easy access to the working face of the landfill. During wet weather, use of a special easy access area for waste disposal may be necessary.

Part (c) of Figure 10.5.3 shows the completed landfill. Five lifts are created, the final cover is installed, vegetation is established, and gas collection wells are installed.

Landfill excavations have sloping bottoms and sides. Excavated side slopes are generally not more than a ratio of 1:1. Their stability must be checked, typically with rotational or sliding-block methods. Bottom slopes are generally 1 to 5%. However, when landfills are built on sloping terrain, bottom slopes can be steeper, requiring stability analysis as well. Operators must also check the stability of the synthetic liner on steeper slopes to ensure that it does not slip or tear. This analysis is based on the friction force between the liner and the material just below the liner. Planners should estimate the bearing capacity of the soil below the landfill and future settlement to ensure that problems associated with differential settling do not ensue after waste is deposited in the landfill. Finally, the pipes used in the leachate collection system must be able to bear the weight of the waste placed on top.

The side slopes of the top of the landfill are generally a ratio of 3:1 or 4:1. Large landfills have benches, or terraces, on the side slope to help reduce erosion by slowing down water as it flows down the sides. The central portion of the top of the landfill is relatively flat because height limitations keep landfills from being pointed cones. However, a slight slope (3 to 6%) is maintained to encourage run-off.

Leachate Control

Water brought in with the waste, precipitation, and surface run-on can increase the amount of water in the landfill, called leachate. Leachate, especially from new landfills, can have high concentrations of COD, BOD, nutrients, heavy metals, and trace organics (Tchobanoglous, Theissen, and Vigil 1993). Leachate that contacts drinking water supplies can result in contamination. For this reason, liners and collection systems are used to minimize the leachate that escapes from landfills. Unless testing indicates that it is not a pollutant, collected leachate is treated before being released in a controlled manner into the environment.

The factors affecting leachate generation are climate, site topography, the final landfill cover material, the vegetative cover, site phasing and operating procedures, and the type of waste material in the fill (O'Leary and Walsh 1991c). Obviously, with all else equal, the more rainfall, the more infiltration into the landfill and the more leachate. Topography can affect the amount of water entering or leaving the landfill site. One purpose of the final cover is to keep water from entering the fill. Current federal regulations require the final cover to have a hydraulic conductivity at least as low as the bottom composite liner. Unless exemptions are made, this requirement means that the final cover must include a geosynthetic layer. If a drainage layer is included in the final cover, this layer further reduces the amount of water infiltrating the fill. Vegetative cover on the final cover reduces infiltration by intercepting precipitation and encouraging transpiration. As already mentioned, proper site phasing keeps the amount of exposed liner area small, thus reducing the collection of rainwater. Finally, the waste deposited in the landfill contains some water, and the resulting moisture content varies with location and waste type. For example, wastewater treatment plant sludges contain significant amounts of moisture. Planners can estimate the amount of leachate generated by a landfill using water balance equations or the EPA's HELP model (Tchobanoglous, Theissen, and Vigil 1993; O'Leary and Walsh 1991c).

Leachate controls are the final cover, the surface water controls that keep water from running onto the landfill, the liner, the leachate collection system, the leak detection system, and the leachate disposal system.

FINAL COVER AND SURFACE WATER CONTROLS

The final cover creates a relatively impermeable barrier over the fill area which keeps rainwater from entering. The

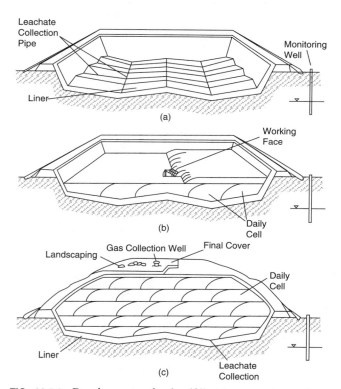

FIG. 10.5.3 Development of a landfill: (a) excavation and installation of landfill liner, (b) placement of solid waste in landfill, (c) cutaway through completed landfill. (Adapted from G. Tchobanoglous, H. Theissen, and S. Vigil, 1993, *Integrated solid waste management: Engineering principles and management issues* [New York: McGraw-Hill].)

slope, soil type, and vegetation determine the surface water run-off characteristics of the site. Planners can determine run-off quantities and peak flows using standard hydrologic run-off techniques, such as the rational method or TR 55. The control of surface run-off generally requires berms to be constructed around the fill area, but drainage ditches can also be used. A detention pond is generally required as well.

LINERS

A landfill can be thought of as a bathtub. Liners make the bottom and sides of the landfill less permeable to the movement of water. Figure 10.5.4 shows a typical liner system. Federal regulations call for a composite liner, consisting of an HDPE geomembrane at least 60 mil thick (1000 mil equals one inch) over 2 ft of compacted soil (clay) with a hydraulic conductivity of less than 1×10^{-7} cm/sec. Equivalent or better alternative liner systems are approved in some cases.

Constructing the soil liner requires spreading and compacting impermeable soil in several lifts, ensuring that the soil contains near optimum moisture content and compaction for minimum permeability. Compaction is usually done with large vehicles with sheepsfoot wheels. Synthetic membranes used in composite landfill liners must be at least 60 mil thick. These membranes can be damaged by heavy equipment and are generally protected with a carefully applied layer of sand, soil, or MSW. Geotextiles can also be used to protect geosynthetic liners.

COLLECTION AND LEAK DETECTION SYSTEMS

Just as a bathtub has a drain to remove bath water, the landfill has a mechanism to remove leachate. The liner in Figure 10.5.3(a) is graded to direct any leachate reaching the liner surface into a leachate collection system. Liner systems are not leak proof. Collecting leachate and removing it from the landfill reduces the hydraulic head on the liner, thus reducing fluid flow through the liner. To speed the lateral flow of leachate once it reaches the bottom of the landfill, a drainage layer is placed over the composite liner (see Figure 10.5.4). The drainage layer can be made of coarse media such as sand or shredded tires, though geonets (high-strength geosynthetic grids less than $\frac{1}{2}$ in thick capable of transmitting high quantities of water) are also common. Geotextiles minimize clogging of the drainage layers by excluding particles. Drainage layers slope toward collection pipes, which direct leachate toward a sump or directly out of the landfill.

Figure 10.5.5 shows a typical leachate collection pipe cross section. The leachate collection pipe is laid in a gravel trench wrapped with a geotextile which allows water to enter the leachate trench but keeps out small particles that could clog the gravel or pipe. Leachate collection trenches lay on top of the liner and travel along local hydraulic low points. The leachate collection system carries leachate out of the landfill cell through the liner or dumps leachate into a sump which is pumped over the side of the liner.

LEACHATE DISPOSAL SYSTEMS

Leachate can be treated by recycling, onsite treatment, or discharge to a municipal wastewater treatment plant. Recycling leachate involves reapplying collected leachate at or near the top of the landfill surface, thus providing additional contact between leachate and landfill microbes. Recycling can reduce BOD and COD and increase pH—with subsequent reduction in heavy metals concentrations. Furthermore, leachate recycling evens the flow of leachate that is removed from the landfill and can enhance the stabilization of the landfill (O'Leary and Walsh, 1991c). Onsite treatment can involve physical, chemical, or biological treatment processes. However, leachate from recently deposited waste is a high-strength wastewater. Furthermore, leachate characteristics change dramatically

Waste
Protective Layer (Soil)
Geotextile
Drainage Layer (Sand, tires, or geonet)
Geosynthetic
Barrier Layer — Clay and Geomembrane (Keeps water out, directs gases toward venting and collection system)

FIG. 10.5.4 Liner system.

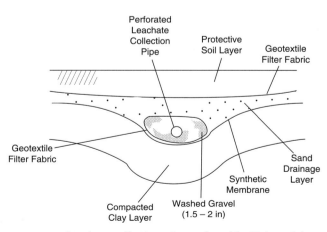

FIG. 10.5.5 Leachate collection pipe and trench. (Adapted from G. Tchobanoglous, H. Theissen, and S. Vigil, 1993, *Integrated solid waste management: Engineering principles and management issues* [New York: McGraw-Hill].)

over the course of the landfill's life. Consequently, treatment processes should be carefully designed and constructed. The most common option is to use a nearby municipal wastewater treatment plant. Leachate is usually transported to the facility by tanker truck, but a pipeline is economic in some cases. Using a municipal wastewater treatment plant to treat a high-strength wastewater may involve extra charges or pretreatment requirements.

LEACHATE MONITORING

The last leachate control element is monitoring. Some landfills use lysimeters, geosynthetic membranes placed in the ground, to detect and collect material directly under the landfill. However, monitoring is most commonly accomplished by collecting groundwater from wells located around the landfill, both upgradient and downgradient of the landfill. Upgradient wells are important in determining whether downgradient contamination is caused by the landfill or some upgradient event. Groundwater is monitored regularly for a number of inorganic and organic constituents (CFR 40 Parts 257–258). Detection of a contaminant at a statistically significant higher concentration than background levels results in increased monitoring requirements. Detection of contaminants at concentrations above groundwater protection levels requires the operator to assess corrective measures. Based on this assessment, a corrective measure is selected that protects human health and the environment, attains the applicable groundwater protection standards, controls the source(s) of release to the maximum extent possible, and complies with the applicable standards for managing any waste produced by the corrective measures. Corrective measures may involve pump and treat, impermeable barriers, or bioremediation.

Gas Control

Waste material deposited in landfills contains organic material. If sufficient moisture is available (more than 20%), indigenous microbes degrade this material. While sufficient external electron acceptors are available, degradation is achieved through respiratory processes that produce primarily carbon dioxide and water. Invariably, oxygen is available at first, entrained in the waste during collection, transport, and unloading. Usually, microbial activity consumes the available oxygen within a short period of time, i.e., days or weeks. If alternative electron acceptors are available to substitute for oxygen, respiratory processes continue (Suflita et al. 1992). The most common alternative electron acceptor in landfills is sulfate, found in gypsum dry wall debris.

Alternative electron acceptors are not available in most of the volume of a typical landfill; subsequently, fermentative processes predominate, ultimately producing landfill gas that is primarily carbon dioxide and methane (see Table 10.5.4). Observed gas yields are less than the theoretical maximum based on stoichiometry and are generally in the range of 4000 cu ft per tn of waste (O'Leary and Walsh 1991b).

Landfill gas must be removed from landfills. The final cover, used to keep out water and support vegetation for erosion control, can trap landfill gases. A build-up of gas in a landfill can rupture the final cover. In addition, the vegetative cover can be killed if the pore space in the final cover topsoil becomes saturated with landfill gases. Finally, methane is explosive if present in sufficient concentration, above 5%. Methane traveling through the landfill or surrounding soils can collect at explosive concentrations in nearby buildings. Migration distances greater than 1500 ft have been observed (O'Leary and Walsh 1991a). For these reasons, landfill gases must be vented or collected.

Gas control can be accomplished in a passive or active manner. Passive landfill gas control relies on natural pressure and convection to vent gas to the atmosphere or flares. Passive systems consist of gas venting trenches or wells, either in the landfill or around it. However, passive systems are not always successful because the pressure generated by gas production in the landfill may not be enough to push landfill gas out.

Active gas control removes landfill gases by applying a vacuum to the landfill. In other words, the landfill gases are pumped out. However, overpumping draws air into the landfill, slowing the production of more methane. After expensive landfill gas extraction equipment is installed, slowing methane production is not desirable. If migration control is the primary purpose of active gas control, recovery wells can be placed near the perimeter of the landfill. However, landfill gas can be an energy source, in which case vertical or horizontal recovery wells are typically placed in the landfill. Landfill gas with 50% methane has a heating value of 505 Btu/standard cu ft, about half that of natural gas (O'Leary and Walsh 1991a).

Collected landfill gas can be vented, burned without energy recovery, or directed to an energy recovery system.

TABLE 10.5.4 TYPICAL LANDFILL GAS COMPONENTS

Component	Percent
Methane	47.4
Carbon dioxide	47.0
Nitrogen	3.7
Oxygen	0.8
Paraffin hydrocarbons	0.1
Aromatic-cyclic hydrocarbons	0.2
Hydrogen	0.1
Hydrogen sulfide	0.01
Carbon monoxide	0.1
Trace compounds	0.5

Source: Data from R. Ham, 1979, Recovery, processing and utilization of gas from sanitary landfills, EPA 600/2-79-001.

When energy is to be recovered, the gas can be piped directly into a boiler, upgraded to pipeline quality, or cleaned and directed to an onsite electricity engine-generator. The first two options are feasible only if a boiler or gas pipeline is located near the landfill, which is not common.

Because of the explosive and suffocative properties of landfill gases, special safety precautions are recommended (O'Leary and Walsh 1991a):

- No person should enter a vault or trench on a landfill without checking for methane gas or wearing a safety harness with a second person standing by to pull him to safety.
- Anyone installing wells should wear a safety rope to prevent falling into the borehole.
- No smoking is allowed while gas wells or collection systems are being drilled or installed or when gas is venting from the landfill.
- Collected gas from an active system should be cleared to minimize air pollution and a potential explosion and fire hazard.

Personnel entering the landfill through gas collection manholes must carry an air supply.

Gas monitoring wells should be placed around the landfill if methane migration could threaten nearby buildings. Gas wells are used to measure gas pressure and to recover gas from soil pore space. The explosive potential of gases can be measured with portable equipment.

Site Preparation and Landfill Operation

Site preparation involves making a site ready to receive MSW and can include (O'Leary and Walsh 1991d):

- Clearing the site
- Removing and stockpiling the soil
- Constructing berms around the landfill for aesthetic purposes and surface water control. Berms are usually constructed around each landfill phase.
- Installing drainage improvement, if necessary. These improvements can include drainage channels and a lagoon.
- Excavating fill areas as phases are built (only for excavated landfills)
- Installing environmental protection facilities, including a liner, leachate collection system, gas control equipment, groundwater monitoring equipment, and gas monitoring equipment
- Preparing access roads
- Constructing support facilities, including a service building, employee facilities, weigh scale, and fueling facilities
- Installing utilities, including electricity, water, sewage, and telephone

- Constructing fencing around the perimeter of the landfill
- Constructing a gate and entrance sign as well as landscaping
- Constructing a convenience center, either for small vehicles to unload waste (to minimize traffic at the working face) or for the collection of recyclables
- Installing litter control fences
- Preparing construction documentation

An efficient landfill is operated so that vectors, litter, and environmental impacts are minimized, compaction is maximized, worker safety is ensured, and regulations are met or exceeded. Regulations control or influence much of the daily landfill operation. For example, regulations require some or all of the following:

- Traffic control
- An operating plan
- Control of public access, unauthorized traffic, litter, dust, disease vectors, and uncontrolled waste dumping
- Measurement of all refuse
- Control of fires
- Minimization of the working face area
- Minimization of litter scatter from the working area
- Frequent cleaning of the site and site approaches
- 6 in of soil cover on exposed waste at the end of the operating day
- Special provisions to handle bulky wastes
- Separation of salvage or recycling operations from the working face
- Exclusion of domestic animals
- Safety training for employees
- Annual reports and daily record keeping

Landfill equipment falls into four groups: site construction; waste movement and compaction; cover movement, placement, and compaction; and support functions (O'Leary and Walsh 1991d). Conventional earth moving equipment is usually used in landfill construction. However, specialized equipment is required for liner installation. The vehicles that bring waste to the landfill dump on the working face. Therefore, operators accomplish waste movement and compaction at the landfill by moving and spreading the waste around the working face and traveling over it several times with heavy equipment, usually compactors or dozers. If soil is used as cover material, it is transported using scrapers or trucks. If trucks are used, additional equipment is needed for loading. Cover soil compaction is done by the same equipment that compacts the waste. The use of an alternative cover material, such as foam or blankets, may require special equipment. A common support vehicle is the water truck, which reduces road dust and controls fires. The selection of landfill equipment depends on budget and the daily capacity of the site.

Closure, Postclosure, and End Use

Both the design and operation must consider the closure and postclosure periods, as well as the end use. Typical end uses include green areas, parks, and golf courses. As phases are closed, the final or intermediate cover may be applied depending on whether the top elevation of the landfill has been reached. Vertical gas vents or recovery wells can be installed as the final elevations are reached. Horizontal gas recovery wells are installed at specified height intervals as the phases are filled. As the side slopes of the landfill are completed, many aspects of final closure can also be completed, including final cover installation and revegetation.

Figure 10.5.6 shows a typical final cover cross section. The surface layer consists of top soil and is used to support vegetation. The vegetation reduces erosion and aesthetically improves the landfill. Grasses are the most common vegetation used, but other plants are used, including trees. Just below the surface layer is the optional drainage layer, used to minimize the hydraulic head on the barrier layer. The drainage layer can be sand or a geonet and is protected from clogging by a geotextile. The next layer is the hydraulic barrier. Current regulations require it to have a hydraulic conductivity at least as low as the bottom liner. Therefore, the barrier layer usually includes a geomembrane. A subbase layer may be necessary to protect the barrier layer.

Final closure involves installing the remaining final cover, planting the remaining vegetation, and adding any fencing required to maintain site security. Revegetation depends on a number of factors (O'Leary and Walsh 1992b). First, the cover soil must be deep enough to sustain the planted species. Grasses require at least 60 cm, while trees require at least 90 cm. The final cover topsoil should be stabilized with vegetation as soon as possible to avoid erosion. Operators should determine the soil characteristics before planting and add lime, fertilizer, or organic matter as required. The bulk density should be measured, and, if too high, amended. Species should be chosen that are landfill tolerant (Gilman, Leone, and Flower 1981; Gilman, Flower, and Leone 1983). Grasses and ground covers should be planted first. If possible, seeds should be embedded in the soil. Trees or shrubs, if used, should be planted only one or two years after grasses are planted. If grasses cannot survive on the landfill, the same will be true of trees and shrubs. The most common problems encountered with revegetation of landfill surfaces are poor soil, root toxicity, low oxygen concentration in the soil pore space, low nutrient value, low moisture content, and high soil temperature. Operators should develop a landfill closure plan which addresses control of leachate and gases, drainage and cover design, and environmental monitoring systems. The postclosure period is currently specified by regulation to be at least thirty years after closure. During this time, surface water drainage control, gas control, leachate control, and monitoring continue. The general problems that must be addressed during this period are the maintenance of required equipment and facilities, the control and repair of erosion, and the repair of problems associated with differential settlement of the landfill surface.

Special Landfills

The distinction between the modern sanitary landfill and hazardous waste landfill is blurred, except the latter usually has two or three liner systems and multiple leachate collection systems (O'Leary and Walsh 1992a). Landfills similar to the sanitary landfill are sometimes built to handle special waste. Special waste is high-volume waste that is not hazardous and can be easily handled separate from the municipal waste stream.

Separate disposal is advantageous if a dedicated disposal facility is required, the waste is perceived to have special associated risks, or the waste carries a lower risk than MSW. An example of a waste with special risks is infectious waste which, though relatively innocuous in the ground, must be handled with special care so that disposal facility workers are not infected. In this case, a dedicated facility may be required for worker safety. An example of a low-risk waste is construction and demolition waste. In this case, using a disposal facility with lower performance standards can reduce disposal cost. Thus, a special landfill is dedicated to one or a few classes of special waste material. Examples of special material include coal-fired electric power plant ash, MSW incinerator ash, construction and demolition debris, infectious waste, asbestos, or any nonhazardous industrial waste subject to subtitle D regulations.

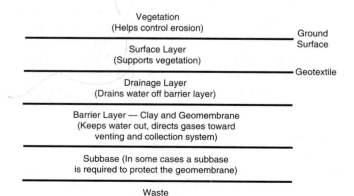

FIG. 10.5.6 Typical final cover.

Conclusion

In the United States, the landfill is the most popular disposal option for MSW. Traditionally, it has been the least-cost disposal option, and it is also a solid waste management necessity because no combination of reduction, recycling, composting, or incineration can currently manage the entire solid waste stream. Developing a new landfill involves site location, landfill design, site preparation, and landfill construction. Locating a new landfill can involve significant public participation. Federal regulations specify many location, design, operation, monitoring, and closure criteria. These regulations reduce the incidence of unacceptable pollution caused by landfills.

—*J.W. Everett*

References

Anandalingham, G. and M. Westfall. 1988–1989. Selection of hazardous waste disposal alternative using multi-attribute theory and fuzzy set analysis. *Journal of Environmental Systems* 18, no. 1: 69–85.

Camp Dresser & McKee Inc. 1984. *Cumberland County landfill siting report*. Edison, N.J.

CFR 40 Parts 257 and 258. *Federal Register* 56, no. 196: 50978–51119.

Erkut, E. and S. Moran. 1991. Locating obnoxious facilities in the public sector: An application of the analytic hierarchy process to the municipal landfill siting decision. *Socio-Economic Planning Sciences* 25, no. 2: 89–102.

Gilman, E., F. Flower, and I. Leone. 1983. Standardized procedures for planting vegetation of completed sanitary landfill. EPA 600/2-83-055.

Gilman, E., I. Leone, and F. Flower. 1981. The adaptability of 19 woody species in vegetating a former sanitary landfill. *Forest Science* 27, no. 1: 13–18.

Morrison, T.H. 1974. Sanitary landfill site selection by the weighted rankings method. Masters thesis, University of Oklahoma, Norman, Okla.

Noble, G. 1992. *Siting landfills and other LULUs*. Lancaster, Pa.: Technomic Publishing Company, Inc.

O'Leary, P. and P. Walsh. 1991a. Landfill gas: Movement, control, and uses. *Waste Age* 22, no. 6: 114–122.

———. 1991b. Landfilling principles. *Waste Age* 22, no. 4: 109–114.

———. 1991c. Leachate control and treatment. *Waste Age* 22, no. 7: 103–118.

———. 1991d. Sanitary landfill operation. *Waste Age* 22, no. 11: 99–106.

———. 1992a. Disposal of hazardous and special waste. *Waste Age* 23, no. 3: 87–94.

———. 1992b. Landfill closure and long-term care. *Waste Age* 23, no. 2: 81–88.

Siddiqui, M. 1994. Municipal solid waste landfill site selection using geographical information systems. Masters thesis, University of Oklahoma, Norman, Okla.

Suflita, J., C. Gerba, R. Ham, A. Palmisano, W. Rathje, and J. Robinson. 1992. The world's largest landfill: A multidisciplinary investigation. *Environmental Science and Technology* 26, no. 8: 1486–1495.

Tchobanoglous, G., H. Theissen, and S. Vigil. 1993. *Integrated solid waste management: Engineering principles and management issues*. New York: McGraw-Hill.

Walsh, P. and P. O'Leary. 1991a. Evaluating a potential sanitary landfill site. *Waste Age* 22, no. 8: 121–134.

———. 1991b. Landfill site plan preparation. *Waste Age* 22, no. 10: 87–92.

———. 1991c. Sanitary landfill design procedures. *Waste Age* 22, no. 9: 97–105.

10.6
COMPOSTING OF MSW

In the United States, 180 million tn, or 4.0 lb per person per day of MSW were generated in 1988 (U.S. EPA 1990). The rate of generation has increased steadily between 1960 and 1988, from 88 million to 180 million tn per day (U.S. EPA 1990). Furthermore, the rate continues to increase (Steuteville and Goldstein 1993). In 1988, 72% of the MSW was landfilled. At the same time, due to strict federal regulations, mainly the RCRA, the number of landfills has decreased (U.S. Congress 1989). For the protection of human health and the environment, old landfills are being closed and new ones must be carefully constructed, operated, and monitored even when the landfill is closed. Thus the cost of disposing MSW by landfilling has greatly increased.

The increasing rate of generation, decreasing landfill capacity, increasing cost of solid waste management, public opposition to all types of management facilities, and concerns for the risks associated with waste management has led to the concept of integrated solid waste management (U.S. EPA 1988). Integrated solid waste management refers to the complementary use of a variety of waste management practices to safely and effectively handle MSW with minimal impact on human health and the environment. An integrated system contains some or all of the following components:

- Source reduction
- Recycling of materials

- Incineration
- Landfilling

The U.S. EPA recommends a hierarchical approach to solve the MSW generation and management problems. Using the four components of integrated solid waste management, the hierarchy favors source reduction, which is aimed at reducing the volume and toxicity of waste. Recycling is the second favored component. Recycling diverts waste from landfills and incinerators and recovers valuable resources. Landfills and incinerators are lower in the hierarchy but are recognized as necessary in the foreseeable future to handle some waste.

Essentially, the goal of integrated solid waste management is to promote source reduction, reuse, and recycling while minimizing the amount of waste going to incinerators and landfills. Composting is included in the recycling component of the hierarchy. This section discusses the composting of MSW.

Aerobic Composting in MSW Management

The organic fraction of MSW includes food waste, paper, cardboard, plastics, textiles, rubber, leather, and yard waste. Organic material makes up about half of the solid waste stream (Henry 1991) (see Section 8.3). Almost all organic components can be biologically converted although the rate at which these components degrade varies. Composting is the biological transformation of the organic fraction of MSW to reduce the volume and weight of the material and produce compost, a humus-like material that can be used as a soil conditioner (Tchobanoglous, Theissen, and Vigil 1993).

Composting is gaining favor for MSW management (Goldstein and Steuteville 1992). It diverts organic matter from landfills, reduces some of the risks associated with landfilling and incineration, and produces a valuable byproduct (compost). At the present time, twenty-one MSW composting plants are operating in the United States (Goldstein and Steuteville 1992). Most of these plants compost a mixed MSW waste stream. This number does not include a larger number of operations which deal solely with organic material, primarily from commercial establishments (grocery stores, restaurants, and institutions) and those facilities composting yard waste. Finstein (1992) states that over 200 such yard waste facilities are in New Jersey alone.

Applications of aerobic composting for MSW management include yard waste, separated MSW, commingled MSW, and cocomposting with sludge.

SEPARATED AND COMMINGLED WASTE

Yard waste composting includes leaves, grass clippings, bush clippings, and brush. This waste is usually collected separately in special containers. Yard waste composting is increasing especially since some states, as a part of their waste diversion goals, are banning yard waste from landfills (Glenn 1992). The U.S. EPA (1989); Strom and Finstein (1985); and Richard, Dickson, and Rowland (1990) provide detailed descriptions of yard waste composting.

Separated MSW refers to the use of mechanical and manual means to separate noncompostable material from compostable material in the MSW stream before composting. The mechanical separation processes involve a series of operations including shredders, magnetic separators, and air classification systems. The sequence is often referred to as front-end processing. Front-end processing prepares the feedstock for efficient composting in terms of homogeneity and particle size. Front-end processing also removes the recyclable components and thus insures a higher-quality compost product since the material which causes product contamination is removed. Still, significant amounts of metals and trace amounts of household hazardous waste are often found after mechanical separation. For this reason, source-separated material is the preferred feedstock to produce the highest quality compost product.

On the other hand, composting partially processed, commingled MSW can divert waste from landfills when the product quality is not too demanding. The compost can also be used as intermediate landfill cover (Tchobanoglous, Theissen, and Vigil 1993). Recently, a planning guide was published for mixed organic composting (Solid Waste Composting Council 1991).

The organic fraction of MSW can be mixed with wastewater treatment plant sludge for composting. This process is commonly known as cocomposting. In general, a 2:1 mixture of compostable MSW to sludge is used as the starting point. Sludge dewatering may not be necessary.

While MSW contains a high percentage of biodegradable material (yard waste, food waste, and paper), one must decide prior to composting whether to keep the organic material separate from the other components of MSW or to begin with mixed MSW and extract the organic material later for composting. For example, yard waste (particularly leaves) is often kept separate from the rest of MSW and composted. This separation allows easier composting (than with mixed MSW) and yields a product with low levels of contamination. The disadvantage is that separate collection of yard waste is necessary.

COCOMPOSTING RETRIEVED ORGANICS WITH SLUDGE

While the fundamentals of sludge composting are applicable to MSW composting, several significant differences exist. The major difference involves preprocessing when MSW is composted. As shown in Figure 10.6.1, receiving, the removal of recoverable material, size reduction, and the adjustment of waste properties (e.g., the C:N ratio and the addition of moisture and nutrients) are essential steps in preparing MSW for composting. Obviously, different preprocessing strategies are needed for source-separated organic MSW and yard waste. Also, the degree of preprocessing depends on the type of composting process used and the specifications for the final compost product (Tchobanoglous, Theissen, and Vigil 1993).

MSW composting employs the same techniques as sludge composting: windrow, aerated static pile, and in-vessel systems. Tchobanoglous, Theissen, and Vigil (1993) note that over the past fifty years, more than fifty types of proprietary commercial systems have been developed and applied worldwide. In general, they are variations of these three basic techniques.

Municipal Composting Strategies

Today, a large degree of public opposition to all types of waste management facilities and concerns for the risks associated with waste management exists. Composting, however, is often perceived as a safer alternative to either landfilling or incineration (Hyatt 1991) and is ranked higher in the integrated solid waste management hierarchy. Nonetheless, composting facilities must be carefully

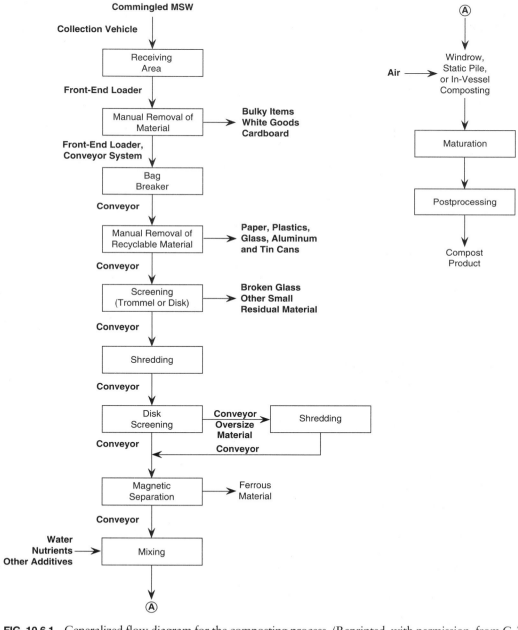

FIG. 10.6.1   Generalized flow diagram for the composting process. (Reprinted, with permission, from G. Tchobanoglous, H. Theissen, and S. Vigil, 1993, *Integrated solid waste management* [New York: McGraw-Hill].)

planned and managed for successful operation. The key elements are elucidated by the Solid Waste Composting Council (1991) and include:

1. Recovery and preparation of compostables
2. Composting
3. Refining
4. Good neighbor planting
5. Positive control of litter, dust, odors, noise, and runoff

The first step involves preprocessing (as previously described). This processing results in the preparation of a good feed stock for composting and the recovery of recyclables. The second step is the composting, which must be properly controlled. Refining involves postcomposting management (e.g., screening) to improve product quality. Good neighbor planting includes a carefully selected site, pleasing appearance, paved access, parking, a secure site, and a clean site. The positive control element includes the treatment of odors and other emissions, pathogen and toxin control, air-borne dust management, noise control, and run-off control.

Compost quality, an important issue, is a function of the physical, chemical, and biological characteristics of the product. In terms of physical aspects, good compost should be dark in color; have uniform particle size; have a pleasant, earthy odor; and be free of clumps and identifiable contaminants, such as glass fragments and pieces of metal and plastic. Chemical characteristics include not only the positive contribution from organic and inorganic nutrients, which are helpful for plant production, but also the detriments associated with heavy metals and toxic organics. Other chemical characteristics include weed seeds, salts, plant pathogens, and possibly human pathogens. Stability and maturity are significant concerns for compost quality and process control.

Quality is a major component of marketing compost, and marketing plays a key role in the effectiveness of any program to compost waste. The primary objective in finding a market for compost is finding an end use of the product. Since composting significantly reduces the volume of MSW, even if the compost is landfilled (as intermediate cover) that use may justify a composting program. However, the value of any program increases when a better end use is secured which further reduces the required landfill space and recovers a resource—a soil conditioner. While compost can be sold, revenue from composting is a secondary objective. While operating a compost facility for profit from compost sales is possible, this situation rarely occurs. Of course, any revenue generated from compost sales can offset the processing cost.

Composting MSW or various portions of the waste stream is an important component of integrated solid waste management. The use of composting is part of the strategy of minimizing incineration and landfilling while promoting source reduction and recycling. At least 50% of the MSW stream is compostable. Composting diverts these materials from less beneficial disposal methods and provides a more environmentally sound MSW program.

A central issue is the tradeoff between collection ease and management concerns. Source separated organics are easier to compost and yield a compost product of higher quality but require separate collection. The use of composting processes and the type of waste to be composted (mixed MSW versus source separation) must be integrated into the overall waste management plan for a given region. In terms of mixed MSW, preprocessing is important to obtain a high-quality product. Regardless of the final compost use or source of feedstock, some degree of preprocessing is necessary to prepare the feedstock for composting. This preprocessing insures proper particle size, moisture content, and nutritional balance.

—*Michael S. Switzenbaum*

References

Finstein, M.S. 1992. Composting in the contest of municipal solid waste management. *Environmental Microbiology* 58: 355–374.

Glenn, J. 1992. The challenge of yard waste composting. *BioCycle* 33, no. 9: 30–32.

Goldstein, N. and R. Steuteville. 1992. Solid waste composting in the United States. *BioCycle* 33, no. 11: 44–47.

Henry, C.L., ed. 1991. *Technical information of the use of organic materials as soil amendments: A literature review.* 2d ed. Solid Waste Composting Council. Washington, D.C.

Hyatt, G.W. 1991. The role of consumer products companies in solid waste management. *Proceedings of the Northeast Solid Waste Composting Conference.* Washington, D.C.: Solid Waste Composting Council.

Richard, T.L., N.M. Dickson, and S.J. Rowland. 1990. *Yard waste management: A planning guide for New York.* Albany, N.Y.: N.Y. State Dept of Environmental Conservation.

Solid Waste Composting Council. 1991. *Compost facility planning guide for municipal solid waste.* 1st ed. Washington, D.C.

Steuteville, R. and N. Goldstein. 1993. The state of garbage in America. *BioCycle* 34, no. 5: 42–50.

Strom, P.F. and M.S. Finstein. 1985. *Leaf composting manual for New Jersey municipalities.* Trenton, N.J.: Rutgers University and the N.J. Dept of Environmental Protection.

Tchobanoglous, G., H. Theissen, and S. Vigil. 1993. *Integrated solid waste management.* New York: McGraw-Hill.

U.S. Congress, Office of Technology Assessment. 1989. *Facing America's trash problem. What next for municipal solid waste?* OTA-0-424. Washington, D.C.

U.S. Environmental Protection Agency (EPA). 1988. *The solid waste dilemma: an agenda for action.* Draft report. EPA/530/SW-88-052.

———. 1989. *Decision makers guide to solid waste management.* EPA/530-SW-89-072.

———. 1990. *Characterization of municipal solid waste in the United States: 1990 update. Executive summary.* EPA/530-SW-90-042A.

Bibliography

American Society for Testing and Materials (ASTM). 1989. Standard specifications for waste glass as a raw material for the manufacture of glass containers. E708-79 (Reapproved 1988). Vol. 11.04 in *1989 Annual book of standards,* 299–300, Philadelphia: ASTM.

Bagchi, A. 1990. Design, construction and monitoring of sanitary landfill. New York: John Wiley & Sons.

Baillie, R.C. and M. Ishida. 1971. Gasification of solid waste materials in fluidized beds. *69th National A.I.Ch.E. Meeting, Cincinnati, Ohio, May 1971.*

Bergvall, G. and J. Hult. 1985. *Technology, economics, and environmental effects of solid waste treatment.* Final report #3033, DRAV Project 85:11. Sweden (July).

Cal Recovery Systems, Inc. 1990. *Waste characterization for San Antonio, Texas.* Richmond, Calif. (June).

California Integrated Waste Management Board. 1991. Unpublished preliminary data from a waste characterization study in Downey and Commerce, CA. Study by CalRecovery, Inc., Hercules, Calif. (Samples collected July 1988; data dated 1991.)

CalRecovery, Inc. 1989. *Waste characterization study for Berkley, California.* (December).

———. 1992. *Conversion factor study—In-vehicle and in-place waste densities.* (March).

Camp Dresser & McKee, Inc. Unpublished data developed by field personnel during waste characterization studies.

———. 1989. *Polk County (FL) waste composition analysis.* (September).

———. 1990. *Cumberland County (NJ) waste weighing and composition analysis.* Edison, N.J. (January).

———. 1990. *Sarasota County (FL) waste stream composition study.* Draft report (March).

———. 1991. *Cape May County (NJ) multi-seasonal solid waste composition study.* Edison, N.J. (August).

———. 1991. *City of Ontario (CA) source reduction and recycling evaluation.* Ontario, Calif. (March).

———. 1991. *City of Wichita integrated solid waste management plan.* Wichita, Kans. (December).

———. 1992. *Atlantic County (NJ) solid waste characterization program.* Edison, N.J. (May).

———. 1992. *Bay County (FL) waste composition analysis report.* (September).

———. 1992. *Frederick County (VA) solid waste composition analysis.* Annandale, Va. (June).

———. 1992. *Jacksonville (FL) waste composition study.* Tallahassee, Fla.

———. 1992. *Prince William County (VA) solid waste supply analysis.* Annandale, Va. (October).

———. 1993. *Berkeley and Dorchester Counties (NC) waste characterization study.* Raleigh, N.C. (April).

———. 1993. *Lake County municipal solid waste characterization study.* Chicago (November).

———. 1993. *Scott Area (IA) municipal solid waste characterization study.* Chicago (February).

———. 1993. *Wake County/City of Raleigh (NC) commercial, institutional, and industrial solid waste characterization study.* Raleigh, N.C. (February).

Cashin Associates, P.C. 1990. *Town of Oyster Bay commercial waste stream analysis.* Plainview, N.Y. (July).

CFR 40 Parts 257 and 258. *Federal Register* 56, no. 196: 50978–51119.

CH2M Hill Engineering, Ltd. 1993. *Waste flow and recycling audit, Greater Vancouver Regional District.* Vancouver (January).

Conrad, E., J. Walsh, J. Atcheson, and R. Gardner. 1981. *Solid waste landfill design and operation practices.* EPA draft report, Contract no. 68-01-3915.

Diaz, L.F. et al. 1993. Composting and recycling municipal solid waste. Chap. 3 in *Waste characterization.* Boca Raton, Fla.: Lewis Publishers.

Glysson, E.A. 1989. Solid waste. In *Standard handbook of environmental engineering.* McGraw-Hill.

Goff, J.A. 1993. Waste from airports. *Waste Age* (January).

———. 1993. Waste from malls. *Waste Age* (February).

Graham, B. 1993. Collection equipment and vehicles. Chap. 27 in *The McGraw-Hill recycling handbook,* edited by H.F. Lund. McGraw-

Hill, Inc.

Ham, R. 1979. *Recovery, processing and utilization of gas from sanitary landfills.* EPA 600/2-79-001.

Harrison, B. and P.A. Vesilind. 1980. Design and management for resource recovery. Vol. 2 of *High technology—A failure analysis.* Ann Arbor, Mich.: Ann Arbor Science.

Hill, R.M. 1986. Three types of low-speed shredder designs. *National Waste Processing Conference, Denver, 1986.* ASME.

Hilton, D., H.G. Rigo, and A.J. Chandler. 1992. Composition and size distribution of a blue-box separated waste stream. Presented at SWANA's Waste-to-Energy Symposium, Minneapolis, MN, January 1992.

Holmes, J.R. 1983. Waste management options and decisions. In *Practical waste management,* edited by J.R. Holmes. Chichester, England: John Wiley & Sons.

Institute for Solid Wastes, American Public Works Association. 1975. *Solid waste collection practice.* 4th ed. Chicago.

Kaiser, E.R. and S.B. Friedman. 1968. *Pyrolysis of refuse component combustion.* (May): 31–36.

Kaminski, D. 1986. Performance of the RDF delivery and boiler-fuel system at Lawrence, Massachusetts facility. *National Waste Processing Conference, Denver, 1986.* ASME.

Killam Associates. 1989; 1991. *Middlesex County (NJ) solid waste weighing, source, and composition study.* Millburn, N.J. (February).

Lipták, B.G. 1991. *Municipal waste disposal in the 1990s.* Radnor, Pa.: Chilton Book Company.

Liu, David H.F. 1974. Solid waste characterization. In *Environmental engineers handbook,* edited by B.G. Lipták. Radnor, Pa.: Chilton Book Company.

Lund, Herbert F. 1993. *The McGraw-Hill recycling handbook.* New York: McGraw-Hill, Inc.

Mallan, G.M. 1971. A total recycling process for municipal solid wastes. Paper 46C, *Nat. A.I.Ch.E., Atlantic City, August 29–September 1, 1971.*

Malloy, M.G. 1993. Waste from hospitals. *Waste Age* (July).

National Solid Wastes Management Association. 1985. Technical Bulletin 85-6. Washington, D.C. (October).

Non-Burn system for total waste stream. 1987. *BioCycle* (April): 30–31.

O'Leary, P. and P. Walsh. 1991. Landfill gas: Movement, control and uses. *Waste Age* 22, no. 6: 114–122.

———. 1991. Landfilling principles. *Waste Age* 22, no. 4: 109–114.

———. 1991. Leachate control and treatment. *Waste Age* 22, no. 7: 103–118.

———. 1991. Sanitary landfill operation. *Waste Age* 22, no. 11: 99–106.

———. 1992. Disposal of hazardous and special waste. *Waste Age* 23, no. 3: 87–94.

———. 1992. Landfill closure and long-term care. *Waste Age* 23, no. 2: 81–88.

Paper Stock Institute. *Guidelines for paper stock.* Washington, D.C.: Institute of Scrap Recycling Institute Inc.

Pfeffer, J. 1992. *Solid waste management engineering.* Englewood Cliffs, N.J.: Prentice-Hall.

Portland Metropolitan Service District. 1993. *Waste stream characterization study.* Results for fall 1993.

Preston, G.T. 1976. Resource recovery and flash pyrolysis of municipal refuse. Presented at Inst. Gas Technol. Symp., Orlando, FL, January 1976.

Rabasca, L. 1993. Waste from restaurants. *Waste Age* (March).

Robinson, W., ed. 1986. *The solid waste handbook.* New York: John Wiley & Sons.

Rugg, M. 1992. *Lead in municipal solid waste in the United States: Sources and forms.* Edison, N.J.: Camp Dresser & McKee Inc. (June).

San Diego, City of, Waste Management Department. 1988. *Request for proposal: Comprehensive solid waste management system.* (4 November).

Sanner, W.S., C. Crtuglio, J.G. Walters, and D.E. Wolfson. 1970. *Conversion of municipal and industrial refuse into useful materials by py-*

rolysis. RI 7428. U.S. Dept. of Interior, Bureau of Mines (August).

Santhanam, C.J. 1974. Flotation techniques. Vol. 3 of *Environmental engineers handbook,* edited by B.G. Lipták. Radnor, Pa.: Chilton Book Company.

Savage, G.M., L.F. Diaz, and C.G. Golueke. 1985. Solid waste characterization. Results of waste composition study in Santa Cruz County, Calif. *BioCycle* (November/December).

Schaper, L.T. and R.C. Brockway. 1993. Transfer stations. In *The McGraw-Hill recycling handbook,* edited by H.F. Lund. McGraw-Hill, Inc.

Scher, J.A. 1971. Solid waste characterization techniques. *Chem. Eng. Prog.* 67 (March).

SCS Engineers. 1991. *Waste characterization study, solid waste management plan, Fairfax County, Virginia.* Reston, Va. (October).

Seattle Engineering Department, Solid Waste Utility. 1988. *Waste reduction, recycling and disposal alternatives: Volume II—Recycling potential assessment and waste stream forecast.* Seattle (May).

Snell, J.R. 1974. Size reduction and compaction equipment. Vol. 3 of *Environmental engineers handbook,* edited by B.G. Lipták. Radnor, Pa.: Chilton Book Company.

Solid waste management: Technology assessment. 1975. Schenectady, N.Y.: General Electric.

Sommerland, R.E., W.R. Seeker, A. Finkelstein, and J.D. Kilgroe. 1988. Environmental characterization of refuse-derived-fuel incinerator technology. *National Waste Processing Conference, Philadelphia, 1988.* New York: ASME.

Steven, W.K. 1989. When the trash leaves the curb: New methods improve recycling. *New York Times,* 2 May.

Surprenant, G. and J. Lemke. 1994. Landfill compaction: Setting a density standard. *Waste Age* (August).

Tchobanoglous, G., H. Theisen, and R. Eliassen. 1977. *Solid wastes: Engineering principles and management issues.* New York: McGraw-Hill.

Tchobanoglous, G., H. Theisen, and S. Vigil. 1993. *Integrated solid waste management.* McGraw-Hill, Inc.

———. 1993. *Integrated solid waste management: Engineering principles and management issues.* New York: McGraw-Hill.

Tuttle, K.L. 1986. Combustion generated particulate emissions. *National Waste Processing Conference, Denver, 1986.* ASME.

U.S. Environmental Protection Agency (EPA). 1976. *Decision makers' guide in solid waste management.* 2d ed. Washington, D.C.: U.S. EPA.

———. 1992. *The consumer's handbook for reducing solid waste.* EPA 530-K-92-003. U.S. EPA (August).

Vesilind, P.A. and A.E. Reimer. 1980. *Unit operations in resource recovery engineering.* Englewood Cliffs, N.J.: Prentice-Hall.

Walsh, P. and P. O'Leary. 1991. Evaluating a potential sanitary landfill site. *Waste Age* 22, no. 8: 121–134.

———. 1991. Landfill site plan preparation. *Waste Age* 22, no. 10: 87–92.

———. 1991. Sanitary landfill design procedures. *Waste Age* 22, no. 9: 97–105.

Index